Horizons of Railway Transport 2018

10th Year of International Scientific Conference

MATEC Web of Conferences Volume 235 (2018)

Strecno, Slovakia
11 - 12 October 2018

Editors:

Jozef Gasparik
Juraj Camaj
Jaroslav Masek
Vladislav Zitricky

ISBN: 978-1-5108-7489-3

Printed from e-media with permission by:

Curran Associates, Inc.
57 Morehouse Lane
Red Hook, NY 12571

Some format issues inherent in the e-media version may also appear in this print version.

No changes have been made to the content of these proceedings. There may be changes to pagination and minor adjustments for aesthetics.

Printed by Curran Associates, Inc. (2019)

For additional information, please contact EDP Sciences – Web of Conferences
at the address below.

EDP Sciences – Web of Conferences
17, Avenue du Hoggar
Parc d'Activité de Courtabœuf
BP 112
F-91944 Les Ulis Cedex A
France

Phone: +33 (0) 1 69 18 75 75
Fax: +33 (0) 1 69 28 84 91

contact-edps@webofconferences.org

Additional copies of this publication are available from:

Curran Associates, Inc.
57 Morehouse Lane
Red Hook, NY 12571 USA
Phone: 845-758-0400
Fax: 845-758-2633
Email: curran@proceedings.com
Web: www.proceedings.com

TABLE OF CONTENTS

University of Žilina

International Scientific Conference

Horizons of Railway Transport 2018

Strečno 11. - 12. October 2018

Scientific Committee

Programme Committee

Organizing Comittee

Ing. Zdenka Záhumenská
Department of Railway Transport, Faculty of Operation and Economics of Transport and Communication, University of Žilina

Ing. Milan Dedík
Department of Railway Transport, Faculty of Operation and Economics of Transport and Communication, University of Žilina

Ing. Matúš Dlugoš
Department of Railway Transport, Faculty of Operation and Economics of Transport and Communication, University of Žilina

Ing. Lukáš Čechovič
Department of Railway Transport, Faculty of Operation and Economics of Transport and Communication, University of Žilina

Ing. Tomáš Čechovič
Department of Railway Transport, Faculty of Operation and Economics of Transport and Communication, University of Žilina

Ing. Vladimíra Štefancová
Department of Railway Transport, Faculty of Operation and Economics of Transport and Communication, University of Žilina

Ing. Michal Petr Hranický
Department of Railway Transport, Faculty of Operation and Economics of Transport and Communication, University of Žilina

General partner of the conference

RTI Wagon RAILTRANS

Conference partners

Slovenská vedeckotechnická spoločnosť dopravy Bratislava

MINISTERSTVO DOPRAVY A VÝSTAVBY SLOVENSKEJ REPUBLIKY

Ministerstvo dopravy a výstavby SR

ŽSR

Železnice Slovenskej republiky

ZSSK CARGO

Železničná spoločnosť Cargo Slovakia, a. s. Bratislava

Železničná spoločnosť Slovensko, a. s. Bratislava

ČD Cargo, a. s. Praha

Budamar Logistics, a. s. Bratislava

WAGON SERVICE travel s.r.o. Bratislava

VSL Software, a. s. Košice

REGIOJET
| STUDENT | AGENCY |

RegioJet Slovesnko a.s.

oltis slovakia

OLTIS Slovakia s.r.o.

ŠPED-TRANS LEVICE, a.s.
International Spedition Ku Bratke 5, 934 05 Levice, SLOVAKIA
PRVÁ SLOVENSKÁ SÚKROMNÁ ŠPEDIČNÁ FIRMA

Šped-Trans Levice, a. s.

azu+

Aktivita zvyšuje úspech, národný rozvojový projekt

Cargo

edp sciences

MATEC Web of Conferences

Statement of Peer review

In submitting conference proceedings to MATEC Web of Conferences, I certify to the Publisher that I adhere to the **Policy on Publishing Integrity** of the journal in order to safeguard good scientific practice in publishing.

1. All articles have been subjected to peer review administered by the proceedings editors.
2. Reviews have been conducted by expert referees, who have been requested to provide unbiased and constructive comments aimed, whenever possible, at improving the work.
3. Proceedings editors have taken all reasonable steps to ensure the quality of the materials they publish and their decision to accept or reject a paper for publication has been based only on the merits of the work and the relevance to the journal.

Title, date and place of the conference

Horizons of Railway Transport 2018, 10th year of international scientific conference, organized by University of Žilina, 11 - 12 October 2018, in Strečno, Slovakia

Proceedings editor(s):

prof. Ing. Jozef Gašparík, PhD.
assoc. prof. Ing. Juraj Čamaj, PhD.
assoc. prof. Ing. Jaroslav Mašek, PhD.
assoc. prof. Ing. Vladislav Zitrický, PhD.

Date and editor's signature

29. 10. 2018

Analysis of the Organisation of Railway Freight Undertaking: A Case Study of HŽ Cargo Ltd.

Borna Abramović[1,*], *Denis* Šipuš[1] and *Martina* Ribarić[1]

[1]University of Zagreb, Faculty of Transport and Traffic Sciences, 10000 Zagreb, Croatia

Abstract. Organisation is a significant segment of every enterprise, and so is the case with companies providing railway freight transport services. The success of a company depends on the organisational structure that makes it much easier to conduct business owing to the clearly defined hierarchy and responsibilities that lead to the ultimate goal – freight transport. This paper defines and describes railway freight transport, as well as the impact of freight transport market liberalisation on the sole organisation of a company. Moreover, various organisational schemes are analysed along with their potential benefits for the railway transport. The organisation of HŽ Cargo Ltd. is shown together with an outline of business activities of certain organisational units of the company.

1 Introduction

Organisation is a willing association of persons whose goal is to use appropriate means to complete certain tasks with the slightest strain possible, in any aspect of the social life [1]. The foundation for successful enterprise lies in effective organisation. A carefully selected and conducted organisation ensures an expedite and thorough fulfillment of goals with minimal expenses. Organisation places important emphasis on human resources – the cadre – as they ensure a successful business. The modern age, in which technology and techniques advance on a daily basis, has in a way forced parties engaged in the railway transport to change and adapt the organisational structure in order for the production process to continue without problems and progress with time. Freight operators need to rise to the various challenges of providing a high quality of service at low cost. The European Union has recently been working to liberalise the market in order to simplify the transport process [2]. Opening the market requires elaborate preparation of the undertakings, and it very often demands a complete restructuring of the enterprise. The future of business strategy, success on the market, and long-term growth all depend on the adaptation of an enterprise to the new changes [3].

2 Organising freight railway transport

An efficient organisation can anticipate issues and their resolution thus ensuring stability of the system, which in turn enables potential progress in terms of abilities, knowledge, and competencies. The organisation has certain elements that determine its quality, such as [4]:

- Established goals: e.g. higher profit, greater capacity for transport, higher productivity,
- Content: the structure of labour, transport, and cost structure,
- Responsible personnel: executives, employees (cadre), means,
- Methodology: the manner in which activities are conducted, and
- Monitoring mechanisms: indicators and collection of feedback.

A transport service is a non-material output and cannot be stored, which asserts the need for the service production and consumption processes to take place simultaneously. The place where that occurs is the market, which is specific due to the creation of freight transport service. The whole point of the organisation of railway transport is ultimately to offer a high quality transport service for its users, which depends on the connectivity between the components of the rail process. The basic components include techniques and technology. [5-6]

The organisational structure is a totality of relations and relationships between all the factors of production and business, as well as those within each individual factor. The organisational structure plays a very important role in the organisation of an enterprise, which ultimately impacts the success of a business.

Types of organisational structures of an enterprise include internal division of labour and forming organisational units on all hierarchical levels of the company. The lowers ones are usually called departments, sectors, divisions, or offices. Each structure is distinct from another because of the different ways of merging similar business activities, and in line with that so is each division of labour. Merging business activities

*Corresponding author: borna.abramovic@fpz.hr

is carried out based on the appropriate functions in the enterprise. Wider and narrower organisational units are able to adapt the organisation more quickly and easily to the changes that occur.

Jermakowicz [7], [8] classified organisational structures based on the degree of functionality and the general degree of decentralisation, as follows:

a) Line structure:
   1) Simple line,
   2) Line-and-staff,
   3) Line with a centralized position,
b) Functional structure:
   1) Functional,
   2) Matrix, and
   3) Project structure
c) Organic structure
   1) Hybrid,
   2) Collegial, and
   3) Amorphous.

3 Railway freight transport market

Liberalisation signifies introducing the 'new' and 'different' legal norms that, in turn, do away with the previous restrictions and bans in various domains of human activity [9]. The chief purpose of the directives and regulations of the EU is to liberalise the transport market for both the cargo and passengers. These regulations also apply to the Republic of Croatia as a full member of the EU, so the first step in the harmonisation of the Croatian legislation with that of the EU was bringing in the Railway Act of 2006, which initiated the adoption of the First railway package of EU directives, the foundation for further harmonisation, creating institutional framework for the railway, that is, reconstructing the railway system in Croatia. The country regularly carries out these harmonisations with the acquis of the EU, and there is a new law currently being drafted which would harmonise legislature with the Fourth railway package.

The liberalisation of the transport market is ultimately what ensures growth and development of transport companies, which depend solely on competitiveness, which implies a reduction of manipulative costs, i.e. rationalizing between creating transport service and increasing safety and commercial speed for freight and passenger transport.

The positive effects of service liberalisation in railway traffic [5] should result in a greater:

- Modal split in favour of railway traffic,
- Number of operators – carriers,
- Number of services for a lower price, and
- Financially sustainable system.

4 A case study of HŽ Cargo d.o.o.

HŽ Cargo d.o.o. is a limited liability company that provides services in domestic and international railway and intermodal freight transport, as well as other activities related to the transport process, such as:

- Loading and unloading freight in railway stations and other locations,
- Accompanying transport activities,
- Public freight transport in domestic and international road transport,
- Storage services,
- Letting property,
- Letting other means of land transport,
- Procuring and distributing goods,
- Brokering on the domestic and international market,
- Shipping,
- Wagon maintenance,
- Wagon cleaning and washing,
- Representing various foreign companies,
- Maneuvering and marshalling,
- Rail security (accompanying freight trains for other companies), and
- Overhaul and maintenance of rail vehicles (minor repairs of wagons for other railway undertakings, as well as the future undertakings in the Republic of Croatia) [10, 11].

HŽ Cargo d.o.o is a hundred-percent owner of its daughter or subsidiary companies, which include AGIT d.o.o, a limited liability company for intermodal transport, Održavanje vagona d.o.o (Wagon maintenance Ltd.), Radionica željezničkih vozila Čakovec d.o.o. (Railway vehicle workshop Čakovec Ltd.), and Remont i proizvodnja željezničkih vozila Slavonski Brod d.o.o. (Overhaul and manufacturing of railway vehicles Slavonski Brod Ltd.). In addition, under HŽ Cargo's shared ownership are Robno-transportni centar Brod d.o.o (Goods Transport Centre Brod Ltd.), 51 per cent, CROKOMBI d.o.o 47.09 %, and Cargo centre Zagreb, 20 %.[12]

HŽ Cargo d.o.o has a staff of 1.608, 168 of whom are in managing positions, with 1422 employees based in various regions [13].

The organisational structure of the company is based on [14] the Railway Act, Railway Safety and Interoperability Act, Companies Act, and Articles of Associations. The organisation of units, their tasks, obligations, authorities, and hierarchy are regulated by Ordinance of the Organisation of HŽ Cargo d.o.o. The organisational structure comprises several organisational units:

- Office of the Management Board,
- IT service,
- RCS customer support centre (IS Cargo),
- Office for Management of Safety, Workplace Safety, Fire Safety and Environmental Protection,
- Office for Controlling and Risk Management,
- Centre for Training of Train Drivers,
- Business departments: Sales, Technology, Finance, HR, Legal affairs, General and commercial affairs,
- Regional areas: West, South, East, Centre, and
- Regional centre, Sub-regional Technical Inspection, Sub-regional Traction.

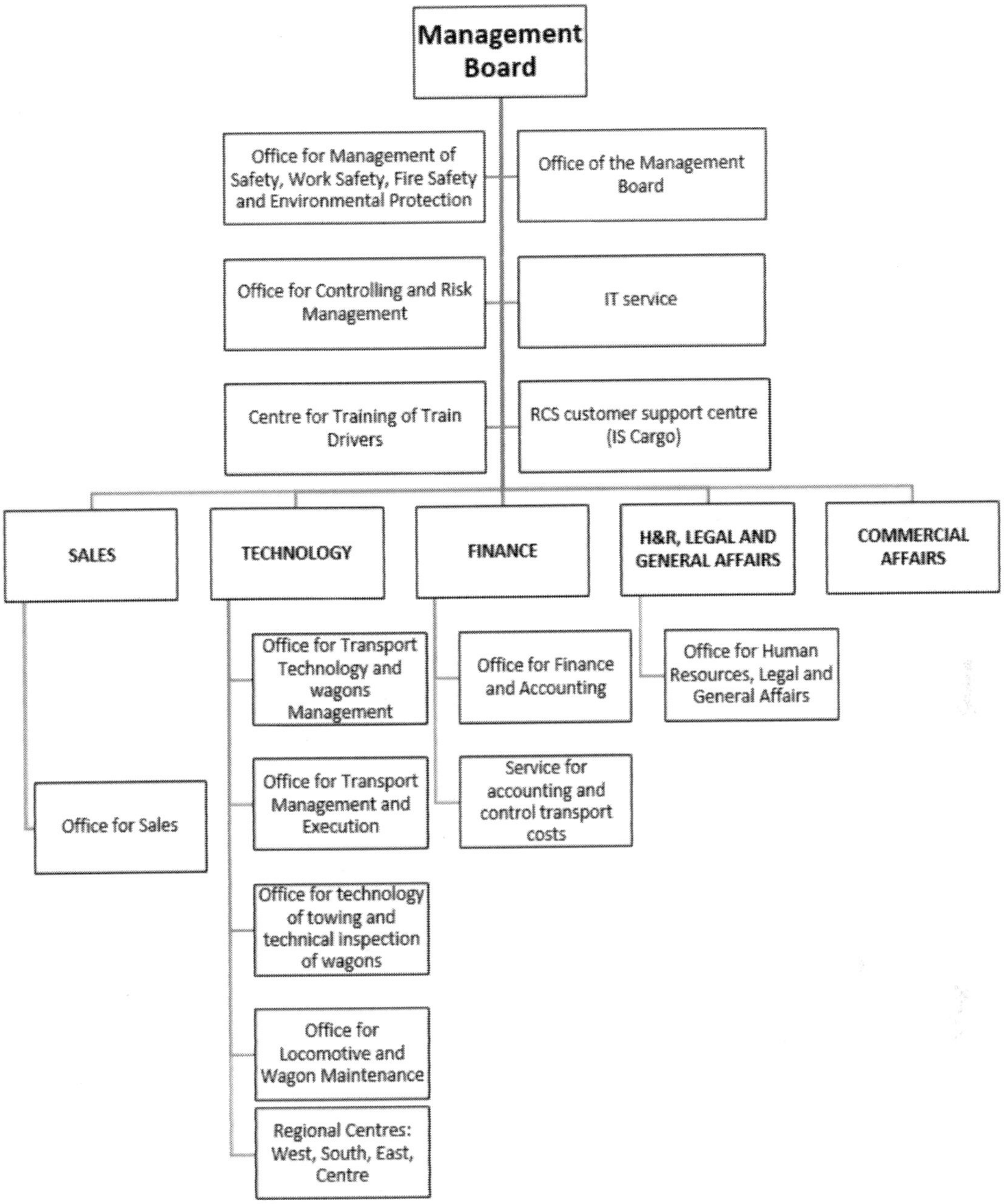

Fig. 1. HŽ Cargo d.o.o organisational structure [11]

The following sections will analyse in detail the following business departments: (1) Sales, (2) Technology, (3) Finance, (4) Human Resources, (5) Legal and General Affairs, and (5) Commercial Affairs.

4.1 Business sector – Sales

Sales department is concerned with activities related to the selling of railway freight transport services on both the domestic and international market, as well as providing intermodal transport services, which is the aim and purpose of every railway undertaking. Sales activities, cost accounting, and billing are done by the Office for Sales and directly in the regional centres of HŽ Cargo d.o.o.

Office for Sales deals with providing railway transport services on the domestic market, transit directions, and intermodal transport at import/export ports, land import/export directions, drafting agreements and user tariffs, processing complaints, providing instructions for service accounting, and monitoring the quantity of transport service. Furthermore, the Office is in charge of the following:[15]

- Laying down the criteria for compiling a list of transport service users,
- Monitoring the execution of transport according to user tariffs for all transport service users,
- Managing service quality and communication with transport service users at the time of accounting and billing,
- Resolving complaints,
- Issuing invoices to transport service users and approving refunds, and
- Balancing accounts for HŽ Cargo d.o.o.

4.2 Business sector – Technology

The technology sector performs a wide array of activities at the Office for Transport Technology and Wagon Management, Office for Transport Management and Execution, Office for Traction Technology and Wagon Inspection, Office for Locomotive and Wagon Maintenance, and at regional centres of the company. [9]

Its business activities include:

- Transport management,
- Work organisation for the HŽ Cargo d.o.o. centres,
- Drafting and interpreting regulations, tariffs, cargo wagon management, monitoring transport and operations,
- Drafting modernisation projects for existing traction vehicles and procuring new ones,
- Preparing working plan period for locomotives and train staff,
- Preparing the work schedule for train staff and locomotives based on the working plan period,
- Maintaining freight wagons and locomotives,
- Planning and monitoring the expenses of wagons maintenance, technical inspection of wagons, operations and technologies of vehicle inspection,
- Planning and monitoring expenses of locomotive technical inspections.

4.3 Business sector – Finance

Finance is an integral part of any business process. Its importance arises from its activity, which is to merge the functions of procuring and utilising capital. The business related to finance takes place at the Office for Financial Operations and Accounting, Office for Accounting and Control of Transport Expenses, and partly also in the regional centres.

The Office of Financial Operations and Accounting is concerned with:

- Billing,
- Paying accounts receivable,
- Controlling accounts receivable and accounts payable,
- Managing local and foreign currency register business activities,
- Liquidating and controlling interests,
- Drafting and controlling a list of short-term and long-term non-monetary assets, and
- Coordinating business with other related companies.

Activities that are conducted at the Office for Accounting and Control of Transport Costs include: [8]

- Evaluating the validity of accounts and accounting documents in receipt and dispatch register,
- Assessing incurred transport costs,
- Entering data from the transport documents into the IT system,
- Compiling reports for invoicing, international accounting, and controlling the accounting from other railway management boards,
- Controlling the train station cash registers, accounting franking and other deposits, controlling accounts payable charged and uncharged in cash, and
- Controlling demurrage and detention in the receipt and dispatch register as well as other charges arising from the transport agreement.

4.4 Business area - Human resources, Legal and General Affairs

Human resources represent the labour force which means it is one of the most vital resources of an enterprise. Employees are essential to every organisation, even more so because their value increases with growth, training, and professional development.

This department's business activities include:

- Introducing acts and articles for the entire Company,
- Offering expert opinions,
- Representing HŽ Cargo d.o.o. In courts and other state bodies,
- Establishing and advocating business policies for human resources management,
- Activities related to European Public Law, methodology and conducting them according to business targets of the company,
- Supporting other organisational units by adopting and complying with regulations, labour and employment relations and legal affairs,

- Permanent shaping and organising of professional training based on the obligatory and other programmes,
- Activities related to social labour, and
- Preparing and payroll accounting.

The aforementioned activities take place within (1) Office for Human Resources, (2) Legal Affairs, and (3) General Affairs.

4.5 Business area - Commercial Affairs

Commercial affairs include conceiving developmental projects, preparing and controlling investments, proposing improvements and rationalisations in business, monitoring the situation of properties of HŽ Cargo d.o.o., investing in them or letting. Purchasing merchandise, works, and services is realised pursuant to Procurement plan, complying thereby with enforced laws related to public procurement. Storage business of HŽ Cargo d.o.o. is conducted in the central storage facilities and accompanying locations in Zagreb.

Business activities carried out by Commercial Affairs are as follows:

- Development, investment, and management of properties owned by HŽ Cargo d.o.o.
- Preparation and implementation of the procurement of merchandise, works, and services, and
- Receipt, storage, and dispatch of goods from the storages.

5 Conclusions

The success of an enterprise comes from a well-organised organisational structure. When opting for the right structure, caution needs to be exercised and the condition in which the company conducts business need to be taken into consideration. The option depends on the company size, number of employees, type of business activity and a variety of factors relevant to business. Therefore, every company must have a unique organisational structure. It is the structure that defines the operation of all employees that are a part of creating the finished product, in this case, transport service. The analysis of the company is what provides insight into the importance of recognising factors that affect the shaping of the company, because, in that way, it is simple to (re)shape the organisational structure.

HŽ Cargo d.o.o. is a company with more than 1.600 employees. Because of its business activities – providing transport services – it is essential that the organisation and division of labour are clearly defined in order to prevent undesirable occurrences. The company is organised based on the principle of line structure, which means that the management board has the upper hand. As mentioned earlier, organisation of the rail freight undertaking is not feasible by utilising functional and organic structures, only line structures. The reason for that lies in the fact that railway transport demands a high level of safety and punctuality, which means there is no room for error since every mistake can cause material or financial loss. Line structures are quite efficient for freight undertakings as they have a clear hierarchy of responsibilities that precisely determine the responsibility of each employee.

The organisation greatly influences the quality of business, and in order to improve it, and thereby the productivity of the company, the enterprise needs restructuring. This enables a different approach of the company to the railway services market, which ultimately very often leads to an increase in quality and the scope of provided services. The issue that is often raised during restructuring is the loss of jobs, one of the greatest challenges. Since the Croatian railway market has been liberalised and now there are several foreign undertakings, restructuring would enable HŽ Cargo d.o.o. to achieve a better position on the liberalised market. Its competitiveness on the market HŽ Cargo d.o.o. should definitely base on an organisation which ensures a quick and efficient process of providing transport services to the satisfaction of end-users.

References

1. D. Badanjak, B. Bogović, and V. Jenić, Organizacija željezničkog prometa. Faculty of Transport and Traffic Sciences, Zagreb, (2006)
2. B. Abramović, V. Zitricky, V. Biskup, European Transport Research Review. 8. (2016)
3. W. Jermakowicz, *Zeitschrift fur Organ.*, **2**, 72–78 (1980)
4. A. Dolinayova, J. Camaj, J. Kanis, Transport Problems, **12**, 139-150, (2017)
5. W. Jermakowicz, *Zeitschrift fur Organ.*, **4**, 191–200, (1980).
6. B. Abramović, *Globalization trends an their impact on the transport system in terms of the EU common market*, (2012).
7. M. Ribarić, "*Analysis of the Organization of Railway Freight Operator*," University of Zagreb, Faculty of Transport and Traffic Sciences, Zagreb, (2017).
8. L. Cerna, A. Dolinayova, J. Danis, **Transport Means**, 414-419, (2016)
9. "www.hzcargo.hr," (2017).
10. HŽ Cargo d.o.o., Pravilnik o organizaciji HŽ Cargo d.o.o. Croatia: Službeni vjesnik, (2016).
11. A. Dolinayova, J. Camaj, J. Danis. *16th International scientific conference proceedings*, 383-390, (2016)
12. HŽ Cargo d.o.o., Operativni plan provedbe politike sigurnosti HŽ Carga d.o.o za 2017. godinu. Croatia: Službeni vjesnik, (2017)
13. R. Kampf, O. Stopka, I. Kubasakova, V. Zitricky. Procedia Engineering, **161**, 1538-1544. (2016)
14. J. Camaj, J. Masek, M. Kendra. *Proceeding of the 20th International scientific conference Transport Means*. 140-144, (2016)
15. E. Nedeliakova, J. Sekulova, I. Nedeliak. Procedia Economics and Finance. **34**. 180-186 (2015)

MATEC Web of Conferences **235**, 00002 (2018) https://doi.org/10.1051/matecconf/201823500002
Horizons of Railway Transport 2018

Empirical Research in Railway Transport Passengers' Interest in the Utilisation of Progressive Methods of Passenger Handling

Eva Brumercikova[1,*] and *Bibiana* Bukova[1]

[1]University of Zilina, Faculty of Operation and Economics of Transport and Communications, Department of Railway Transport, Univerzitna 8215/1, 010 26 Zilina, Slovakia

Abstract. New technologies utilised abroad offer passengers a wide variety of options that can be adapted to passengers' needs. The application of progressive ways of passenger handling requires the provision of integrated transport systems. Thanks to innovative technologies carriers can extend their offering of transport services with brand new services enabled with these technologies. The article presents the results of empirical research in railway transport passengers' interest in the utilisation of progressive methods of passenger handling in the Slovak Republic.

1 Introduction

Many countries not only in Europe but also in America and Asia utilise there in the sphere of public passenger transport some new innovative technologies and systems which improve the quality of transport services. Innovations in transport technologies, namely in the area of passenger handling, introduce a wide spectrum of options to adapt the infrastructure to passengers' needs. Technologies used abroad provide passengers with comfort and convenience [1,2].

The following technologies are commonly applied in foreign countries: NFC (Near Field Communication), and EMV (Europay/MasterCard/VISA). The newest technology which for example utilises electronic identification cards of citizens (ID cards) to identify a passenger and to calculate a fare is the Account Based Ticketing ("ABT") technology

2 Empirical Research in the Problem

The analysis of passengers' interest in new ways of handling passengers in railway transport stems from empirical research.

The Selection of a Basic Group Size

The minimum count of questionnaires was calculated with the formula (1) [3-5].

$$n \geq \frac{p * (1-p)}{\frac{e^2}{z^2} + \frac{p * (1-p)}{N}} \tag{1}$$

where:

n – minimum count of questionnaires from respondents,
z – value corresponding to a selected significance level,
p – relative percent (%) occurrence,
e – permissible measurement error expressing the percentage in which the found value can differ from the actual value of the basic group,
N – the size of the basic group.

3 The Evaluation of Empirical Research

The questionnaire was aimed at existing and potential customers of railway companies in Slovakia. The goal of the questionnaire was to find out what method of payment was most frequently used by respondents when buying a travel document, together with providing a reason why they used that method most frequently. [6]Another goal was to find out if respondents knew some other methods of payment for a travel document which are used abroad, and if they were interested in utilising one of technologies mentioned above. [7]

Total number of respondents was 519, out of which 269 ones responded using a personal questionnaire and 250 ones responded using an electronic questionnaire provided by Google. Return of questionnaires (out of total 519 respondents) was 100%. Based on the Table 1 a permissible measurement error calculated using the formula (1) was expressed; it tells about the percentage in which the found value can differ from the actual value of the basic group.

Table 1. Permissible Error of Measurement.

Permissible Error	Number of Questionnaires	Permissible Error	Number of Questionnaires
2	2339	6	264
3	1049	7	194
4	592	8	149
5	380	9	118
		10	96

4 Identification Data about Respondents

The base for empirical research evaluation is formed with the analysis of respondents' structure according to

* Corresponding author: eva.brumercikova@fpedas.uniza.sk

demographic criteria. With regard to the article scope we present only some data. 58% of men participated in the empirical research; the most frequent age group is expressed in the interval 26 - 40 years (almost 50%). With regard to the achieved education of respondents the following grades were present: primary education: 9%, secondary education without school leaving examination (A Levels): 14%, secondary education with school leaving examination (A Levels): 33%, and university education: 44% respondents. The biggest group of respondents (61%) were the employees.

5 Current Method of Payment for a Travel Document Used by Respondents

Currently there in passenger railway transport it is possible to use 5 payment options to buy a travel document: cash, cashless with authorisation (PIN code), cashless contactless, Internet banking and purchase of an SMS ticket. Respondents, who participated in the empirical research, use only the following 4 payment options:

- cash payment system: 44% - mostly passengers utilising free transport,
- cashless payment system with authorisation: 12%,
- cashless payment system contactless: 19%,
- internet banking: 25%.

Respondents expressed their opinion on cashless payment where better part of responses was positive, however, some respondents also had a negative opinion; they prefer cash payment system and they do not trust in cashless payment; they consider it dangerous and untrustworthy in some cases. [8]

6 The Respondents´ Interest in Progressive Methods of Passenger Handling

Empirical research was used to find out if respondents knew technologies used in progressive methods of passenger handling, namely the acronym NFC (Near Field Communication - this technology is part of all smartphones and tablets), EMV (Europay/MasterCard/VISA - a global standard for credit and debit cards which is used in each cashless payment), and ABT (Account Based Ticketing - it uses electronic ID cards, debit and credit cards to identify a passenger and subsequently to calculate a fare; no other chip card is needed)[9,10]. Respondents could have marked several responses, not only one of them. More than half of respondents (53.90%) stated they did not know any of the acronyms. The second group of respondents replied they knew the acronym NFC (33%). The EMV technology was known only by 16.5% respondents. At least 10.30% respondents were aware of the acronym ABT.

After that respondents were given explanations of the acronyms and technologies, so the interest in utilising these technologies in passenger transport could be determined. Based on these explanations respondents should have replied if they were interested in using some of those technologies. Again several responses could have been marked in this question. The majority of respondents replied they would be interested in using the NFC technology. The EMV technology, which is based on paying for a travel document by means of a payment card (which is currently used in cashless payment), would be interesting for 36.10% respondents. The ABT technology would be interesting for the least respondents (30.80%); although lately this technology has been promoted abroad, which is related to electronic identification cards (ID cards) as well as the public equipped with credit and debit cards. Other respondents (12.90%) have no interest in using any of the given technologies.

The above mentioned technologies enable variations in consideration of logging in and logging off the system for passenger handling, namely the connection:

- check in/check out - a passenger is required to hold their ID card or active NFC mobile phone near a scanning device before each beginning of a journey and after completing the journey. After holding the card near the scanning device upon the journey completion the fare is deducted. [11-13]:
- walk in/walk out - scanning devices are placed near the door of transport means within public passenger transport. Passengers identify themselves using their ID card or active mobile phone with a carrier's mobile application enabled; the phone is hold near the scanning device at each walk in/walk out of the transport means.[14]
- be in/be out - identification of a contactless chip card or active NFC mobile phone during the journey. This way, passengers register themselves during the entire journey. [15]

Empirical research was used to find out a preferred kind of connection into and out of the system. The total number 519 of respondents was decreased in 66 respondents due to lack of interest in using cashless payment and new technologies. There was an assumption that not all respondents would understand the functioning of given connections, thus they were explained. Thanks to these explanations respondents would mostly utilise the connection of walk in/walk out type (46%) and the connection of check in/check out type (40%). There are three login options mentioned above. The connection of be in/be out type was not offered in the research due to its investment intensity; instead respondents were offered the option to log into the system using a stationary automatic machine. This variant was marked by the least respondents (14%).

7 Conclusion

On the basis of the realised empirical research in passengers' interest in new methods of passenger handling in railway transport in the territory of the Slovak Republic it may be concluded that passengers are mostly interested in the NFC technology which uses a contactless chip cards and mobile phones to identify passengers and pay for travel documents. Moreover, results of the realised empirical research imply that passengers would utilise the connection of walk in/walk out type between media and scanning devices which would be placed in the means of transport.

The NFC technology has been occurring on the transport market abroad for more than 10 years; it has been undergoing an innovation in the area of a cashless payment, so passengers can (besides a contactless chip card and mobile phone) use their payment card, too, to pay for a travel document.

The way of a cashless payment is also used in the EMV technology and the newest technology ABT which is currently applied also in the capital city of Hungary, Budapest, where (besides a payment card) electronic identification cards (ID cards) and national cards of carriers are used in order to identify a passenger and to pay a fare [14,16].

The paper is supported by the VEGA Agency by the Project 1/0791/18"The Assessment of Economic and Technological Aspects in the Provision of Competitive Public Transport Services in Integrated Transport Systems" that is solved at Faculty of Operations and Economics of Transport and Communication, University of Žilina.

References

1. L. Jedlinsky, J. Caban, L. Krzywonos, S. Wierzbicki, F. Brumercik, Journal of Vibroengineering, **17**. 175-187. (2015)
2. N. Nestorova, S. Goncharuk, V. Anisimov, A. Anisimov, V. Shvartcfel, Procedia Engineering. **165**. 1547-1555. (2016)
3. D. Barta, M. Mruzek, M. Kendra, P. Kordos, L. Krzywonos, Advances in Science and Technology, **10**. 240-247. (2016)
4. J. Danis, A. Dolinayova, J. Camaj, AER – Advances in Engineering Research. **62**. 187-190. (2016)
5. L. Cerna, V. Zitricky, J. Danis, Open Engineering. **7**. 6-13. (2017)
6. J. Masek, M. Kendra, Regulated and Unregulated Competition on Rails. 117-129. (2013)
7. Z. Krzysiak, G. B. Artnik, W. Samociuk, J. Zarajczyk, K. Plizga, B. Rachwal, S. Wierzbicki, L. Krzywonos, F. Brumercik, Przemysl Chemiczny. **96**. 279-282. (2017)
8. L. Jedlinsky, J. Caban, L. Krzywonos, S. Wierzbicki, F. Brumercik, Journal of Vibroengineering. **17**. 175-187. (2015)
9. B. Abramovic, V. Zitricky, V. Biskup, European Transport Research Review. **8**. (2016)
10. J. Dolinayova, J. Camaj, J. Kanis, Transport Problems, **12**, 139-150, (2017)
11. J. Lalinska, J. Camaj, E. Nedeliakova. *Transport Means – proceedings of the International Conference*. 729-733. (2015)
12. J. Ližbetin, O. Stopka. Open Engineering. **6**, 1. 441-445. (2016)
13. V. Cempirek, P. Nachtigall, J. Siroky. Open Engineering, **6**, 637-641, (2016)
14. D. Seidenglanz, M. Kvizda, T. Nigrin, Z. Tomes, J. Dujka. Journal of Transport Geography. **54**. 414-429. (2016)
15. M. Poliak. Ekonomicky casopis. **61,** 2. 206-220. (2013)
16. M. Kvizda, Rewiew Of Economic Perspectives. **17**, 3. 331-334. (2017)

Research of the strength of the bearing structure of the flat wagon body from round pipes during transportation on the railway ferry

Oleksij Fomin[1], *Juraj* Gerlici[2], *Alyona* Lovskaya[2], *Kateryna* Kravchenko[2,*], *Pavlo* Prokopenko[4], *Anna* Fomina[5] and *Vladimír* Hauser[2]
[1]State University of Infrastructure and Technology, Cars and Carriage Facilities Department, 04071, Kyrylivska str., 9, Kyiv, Ukraine
[2]University of Zilina, Transport and handling machines department, 01026, Univerzitná 8215/1, Zilina, Slovak Republic
[3]Ukrainian State University of Railway Transport, Department of Wagons, 61050, Feuerbach Square 7, Kharkov, Ukraine
[4]Branch "Research and design and technological Institute of Railway Transport »PJSC« Ukrzaliznytsya»,
03038, Ivana Fedorova, 39, Kyiv, Ukraine
[5]Volodymyr Dahl East Ukrainian National University, Department of railway, road transport and handling machines, 93400, Central Avenue 59a, Sewerodonetsk, Ukraine

Abstract. The dynamic loading of the bearing structure of the flat wagon from round pipes was investigated during transportation on the railway ferry. The main types of movements of the railway ferry with wagons located on its decks are taken into account. The research has been carried out on the strength of the bearing structure of the flat wagon by the finite element method. The coefficient of the reserve for the fatigue strength of the bearing structure is calculated, taking into account the action of cyclic loads in the conditions of sea disturbance.

1 Formulation of the problem

The active competition of railway transport with other modes of transport, both on the internal market of goods transportation and through international transport corridors, causes the necessity of providing the transport industry with wagons of the new generation. The construction of such wagons at the present stage of development should ensure that they can be used not only for railways, but also for international rail and water connections [1-9]. Therefore, the issue of designing and putting into operation of fundamentally new structures of wagons with improved technical and economic indicators and an expanded spectrum of functional capabilities is a topical issue.

2 Analysis of recent research

The issue of designing the rolling stock for the carriage of heavy goods is considered in [4]. Dynamics and strength research was carried out using modern software tools such as ProMechanica and CosmosWorks. When designing the bearing structure of the transporter, research was conducted on the possibility of its execution from various types of materials.

Comparison of analytical stress analysis and numerical calculation of mobile work machine part is considered in [5].

The study of the dynamics of a railway wagon with an open loading platform is given in [7]. The calculation is carried out in the software environment of MSC Adams. Investigation of stability against the rollover of a wagon was carried out when it is entered into a curve with a radius of 250 m, taking into account different speeds of motion.

Determination of dynamic parameters of wagons during transportation on a railway ferry in works is not carried out.

The structural features of the wagon for intermodal transportation are considered in [8]. The wagon has a reduced middle part, and the presence of the back part makes it possible to load / unload the vehicle on / off it with a scooter. An analysis of the design of a new-generation flat wagon is given in [9]. The feature of a flat wagon is the possibility of adjusting the useful length depending on the dimensions of the cargo being transported.

In the designs of the flat wagons considered, there are no nodes for fixing of chain binders for the purpose of transportation on railway ferries.

Studies on the expediency of using round pipes as elements of bearing systems of flat wagons are given in [10]. On the basis of theoretical studies of the stress-strain state of the bearing structure of the flat wagon model 13-401, the reserves of the strength of its components were determined and the use of a round section pipe frame as a bearing element was proposed. The study of the dynamics of the improved bearing structure of the flat wagon in the work is not carried out.

* Corresponding author: kateryna.kravchenko@fstroj.uniza.sk

Features of creation of ideal bearing structures of railway vehicles with optimal technical and economic indicators are covered in [11, 12]. The question of studying the dynamic loading of bearing structures of wagons during transportation on railway ferries in the considered works was not investigated.

Illumination of the features of the study of the structural strength of the flat wagon from round pipes when transporting on a railway ferry.

To achieve this purpose, the dynamic loading of the bearing structure of the flat wagon of round pipes was investigated during transportation on the railway ferry. The main types of movements of the railway ferry with wagons located on its decks are taken into account: forward movements relative to the vertical axis, angular displacements around the transverse axis, and angular displacements around the longitudinal axis. The solution of differential equations is carried out by the Runge–Kutta method. The research has been carried out on the strength of the bearing structure of the flat wagon by the finite element method implemented in the CosmosWorks software environment. The coefficient of the reserve for the fatigue strength of the bearing structure of the flat wagon during transportation on the railway ferry is calculated, taking into account the action of cyclic loads in the conditions of sea disturbance.

3 Presentation of the main material

The structural features of a new generation flat wagon, bearing elements of which are made of pipes of circular cross section, are given in [10]. For the possibility of transportation of a flat wagon on a railway ferry, it is proposed to install on its bolster beams the nodes for fixing the chain binders (Figure 1) [13].

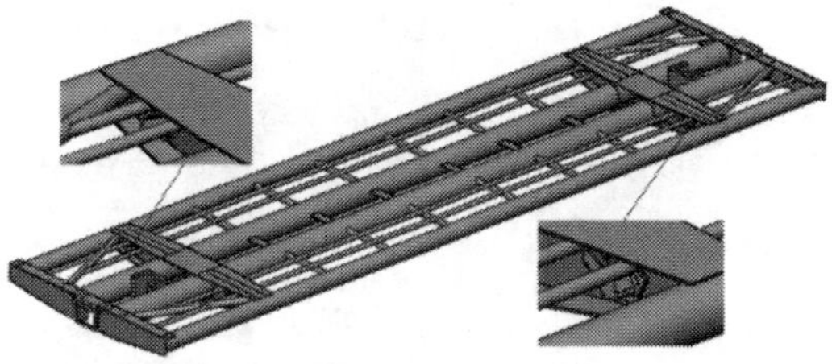

Fig. 1. Improved bearing structure of a flat wagon with nodes for fixing relative to the decks of the railway ferry.

The study of the dynamic loading of the bearing structure of the flat wagon was carried out using the mathematical model presented in [13, 14].

The main types of movements of a railway ferry with wagons located on its decks are taken into account: forward movements relative to the vertical axis Z (1), angular displacements around the transverse axis Y (2), angular displacements around the longitudinal axis X (3).

The diagram of the movements of the flat wagon fixed to the deck of the railway ferry is shown in Figure 2.

$$D' \cdot \ddot{q}_1 + \Lambda_z \cdot \dot{q}_1 = \Lambda_z \cdot \dot{F}(t); \quad (1)$$

$$I_\varphi \cdot \ddot{q}_2 + \left(\Lambda_\varphi \cdot \frac{L}{2}\right)\dot{q}_2 = p' \cdot \frac{h}{2} + \Lambda_\varphi \cdot \frac{L}{2} \cdot \dot{F}(t); \quad (2)$$

$$I_\theta \cdot \ddot{q}_3 + \left(\Lambda_\theta \cdot \frac{B}{2}\right)\dot{q}_3 = p' \cdot \frac{h}{2} + \Lambda_\theta \cdot \frac{B}{2} \cdot \dot{F}(t), \quad (3)$$

where $q_1 = z$, $q_2 = \varphi$, $q_3 = \theta$ – generalized coordinates corresponding to: z – the displacement of the body relative to the vertical axis passing through its center of mass, φ – the angular displacement around the transverse axis, θ – the angular displacement around the longitudinal axis. The origin of the coordinate system is located at the center of the mass of the railway ferry.

D' – massive water extrusion, L, B – the length and width of the railway ferry; h – the height of the board of the railway ferry; Λ_i – coefficient of resistance fluctuations; p' – wind load; and $F(t)$ – the law of the force of efforts that disturbs the movement of a railway ferry with the bodies of cars located on its decks.

The resistance coefficient of movement of a railway ferry is determined by the following formula:

$$\Lambda_i = \int_{-L/2}^{L/2} \rho \omega T^2 \overline{\Lambda}_i dL, \quad (4)$$

where $\overline{\Lambda}_i$ – dimensionless damping factor, which is determined by the graphical dependencies of the reference literature, in accordance with the characteristics of the railway ferry [15].

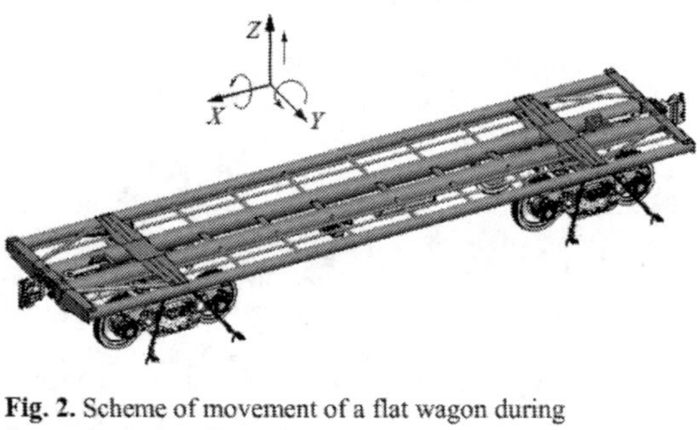

Fig. 2. Scheme of movement of a flat wagon during fluctuations of a railway ferry.

The model takes into account that the wagon is rigidly fixed relative to the deck and carries with it the movement, that is, it does not have its own degree of freedom. The impact of sea waves on the building of a railway ferry with wagons placed on its board was not taken into account. It is taken into account that the railway ferry is loaded with flat wagons using their full load-carrying capacity.

The research was conducted for the case of the carriage of wagons on the ferry "Heroes of Shipka" by the Black Sea.

The solution of differential equations (1-3) was carried out in the software environment of Mathcad [16, 17] by the Runge-Kutta method.

It was established that when moving the railway ferry in the vertical direction, the maximum accelerations act on the flat wagons located on the upper deck and make up 0.82 g, with angular displacement around the transverse axis for the extreme of the open stern of the wagon - 0,1 g, with the angular displacement around the longitudinal the axis for the extreme from the bulwark of the wagon - 0,23g.

The resulting numerical values of acceleration are taken into account when calculating the strength of the bearing structure of the flat wagon. The strength study was conducted using the finite element method implemented in the CosmosWorks software environment.

When creating a finite-element model, isoparametric tetrahedrons was used, the optimal number of which was determined by the graph-analytical method. The number of elements of the grid was 3056795, the nodes - 849631, the maximum dimension of the element - 15 mm, the minimum - 3 mm, the maximum aspect ratio of the elements - 584,93 the percentage of elements with a side relation of less than 3 - 89,3, more than 10 - 0,228. The minimum number of elements in the circle was 8, the ratio of increase in the size of the element - 1.6.

The model of the structural strength of the flat wagon at angular displacements relative to the longitudinal axis, as in the case of the greatest loading of the bearing structure, is shown in Figure 3.

When compiling the strength model, the following loads are taken into account: vertical static P_v^{st} and dynamic load, as well as loads that act on the body of the wagon through the chain binders P_{chb}. Since the chain binders have a spatial arrangement, the forces that will be transmitted to the supporting structure through them are decomposed into three components, taking into account the angles of their placement in space.

The fixing of the model was carried out in the areas of buckling of the bearing structure of the flat wagon on the bogies, as well as the supporting surfaces of mechanical stop-jacks. Steel grade 09G2S was used as the material of the bearing structure of the flat wagon.

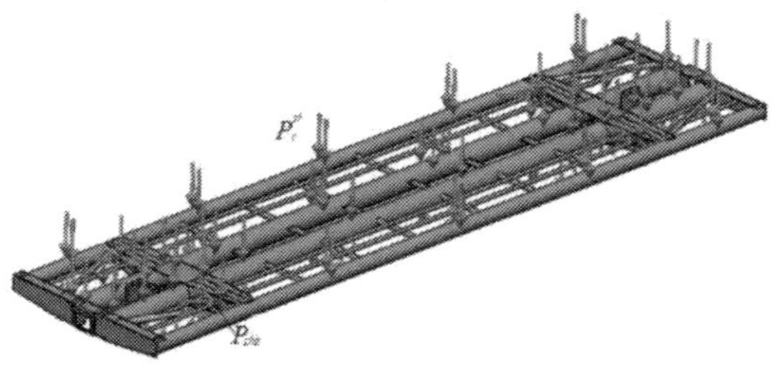

Fig. 3. Model of strength of the bearing structure of the flat wagon.

The results of the calculation are shown in Figures 4, 5. The conducted studies allowed us to conclude that the maximum equivalent stresses arise in the zones of interaction of the bolster beams with the ridge beam and make up about 280 MPa, that is, they do not exceed the permissible values [18, 19]. The distribution of maximum equivalent stresses along the length of the ridge of the flat wagon is shown in Figure 6.

Maximum displacements arise in the middle part of the main longitudinal beam of the frame by placing the chain binders that are tensioned at the angular displacement of the bearing structure of the flat wagon and make up 27.7 mm (Figure 7). The maximum deformations were $3{,}19 \cdot 10^{-3}$.

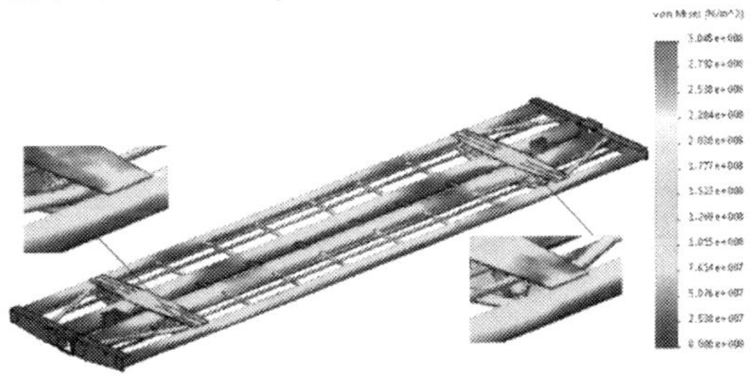

Fig. 4. The stressed state of the bearing structure of the flat wagon at angular movements around the longitudinal axis (top view).

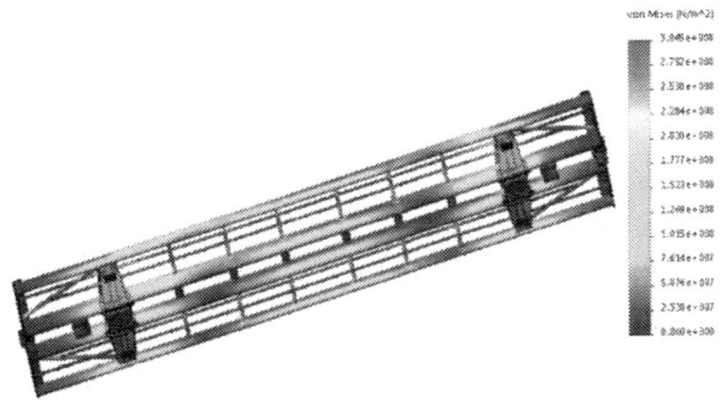

Fig. 5. The stressed state of the bearing structure of the flat wagon at angular movements around the longitudinal axis (bottom view).

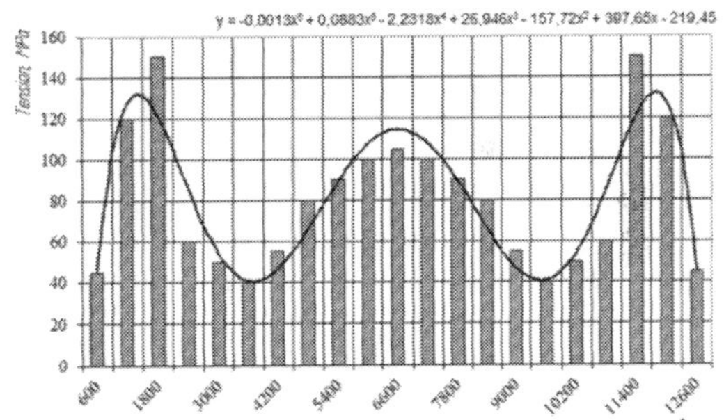

Fig. 6. Distribution of maximum equivalent stresses over the length of the ridge beam of the flat wagon.

The results of calculations at other schemes of displacements of the flat wagon allowed to conclude that the maximum equivalent stresses in the bearing structure do not exceed the permissible values.

To ensure the strength of the bearing structure of the flat wagon under the influence of cyclic loads in conditions of sea disturbance, the calculation of the reserve factor for fatigue strength was performed. The calculation is carried out according to the method shown in [20]. At the same time, the coefficient of reserve for fatigue strength in a multifaceted tense state is determined:

$$n = \sigma_{-1} / \sigma_{equ} \cdot K_m \quad (5)$$

where σ_{-1} – limit of endurance; σ_{equ} – equivalent stresses; K_m – a coefficient that takes into account the characteristics of the material.

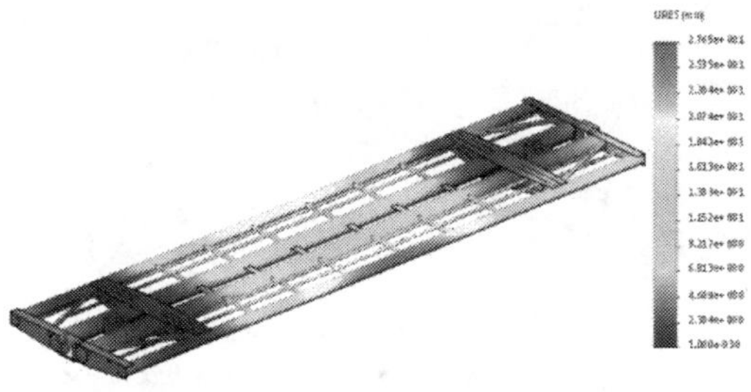

Fig. 7. Moving in the nodes of the bearing structure of the flat wagon.

$$K_m = \frac{1-R}{2}\left[\frac{1}{n_{-1} \cdot \varepsilon_\sigma} + \left(\frac{1}{\beta} - \beta\right)\frac{1}{\alpha_\sigma}\right]\frac{1}{\beta_y} + \frac{1+R}{2\alpha_\sigma}\psi_\sigma, \quad (6)$$

where R – coefficient of asymmetry of the cycle of stresses; n_{-1} – coefficient, which characterizes the sensitivity of the material to stress concentration; ε_σ – scale factor; β – coefficient of surface quality; β_y – coefficient of strengthening the surface; α_σ – theoretical concentration of stresses; ψ_σ – coefficient taking into account the influence of the asymmetry of the cycle.

The calculations made allow us to conclude that the coefficient of reserve for fatigue strength is $n \approx 1.7$, that is, within the limits of permissible [18, 19].

Conclusions

1. In order to increase the efficiency of combined transport, the construction of a new generation of flat wagon has been proposed. The peculiarity of the flat wagon is that its load-bearing elements are made of round pipes, which allows reducing the carriage tare compared with the prototype wagon by almost 5%. For the possibility of carriage a flat wagon on a railway ferry, it is proposed to set up the nodes on the bolster beams for chain binders fixing;
2. The maximal accelerations, acting on the supporting structure of the flat wagon of round pipes, are determined at transportation on the railway ferry. When moving the railway ferry in the vertical direction for the flat wagon, placed on the upper deck, the maximum acceleration value is 0.82 g, with the angular displacement around the transverse axis for the extreme of the open stern of the wagon - 0,1 g, with the angular displacement around the longitudinal the axis for the extreme from the bulwark of the wagon – 0.23 g.
3. Researches on the strength of the bearing structure of the flat wagon during transportation on the railway ferry allowed to conclude that the maximum equivalent stresses do not exceed the permissible and make up about 280 MPa;
4. Conducted calculation of the reserve factor for the fatigue strength of the bearing structure of the flat wagon during transportation on the railway ferry. The results of the calculation showed that the value of the reserve factor for fatigue strength is within the permissible limit;
5. The conducted researches will allow to increase efficiency of operation of flat wagons through international transport corridors.

References

1. J. Dizo, J. Harusinec, M. Blatnicky, MATEC Web of Conferences, **134** Article Number: 00010 (2017).
2. P. Šťastniak, M. Moravčík, Manufacturing technology: journal for science, research and production. Vol. **17**, **3** (2017).
3. V. Hauser, O. Nozhenko, K. Kravchenko, M. Loulova, J. Gerlici, T. Lack, Procedia Engineering **192**, 289-294 (2017).
4. V. Hauser, O. S. Nozhenko, K. O. Kravchenko, M. Loulová, J. Gerlici, T. MATEC Web of Conferences, **157**, Article number 03010 (2018).
5. G. Divya Priya, A. Swarnakumari, Intern. J. of Innovative Technology and Research Vol. 2, **6** (2014).
6. M. Blatnická, M. Blatnický, J. Dižo, M. Sága, Manufacturing Technology, Vol.18, **2**, 1 (2018).
7. M. Nader, M. Sala, J. Korzeb, A. Kostrzewski, Logistyka **4** (2014).
8. W. Krason, T. Niezgoda, *Bulletin of the Polish Academy of Sciences Technical Sciences* Vol. 62, **4** (2014).
9. WBN Waggonbau Niesky GmbH, *Developing a flexible platform of freight wagons*, Intern. Edition, **1** (2016).
10. O. Fomin, A. Lovskaya, A. Plakhtiy, V. Nerubatsky, *Scientific Bulletin of National Mining University*, **6 (162)**, (2017).
11. O. V. Fomin, O. V. Burlutsky, Yu. V. Fomina, Scientific and technical journal «Metallurgical and Mining Industry», **2** (2015).
12. O. Fomin, I. Kulbovskiy, E. Sorochinska, S. Sapronova, O. Bambura, Eastern-European Journal of Enterprise Technologies, Vol. 5, **1 (89)**, (2017). doi: 10.15587/1729-4061.2017.109588.
13. A. A. Lovskaya, *The Manuscript. Thesis for the degree of candidate of sciences*, Kiev, (2013).
14. A. A. Lovskaya, *Collection of scientific works.* Donetsk, 37, (2014).
15. S. N. Blagoveshchensky, A. N. Kholodilin, *Handbook of the statics and dynamics of the ship*, 2, Dynamics (pitching) of the ship, (1975)
16. D. V. Kiryanov, *Mathcad* 13. (2006).

17. A. A. Alyamovsky, SolidWorks/COSMOSWorks 2006 – 2007. Engineering analysis by the finite element method, (2007).
18. Norms for the calculation and design of railway wagons of the Ministry of Railways of gauge 1520 mm, M., VNIIZhT, (1996).
19. Freight wagons. Requirements for strength and dynamic qualities. GOST 33211-2014, (2014).
20. A. A. Shirshov, Mechanical engineering **8** (2013)

Transport of dangerous goods by rail

Andrea Galieriková[1]*, *Jarmila* Sosedová[1], *Andrej* Dávid[1] and *Miroslav* Bariak[1]

[1]Department of Water Transport, Faculty of Operation and Economics of Transport and Communications, University of Zilina, Univerzitna 8215/1, 010 26 Zilina, Slovakia

Abstract. Every day, products identified as dangerous goods are transported across the EU by road, rail, water and air. Currently, large quantities of different types of dangerous goods are transported by rail. The influence of random factors and events can lead to an accident resulting in a leakage of hazardous substances. These types of incidents not only threaten the safety of rail transport, but also life, environment and property. The paper defines legal framework of DGT (dangerous good transport), risks during the carriage of hazardous materials and the main advantages of the rail transport of dangerous goods, compared with other transport modes.

1 Introduction

In the last few years, traffic volume, especially of hazardous substances, transported by rail has increased. Safety is becoming an issue of major concern. Regulations and preventive measures present powerful tool for increasing the safety and efficiency of "dangerous transports". Currently, each mode of transport, used for carriage of dangerous goods, has its own regulations for proper and correct ways of manipulations, package, labelling and transportation of hazardous substances. [1]

These rules have to be primarily respected by the shipper and the transporter. It is not possible to ensure enough protection of the population and the environment without the population's active participation and responsibility.

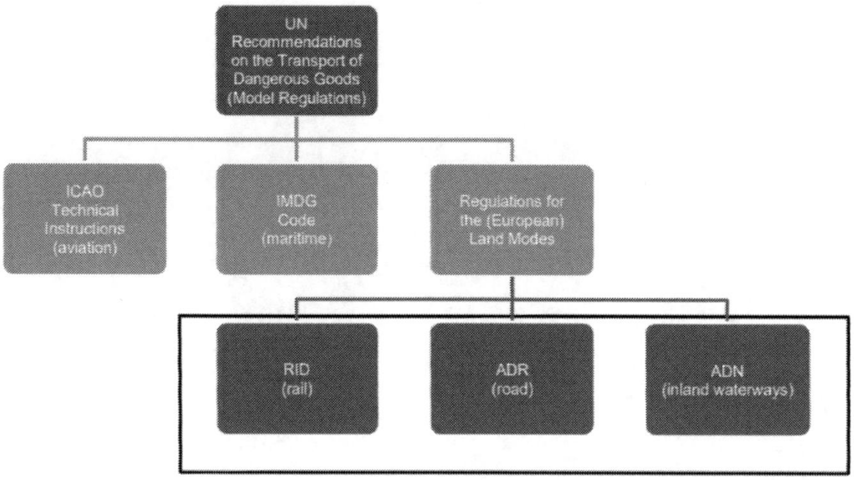

Fig. 1. Regulations for the international carriage of dangerous goods (Authors according to the Orange book)

2 Regulations for international transport of dangerous goods

2.1 International DGT by road

The international regulation of the carriage of hazardous substances by roads within the EU is achieved via the 'European Agreement Concerning the International Carriage of Dangerous Goods by Road' (known as ADR), drawn up by the United Nations Economic Commission for Europe (UNECE) in Geneva [2,3].

The provisions of ADR set out how producers, consignors and carriers should classify, package, label and transport dangerous goods. Specific vehicle and tank requirements and

* Corresponding author: andrea.galierikova@fpedas.uniza.sk

various other operational requirements such as driver training are also included.

2.2 International DGT by inland waterways

Carriage and manipulation of the hazardous substances in the terms of inland navigation are regulated by the AND Agreement. The agreement divides substances and materials into the classes (classification) according to the hazardous properties of the substances. It also sets uniform conditions for the labeling of dangerous goods and means of transport carried on waterways within the EU [4-6]. ADN was formed by grouping of two agreements – ADN-R (Agreement for the Rhine river) and ADN-D (for the Danube river). The agreement was established in 2006 in Geneva. The ADN consists of nine parts, which deal with the transport, handling and labelling of dangerous goods. [8,9]

2.3 International DGT by sea

The principal international rules for proper carriage of dangerous goods by sea are published in the IMDG code - 'International maritime dangerous goods code', which closely reflects the UN Model Regulations. Modifications of the IMDG code are the responsibility of IMO sub-committees. The carriage of goods in ships is covered by separate codes such as the 'International bulk chemical code' and the 'International gas carrier code'. The IMO is based in London; the UK's Maritime and Coastguard Agency, an agency of the Department for Transport (DfT), plays (in consultation with industry) a leading part in its deliberations [10].

2.4 International DGT by air

The International Civil Aviation Organization (ICAO) established the Technical Instructions for the Safe Transport of Dangerous Goods by Air (ICAO technical instructions), which are closely aligned with the UN Model Regulations, and cover the following: [11]

- classification of substances,
- proper packing,
- labelling and marking of packages, the documentation of consignments,
- the acceptance procedures to be used by air operators,
- the loading of dangerous goods on aircraft and the training of personnel involved in the transport of dangerous goods by air [12].

2.5 International DGT by rail

The international carriage of dangerous goods by rail within the EU is governed by annex I of the Convention Concerning International Carriage by Rail (COTIF).

Annex I is the 'Regulations Concerning the International Carriage of Dangerous Goods by Rail', more commonly known as 'RID'. RID is published by OTIF and updated every 2 years. Any amendments to the regulations are discussed and agreed by the RID Committee of Experts [11].

The principal international rules for proper carriage of dangerous goods by sea are published in the IMDG code - 'International maritime dangerous goods code', which closely reflects the UN Model Regulations. Modifications of the IMDG code are the responsibility of IMO sub-committees. The carriage of goods in ships is covered by separate codes such as the 'International bulk chemical code' and the 'International gas carrier code'. The IMO is based in London; the UK's Maritime and Coastguard Agency, an agency of the Department for Transport (DfT), plays (in consultation with industry) a leading part in its deliberations [11].

The structure of Regulations RID is divided into seven parts, each of which is divided into chapters and then sections and paragraphs. In accordance with the conditions laid down in the RID dangerous goods are divided into nine hazard classes (table 1). [12]

Table 1. Classification of dangerous goods.

Class	Substance
1	Explosives
2	Gases
3	Flammable Liquids
4.1	Flammable Solids
4.2	Spontaneously Combustible Solids
4.3	Dangerous when Wet
5.1	Normal Oxidizing Agent
5.2	Organic Peroxide Oxidizing Agent
6.1	Poisons
6.2	Biohazard
7	Radioactive
8	Corrosive
9	Miscellaneous

3 Regulations for international transport of dangerous goods

Risk is commonly understood as the combination of the magnitude and likelihood of potentially harmful events (hazard), and the severity of their consequences (1). In this context risk is estimated as the product of hazard and consequences:

$$Risk = Hazard \times Consequences \qquad (1)$$

The risk assessment process consists of several steps:

- identification of risks,
- risk analysis,
- risk evaluation,
- risk treatment [4].

The risk identification process includes the identification of main factors that may affect the occurrence of an emergency or accident in rail transport:

- **the state of the track** – level of rail quality,
- **line characteristic** – can depend on the interference deriving from the trains themselves (single or double track) and on the line alignment,
- **traffic system** – using of the automatic block signal (safer) or the telephonic signal (the least efficient), both require a visual check by or intervention on the part of the driver,
- **railway operation** – human error (accidental/no accidental),
- **wagon characteristics** – wagon defects (condition of a wagon – its age, the last maintenance, check),
- **probability of hazardous material release** – type of material transported, quantity of material, modality of transport, train speed [5].

Based on historical data, the most possible types of accidental events that should be considered in a Risk Assessment are: [13]

- **collision,**
- **derailment,**

- **grade crossing** .

Most of these accidents are caused by events resulting in impacts external to the mode of transport.

Monitoring the transport situation and subsequent communication and consultation may lead to risk treatment process (in the form of preventive measures and procedures that may eliminate the emergencies and risks to acceptable level). [14]

In accordance with the risk management the Commission of Experts OTIF (Working Group on Standardization of Risk Analysis) developed Generic Guideline for the Calculation of Risk due to Railway Transport of Dangerous Goods. This guideline is intended to assist in the implementation of basic principles that are used for risk assessment in Chapter 1.9 of the RID Regulations, "Transport restrictions by the competent authorities". [15] At the present guideline uses widely known and utilised procedures for analysis and risk evaluation (for example method Fault Tree Analysis FTA, Event Tree Analysis ETA or Quantitative Risk Analysis QRA)

4 Benefits of railway transport of hazardous materials

4.1 Safety

Reliable freight transport is the primary precondition for transport quality. In the comparison with other transport modes, the rail transport is certainly the most reliable traffic mode. The reliability does not result from a lower number of traffic accidents, but also lesser losses incurred from the accidents [16]. Accident rate of rail transport parties is almost zero, which is not true for car, neither air transport. The road transport carries along a direct danger for the safety of participants in every accident. In the rail transport, according to available statistics, the results of the investigation of accidents, the most common causes of the accidents are human error and equipment failure. The surroundings are also not negligible. Figure 2 shows the internal and external sources of threats, which are the causes of accident in rail transport.

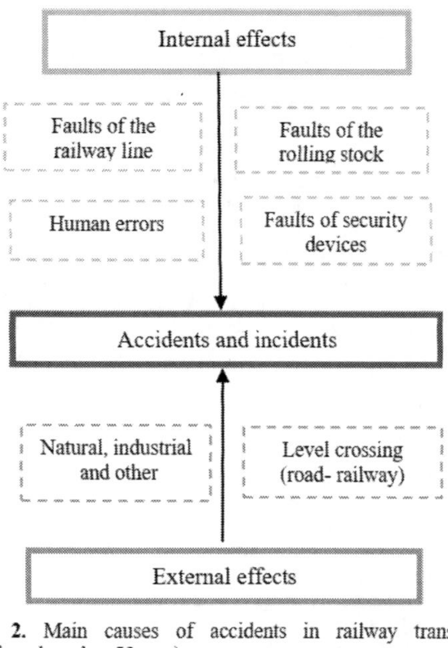

Fig. 2. Main causes of accidents in railway transport (Authors based on Verma)

Figure 2 shows that the causes of accidents can be defined according to the identified threats. Railway transport also disburdens overburdened network of highways and roads.

4.2 Ecology and environment

The high amount of transports and requirements on speed, reliability, punctuality and flexibility of transports increase every day. Millions of tons of goods transported each day by various modes of transport have an impact on quality of environment. Ability of meeting customer's requirements in an environment-friendly manner has become therefore one of the basic criteria for quality of transport services, as well as a significant competitive advantage of railway carriers. Railway transport is an efficient and environment-friendly transport system, as the large volume of goods can be transported on long distances quickly and with a minor impact on environment. Compared with road and air transport, railway transport produces the lowest amount of emissions and requires much lower costs on regeneration of damaged environment. From the total amount of costs on reduction of negative impacts of transport industry on environment, only 8% comes from railway transport, while up to 90% comes from road transport, even though its traffic performance is by 50% lower than the traffic performance of railway transport [9].

The most serious problem near railway lines is noise [10]. Noise produced by the railway transport on environment is lower than by the road transport. The intensity of the noise is influenced by several factors. Among the most significant factors there are means of propulsion of locomotives, inequalities of rails, base of ballast, speed, and different kinds of trains.

Measures to reduce noise can be implemented by means of suitable design of railway track and bearable construction of railway subgrade, including the imposition of tracks or construction of noise barriers. A very good solution is to lead the railway tracks parallel with the roads, motorways, i.e. to create transport corridors, and to protect the surrounding environment by ground mounds with higher vegetation [13].

4.3 Less restrictions

In the terms of transport of dangerous goods, road transport has many restrictions according to ADR (transport through the tunnels, amount of shipments etc.) In general, compared with road carriers, the important advantage of railway transport can also be lower transport restrictions, such as limitations in utilizing highways and primary roads, limitations in traffic peaks, or obligations to keep safety breaks. Advantage of railway transport out of the EU countries is also shorter waiting times on border crossings [8].

4.4 Optimisation of transport processes

Considering all transport claims a new trend of a combined transport is coming to the sphere of transports. The transport system lays stress on utilizing of railway transport supplemented by advantages of other traffic modes. It combines flexibility and fast transposition of goods with a reasonable consumption of energy and positive ecological aspect, minimizes impacts on environment and that way charges the society with rational ecological costs [3], [12].

5 Procedures to improve the safety of railway dangerous goods transport

Strict qualification, strengthening the training of practitioners
This objective can be reached by improving the professional trainings by qualified training institutions. The institutions should prepare every participant of the dangerous goods transport to guarantee the quality of training, so the operators and managers have a knowledge of proper respond on each type of emergency.

Implementation of security measures
Handling of dangerous goods must be strictly monitoring, and all the security procedures must be respected. Requirements on the proper labelling and marking of shipments, tanker filling and handling must be complied [7]. The loading operations of the hazardous materials must be monitor during the whole process. In the case of carriage of larger transport volume, the second and third category of dangerous goods, the measurements standards must be strictly enforcing to avoid the occurrence of the leakage. Mainly forwarders should conscientiously implement the system of tanker inspection according to current standards and procedures.

Contingency plans
Transport of dangerous goods represents big responsibility. In the case of an accident, not only significant economic losses, but also serious social impact can cause. To ensure the safety, in the case of leakage, fire or other accidents, prompt management should be made. This can be reached by establishing and improving the railway dangerous goods transport contingency plans. Plans should contain emergency procedures, methods, proper responding, information network etc [5]. Emergency plans should be situated at all handling stations, intermodal hubs and at the port terminals.

6 Design of minimising-risk model for railway transport of dangerous goods

The model of risk minimising during the transport of dangerous goods by rail consists of 3 steps. Hazard identification and risk estimation by using expert estimates; the second step is to reveal the risk sources based on the selection from the threat library; and the last part represents the decision-making process for the transport of dangerous goods by rail.

6.1 Hazard identification and risk estimation

In 2009, the European Railway Agency approved the Common Safety Method (CSM) as the basic model for managing rail transport risks. The first step of the proposal is to carry out the identification of risk sources according to the standard with respect to the CSM method. First, a system and the context have to be defined. The rail transport system can be divided into:

- transport infrastructure,
- means of transport and technology,
- management and information systems,
- human resource,
- repair capacities.

In the paper, the attention was paid in particular to the first three sections of the rail transport system. The context was mainly determined in relation to the likely occurrence of emergencies. Depending on the threat location, it may be:

- on the train,
- on the track and on the bridge,
- in the station,
- in the tunnel,
- in the vicinity of the track.

The second step of the model is to find out the threats that can endanger safe railway transport of dangerous goods. Based on data from the KISDIS Crisis Information Project, it is possible to divide the reactions to:

- human factor errors (fatigue, error, inattention, lack of communication, inadequacies in the organization, inappropriate repair, operator error, non-compliance with work procedures, non-compliance with occupational safety rules, etc.),
- extreme natural phenomena (avalanche, frost, constant frost, whirlwind, extreme drought, floods, long-lasting rain, torrential rain, snow calamities, etc.),
- technical failures (failure of safety systems, mechanical failure, failure of auxiliary equipment, failures of safety systems),
- criminal activity (individual and organized crime, sabotage, terrorism),
- Other (racial and religious conflicts, unauthorized demonstrations, unprofessional handling of dangerous goods, etc.).

The examination of the likelihood of risk consists of:

- operational failures,
- staff failure,
- threatening a person outside the system,
- the impact of floods,
- the influence of the vortex,
- impact of storm (lightning),
- the influence of high temperatures,
- impact of earthquake.

The risk categorization has been chosen according to the expert opinions on the transport of dangerous goods by rail.

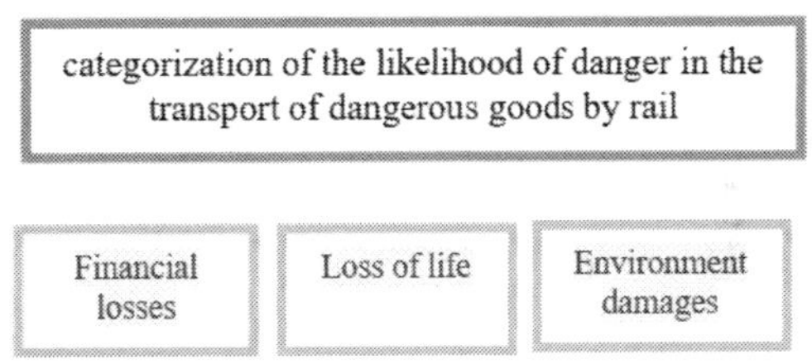

Fig. 3. Main causes of accidents in railway transport (Authors)

The second part of the model consists of three steps that help in the process of dealing with the crisis (Figure 3).

1. **activation of the crisis staff**

- receiving a report on the occurrence of an emergency and verifying the report
- initial evaluation of an emergency
- convocation of the crisis staff
- crisis staff starts up the realisation activities [13].

2. **starting rescue work**

- initiating activities under the crisis plan
- getting more detailed information from the place
- of emergency
- monitoring and evaluating the fulfilment of tasks
- according to the crisis plan

- convocation of the crisis staff
- performing rescue work
- logistics support on the place of emergency
- informing superiors about the progress of the emergency solution
- informing own employees
- informing customers
- informing the media
- conducting a crisis management meeting
- inviting experts to deal with emergency
- ending rescue work
- update the crisis plan
- the selection of workers and means for dealing with the emergency
- termination of rescue activities
- informing own employees about the current working regime.
- informing customers of the current provision of services
- informing the media about the state of the emergency response and its impact on society, the population and the environment
- calculating damages and losses
- termination of the crisis staff [17].

3. finishing rescue work and system recovery
The third part of the model is a sub model of the decision-making process for the transport of dangerous goods by rail. An important element is the establishment of a National Transport Information Centre (NTIC) to collect all transport data. In the case of emergencies, it provides information as required to individual agendas. NTIC has information about the railway lines and rail traffic. Among the components of the agenda are (in Slovakia): firefighters, police, emergency medical services, crisis management, railway administrations, Slovak hydrometeorological institute, infrastructure manager (ZSR), carriers and shippers. The information from the system should be available in addition to rescue services and distribution publishing systems such as systems of the Ministry of Transport and Construction, traffic portal, carriers, shippers, and to subjects of crisis management.

Adequate decision-making process for the occurrence of extraordinary events and accidents during the transport of dangerous goods by rail is aimed at implementing telematics applications [18].

7 Conclusion

In the rail transport there are several risks and threats which should be taken into consideration. Special attention must be paid to the transport of dangerous goods, which is very risky. Transport of hazardous substances could have disastrous consequences on people, as well as the environment and the property. Safety, as the most important element in the terms of the transport of dangerous goods, is reduced by the increasing the level of risk. Risk management allows to know the areas of greatest risk, identify individual risks, express their value and assess them using different methods and procedures. They enable proposing measures to minimize the risks to acceptable levels. It is crucial to recognize all potential threats and eliminate them to the acceptable level to ensure the safety and efficiency of railway transport in general.

References

1. R. Anderson, C. Barkan. Transportation Research Record. Pp. 88-98. (2004)
2. European Agreement concerning the International Carriage of Dangerous Goods by Road. UNECE (2017)
3. European Agreement concerning the International Carriage of Dangerous Goods by Inland Waterways. UNECE (2017)
4. A. Galierikova, J. Sosedova.*Communications : scientific letters of the University of Žilina*. **19**. **2** 80-85. (2017)
5. Gheorghe et al. Reliability Engineering & System Safety. **88**, **3**. 247-272. (2005)
6. International maritime dangerous goods code (IMDG Code). International Maritime Organisation. (2016)
7. O. Stopka, R. Kampf. Transport, 33, 280-290. (2018)
8. T. Kalina, et al. ALTA-Journal of Interdisciplinary Research.. **7**, **2**. 274-281 (2017)
9. Orange book: *Recommendations on the transport of dangerous goods*. United Nations. 17th edition. Vol. **I**. (2015)
10. I.A. Papazoglou *Final report to KOVERS, unpublished working paper*, Swiss Federal Institute of Technology, ETH, Zurich (2000)
11. P. Piala, et. al. *Communications : scientific letters of the University of Žilina*. **19**, **2**. 116-119. (2017)
12. P. Drizdziel, B. Bukova, E. Brumercikova. Transport problems **10**, **4**, 5-13. 2015
13. *Regulation concerning the International Carriage of Dangerous Goods by Rail* (RID). Intergovernmental Organisation for International Carriage by Rail (2017)
14. *Technical Instructions for the Safe Transport of Dangerous Goods by Air* (ICAO technical instructions). Council of ICAO. (2016)
15. L. Černá, A. Dolinayová, J. Daniš. In: *Proceedings of the 20th International Scientific Conference Transport Means 2016.*, Pages: 414-419. (2016).
16. L. Černá, J. Mašek. *Tranport Means – Proceedings of the International Conference*. 567-570. (2015)
17. M. Verma. Transportation Research Part C : Emerging technologies. **19**, **5**. 790-802 (2011)
18. Project KISDIS, 2012. Online. Available at <http://anakan.cz/soubory/**kisdis**/metodika-krizoveho-rizeni.pdf>

Possibilities of Using Tracking Methods for Trains in the Czech Republic

Jiří Hanzl[1], *Ladislav* Bartuška[1,*], *Josef* Šedivý[1], *Tomáš* Kůs[1], *Martin* Kůs[1] and *Jiří* Novotný[1]

[1]Institute of Technology and Business in České Budějovice, Department of Transport and Logistics, Okružní 517/10, 370 01 České Budějovice, Czech Republic

Abstract. The aim of this paper is to familiarize readers with several telematic systems that may be observed on the Czech railway network. In order to function correctly, they use special technical devices for detecting positions of trains and their tracking. In the introductory part, a brief overview of individual ways of train detecting and tracking in the Czech Republic is given, which is then followed by a detailed description of the related telematic systems, including the principles of their functioning and graphic presentation of their practical use. The paper also contains demonstrative photo documentation and schematic illustrations of principles of functioning of the discussed systems.

1 Introduction

Devices for detecting positions of trains and their tracking are one of the basic sources of information for different systems used on railways. These may include, for instance, various information and navigation systems that inform passengers at stations or cargo carriers about the current positions of trains on railway tracks and their delays or automated train management systems that are capable (based on the data obtained from detectors about the positions of trains) of automating and optimizing their journey with regard to minimum traction power consumption, adhering to the planned journey time without delays and indeed safe railway vehicle stopping. Within the Czech railway network, one may come across four basic ways of train detecting and tracking – axle counters, track circuits, magnetic information points and eurobalises. In the next part, the previously mentioned ways are briefly described and then the related telematic systems used on the Czech railway network are introduced.

2 Devices for detecting positions of trains and their tracking

Certain ways of detecting positions of trains have been applied particularly in the context of increasing requirements for railway safety and increasing the capacity of the railway infrastructure. At present, automatic or semiautomatic devices are largely used for train tracking, and depending on their technical design and function, they are divided into the point devices and the continuous devices. In the Czech Republic, it is possible to observe four basic ways of detecting positions of trains – axle counters, track circuits, eurobalises and magnetic information points.

Axle counters are devices for detecting railway vehicles on particular sections of railway tracks using point units, such as wheel detectors. They are therefore point-tracking devices made up of two parts, i.e. an outer part and an inner part. The outer part consists of the wheel detectors, whilst the inner part comprises a device which (on the basis of information from the outer part) is able to recognize whether railway vehicles are or are not located on the given sections of railway tracks. On the other hand, track circuits are used to detect railway vehicles on particular sections of railway tracks by means of the electric circuit and changes in ratios therein [1]. Thus, these are continuous tracking devices and each track circuit consists of a track line and connected equipment, with the track line encompassing a railway upper, where rails form wires and sleepers and gravel serve as insulation. Track circuits can be divided into two basic types, i.e. series and parallel wiring circuits. Both of these types work on the principle of changing the magnitude of the electric current that takes place in the circuit. The track circuit components are constituted by an electric power source, an adjustable resistor and an intrinsically safe relay (with an armature in the relay dropping off when the power is interrupted). As for the so-called eurobalises, positions of trains are detected through point units intended for an inductive transmission of information between railway vehicles and railway tracks within the ETCS (European Train Controlling System, see Subchapter 3.2). Some eurobalises contain information about their exact position and serve as reference points for passing railway vehicles. On their passing over the balises, information about the position is then transferred to the vehicles via a transmission module placed on them. Lastly, magnetic information points are used to detect positions of railway vehicles within the CRV and AVV systems. The

* Corresponding author: bartuska.vste@seznam.cz

 https://doi.org/10.1051/matecconf/201823500005

information about positions obtained from such points is intended solely for the purposes of these systems. Hence, there is only way of how to use them – magnetic information points have data about their exact positions encoded through combining differently placed magnets and this information is transferred to railway vehicles on their passing over the magnetic information points via sensors placed on the vehicles [2].

3 Possibilities of using tracking methods for trains

As stated in the previous chapter, technical devices for train tracking represent an important source of information for all modern security equipment that (based on train positioning data) ensure safe and continuous operating on the rail network. These are not only station, rail line and crossing security devices, but mainly devices designed to check compliance with driving parameters, and (in the event of their non-observance) to directly intervene in train control. The following subchapters (3.1 and 3.2) are focused on two train security systems (LS and ETCS) which are linked to Czech trains.

3.1 LS system

The LS system is seen as a continuous train security system that was developed in former Czechoslovakia back in the 1960´s and has since been modified several times. Currently, the LS 90 version is used [3]. However, one may not rely too much on further development of the system, particularly due to gradual deployment of the ETCS (see Subchapter 3.2), which is to sequentially replace different train security systems in other countries as well. The main function of the LS system is to transfer a signal via track circuits to a railway vehicle´s cabin (cab), where the signal is displayed on a signal repeater (Fig. 1.).

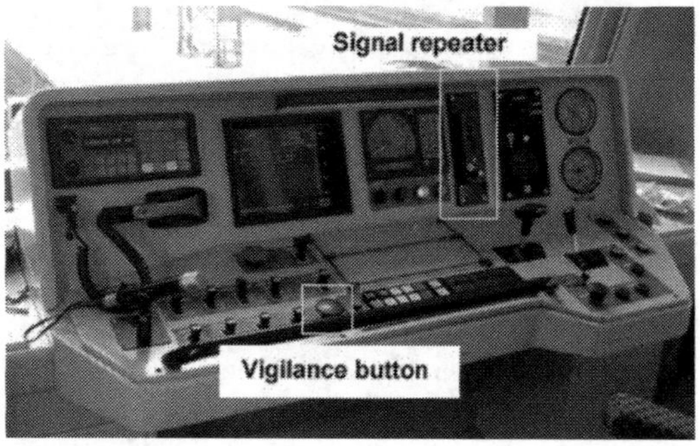

Fig. 1. Control panel with the LS system. [4]

Thus, train drivers are able to see the signal on the following signal device at the beginning of a particular section. The system also works with the so-called vigilance device (button), and while driving on railway tracks not having the LS, train drivers must press the button every 20 seconds. On the contrary, its pressing on the LS railway tracks is only required when the next signal device indicates a red "Stop" light or a yellow "Deceleration" sign (Fig. 2.).

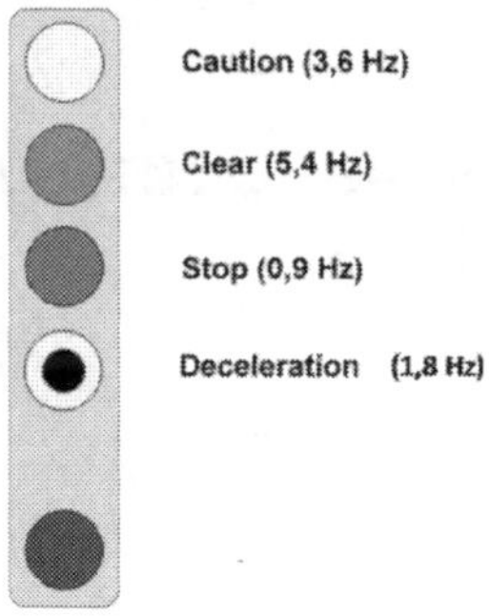

Fig. 2. Signal device – frequencies for individual signals.

The LS system consists of a trackside part and an on-board part. The former comprises independent track circuits with an encoder that transmits a code with information about the current signal via track circuits. On the other hand, the latter allows to transfer information from a signal on a given railway track to a driving vehicle and to check its driver´s vigilance, and is made up of scanning coils, a signal repeater and a vigilance device (button). The coils are located at the front of the vehicle and are used to capture the code transmitted by the encoder via track circuits. The signal repeater then visually communicates information about the upcoming signal to the driver [5].

The signal transmitting encoder is usually located at a respective signal device from which it receives data and current signals. The data are then converted by the encoder into a code of a certain frequency corresponding to a particular signal. The code is subsequently transmitted by track circuits, and upon a railway vehicle entering a given railway track section, the circuit is connected via conductive axles by which there are attached coils able to read the transmitted code [6]. Afterwards, this code is processed and the information about the current signal is transferred to the signal repeater. The LS system functioning is schematically shown in Fig. 3.

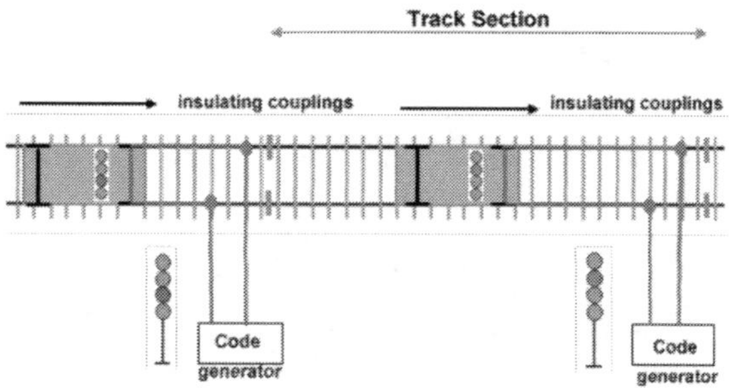

Fig. 3. Principle of the LS system – schematic illustration.

3.2 ETCS

The European Train Control System (ETCS) is part of the ERTMS project. This project was established to

https://doi.org/10.1051/matecconf/201823500005

develop a unified security system that would replace various security systems in individual European countries, thereby achieving interoperability on the European rail network and increasing its competitiveness. The ERTMS has two basic subsystems, i.e. the aforementioned ETCS and the international GSM-R (Global System for Mobile Communications-Railway) for wireless voice and data communication between railway vehicles and dispatch centers [7].

To put it simply, the main principle of fully-fledged ETCS functioning resides in trains knowing their exact position on railway tracks, which is continually reported to dispatch centers. They are then aware of positions of all trains within the network and issue the so-called driving authorization for each train, which is the permission for the trains to be driven to precisely defined destinations (complemented by possible restrictions on driving in particular sections). Based on the permission, individual trains control their speed or stop to avoid collision with other trains. Thus, the main functions of the ETCS include controls of driving authorizations, controls of perceiving and respecting the restrictions by drivers and full controls of drivers and their work. The ETCS consists of two parts - a trackside part and an on-board part.

The trackside part involves eurobalises transferring information between railway tracks and railway vehicles, LEUs (Lineside Electronic Units), RBC (Radio Block Center), euroloops and additional radio circuit. The on-board part encompasses EVC (European Vital Computer), JRU (Juridical Recording Unit), the DMI imaging unit (see Fig. 4), BTM (Balise Transmission Module) and a set of equipment for accurate detecting of the speed of vehicles, including their odometry.

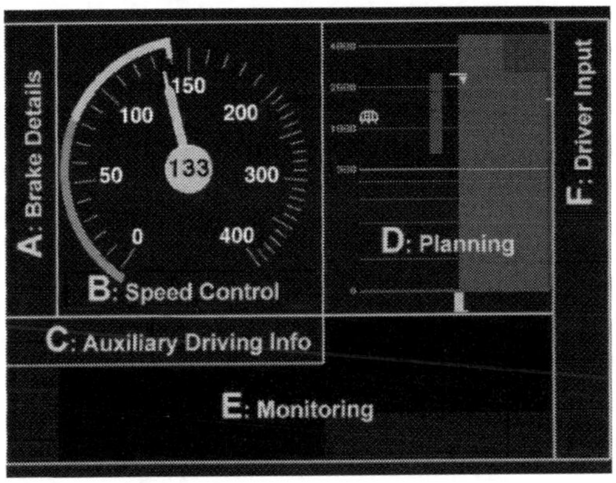

Fig. 4. DMI imaging unit. [8]

At present, there are three ETCS levels differing from each other by the extent of the ETCS involvement as train security equipment. Dividing the system into individual levels facilitates its gradual implementation and replacement of various security systems in other countries. At lower levels, the ETCS rather complements the already existing devices, yet it is gradually taking over their functions with its increasing level.

3.2.1 Level 1 (L1)

It is the lowest level of the ETCS involvement complementing the already used security devices. Controlling the availability of rail sections is still ensured by standard ways, such as the above mentioned track circuits or axle counters. However, railway tracks are also fitted with fixed and switchable balises. The fixed balises serve as reference points passing information about positions to railway vehicles as well as a source of information about invariable railway track characteristics. The switchable balises are connected through cables with the LEUs that are interconnected with security devices, thereby obtaining information about the railway track availability and the current signal of signal devices that rail vehicles are approaching [9].

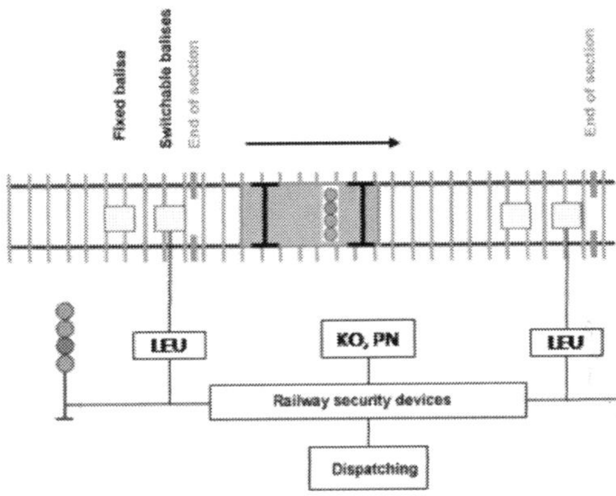

Fig. 5. Principle of the ETCS L1 – schematic illustration.

Subsequently, all the data are transferred via the switchable balises to railway vehicles, whereby their EVCs calculate the maximum speed and braking points. The signal data are transferred to railway vehicles on their passing over the balises and are not changed since then. Thus, in case there are any changes in the signal after the vehicles have passed over the balises, this will be detected by their drivers from the respective signal device. Solving this issue may reside in using the aforementioned euroloops (instead of the balises) between the "Caution" signal and the main signal device as these (unlike the balises) pass information to vehicles continuously [9].

Trackside functions of L1:

- authorization to drive (driving authorization) over switchable balises according to block sections based on data from security devices (Fig. 6),
- transfer of driving authorization and railway track description to railway vehicles.

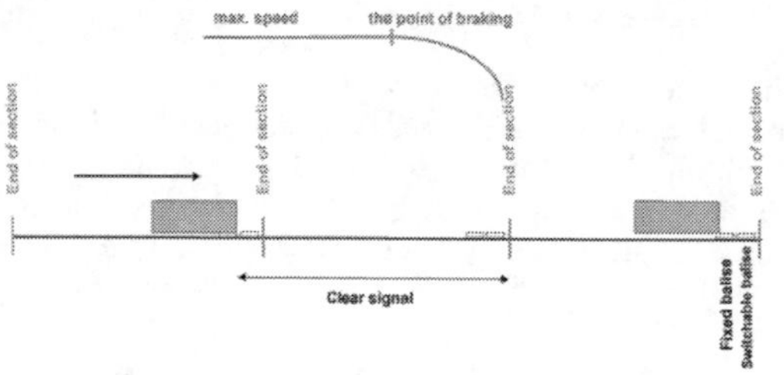

Fig. 6. Driving authorization in ETCS L1 – schematic illustration.

On-board functions of L1:

- receiving driving authorization and railway track description from balises,
- calculation of dynamic speed profile,
- comparing the current speed of railway vehicles with the permitted speed, possible application of brakes,
- on-board signaling for train drivers [10].

3.2.2 Level 2 (L2)

The L2 uses only the fixed eurobalises serving as reference points. Using the switchable balises is not required since all the necessary information (driving authorization, railway track description) is continuously transferred to railway vehicles via the GSM-R from RBC which continuously observes positions of railway vehicles and is in communication with a security device monitoring the availability of railway track sections. Based on these data, RBC (through the GSM-R) issues driving authorization to railway vehicles. Owing to the continuous communication between RBC and the railway vehicles, it is no longer necessary to use signal devices on railway tracks.

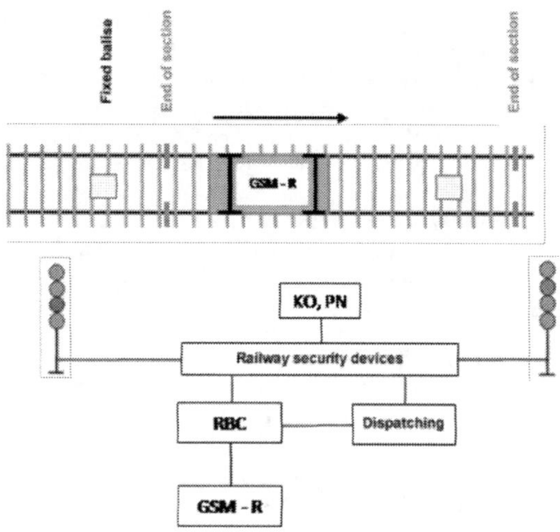

Fig. 7. Principle of ETCS L2 functioning – schematic illustration.

Trackside functions of L2:

- registration of all trains equipped with the ETCS in RBC,
- tracking of all trains equipped with the ETCS in RBC,
- authorization to drive issued by RBC according to block sections based on data from security devices (Fig. 8),
- continuous transfer of driving authorization and railway track description to railway vehicles via the GSM-R.

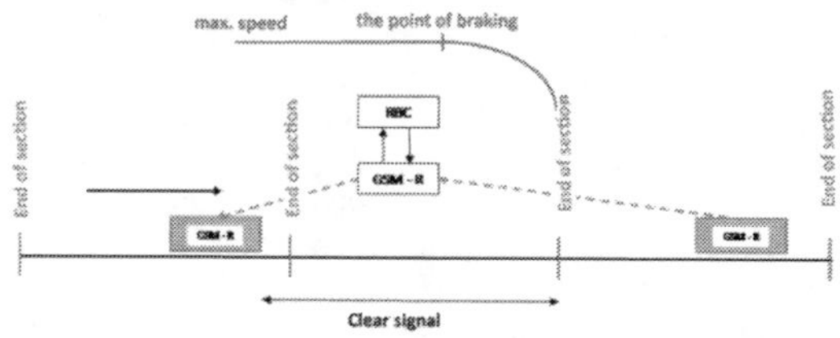

Fig. 8. Driving authorization in ETCS L2 – schematic illustration.

On-board functions of L2:

- transmitting positions to RBC via the GSM–R,
- calculation of dynamic speed profile,
- comparing the current speed of railway vehicles with the permitted speed, possible application of brakes,
- on-board signaling for train drivers [10].

3.2.3 Level 3 (L3)

It is the highest level of the ETCS where train security is provided exclusively by the ETCS as such without additional security devices and without fixed units that are replaced by virtual units. Railway tracks are equipped with fixed balises, again serving as reference points for the calibration of data about positions of railway vehicles. The RBC system continuously monitors the movement of all railway vehicles on railway tracks and issues driving authorization and other necessary information [11-13].

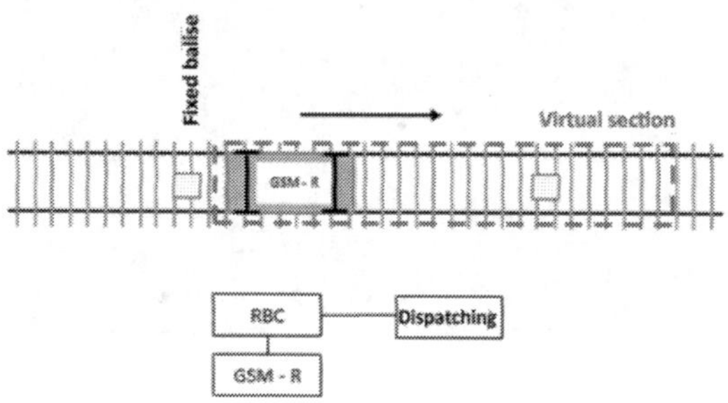

Fig. 9. Principle of ETCS L3 functioning – schematic illustration

Trackside functions of L3:

- registration of all trains equipped with the ETCS in RBC,
- tracking of all trains equipped with the ETCS in RBC,
- determining the end of a journey and its cancelling in accordance with information from railway vehicles,

- driving authorization by RBC up to the last reported position of the previous train based on tracking positions of all railway vehicles on railway tracks (Fig. 10),
- continuous transfer of driving authorization and railway track description to railway vehicles via the GSM-R.

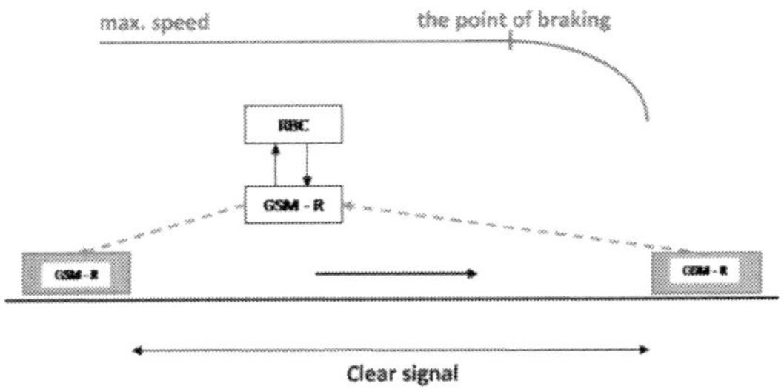

Fig. 10. Driving authorization in ETCS L3 – schematic illustration.

On-board functions of L3:
- transmitting positions to RBC via the GSM–R,
- trains monitor their integrity and transmit it to RBC,
- calculation of dynamic speed profile,
- comparing the current speed of railway vehicles with the permitted speed, possible application of brakes,
- on-board signaling for train drivers [14].

4 Conclusion

The aim of the paper was to acquaint readers with train tracking devices and describe selected telematics systems used on the Czech railway network. In the introductory part, particular ways of detecting positions of trains on railway tracks were briefly described, whilst the following chapters were related to train security devices – the LS system and the European Train Control System (ETCS). Both systems differ from each other by security levels and ways of functioning. The main function of the LS system is to transfer signals to cabins of railway vehicles and to check their drivers' vigilance by means of a special device (button). In order to transfer information about the status of signals to railway vehicles, the system uses track circuits. The main function of the ETCS is to issue driving authorization and to check the security of the trains being operated – based on the input data from railway tracks, the system is able to control the speed of trains or possibly even stop them. The system comes in three levels depending on the extent of its involvement in train security, yet only the Level 1 and the Level 2 systems are used in the Czech Republic [13]. The Level 3 system works completely independently without fixed units that are replaced by virtual units the ETCS itself creates. Due to technically more demanding solutions, however, this is mainly reflected in high acquisition and operating costs of this system.

The paper originated within the project "Research for applying practical skills on software specific for students of logistics disciplines" as a Specific university research on the Institute of Technology and Business in Ceske Budejovice.

References

1. J. Gašparík, B. Abramović, V. Zitrický. Tehnicki Vjesnik,. **25**, **4**. 1190-1195, (2018)
2. V. Chudáček a kol., *Detection of trains in railway security technology (in Czech)*, II. Edition (2015). Available on: http://zabzar.cz/sites/default/files/KO_n.pdf (2005)
3. L. Ferfecki, *ETCS Train Alarm System: bachelor thesis*. Pardubice: University of Pardubice. (2010)
4. Atlas of locomotives (in Czech). 471 (2018). Available on: http://www.atlaslokomotiv.net/loko-471.html
5. T. Dlask, *Design of a cooperative system for communication at railway crossings: Diploma thesis*. Prague: Czech Technical University in Prague. (2005)
6. V. Luptak, J. Gasparik, J., M. Chovancova, MATEC Web of Conferences. **134**. Article number 0003318th. International Scientific Conference, LOGI 2017. (2017)
7. P. Pinkas. Journal Automa **1** (2014). Available on: http://automa.cz/res/pdf/52955.pdf
8. J. Klimkova, J. Camaj, J. Gasparík. *Transport means 2016: proceedings of the 20th international scientific conference*. 891-896. (2016)
9. TORC2U [online]. *TRAIN ELECTRONICS*. Available on: http://torc2u.com.au/electronics/
10. T. Konopac, *ERTMS/ETCS: semester work*. Prague: Czech Technical University in Prague. (2010)
11. J. Sedivy, *Possibilities of Using Tracking Methods for Trains in the Czech Republic: semester work*. České Budějovice: Vysoká škola technická a ekonomická v Českých Budějovicích. (2017)
12. V. Luptak, O. Stopka, K. Jerabek, MATEC Web of Conferences, **134**, Article number 00034, 18th International Scientific Conference, LOGI 2017; (2017)
13. J. Masek, P. Kolarovszki, J. Camaj, Procedia Engineering, **134**. 231-236 (2016)
14. A. Dolinayova, L. Cerna, V. Zitricky, *The role of railway transport in east-west traffic flow conditions of the Slovak Republic*. 121-169 (2018)

Analysis of the effectiveness of noise abatement measures in relation to the type of train

Michal Petr Hranický[1*], *Vladimíra* Štefancová[1] and *Martin* Kendra[1]

[1] University of Zilina, Faculty of Operation and Economics of Transport and Communications, Department of Railway Transport, Univerzitna 8215/1, 010 26 Zilina, Slovakia

Abstract. The paper is focused on the comprehensive analysis of the railway noise issues. Not only does it deal with the influence on the environment, but also with the Czech laws in this field. Moreover, it presents all the options how the noise from the railway transport might be reduced and several proposals how to increase effectiveness of noise protection. In conclusion, the theoretical part is complemented by a practical one in the form of measuring and subsequent analysis.

1 Introduction

Sound is one of the fundamental and inherent parts of the environment and the world we live in. People get a remarkable piece of information thanks to it. In essence, this is a gradual longitudinal mechanical waving of the elastic environment that is perceived by the hearing. It is formed by the vibration of liquids, bodies or air, either on the basis of natural phenomena occurring in the environment or human activities. Human hearing is capable of capturing sound ranging from 20 to 20,000 Hz. [1]

For the purposes of this paper the noise is considered as any sound that is unwanted or has adverse or unpleasant effects on our organism. There are a number of professional definitions, but in general it is a sound that disturbs, harms, or damages (health, property, or the environment). It is important to note that it is not decided on the basis of physical parameters whether the sound is noise or not but purely by considering subjective feelings.

Rail transport is much more environmentally friendly than road transport, but noise can pose both a health risk for people living near the track and an obstacle to building new routes. Therefore, it is necessary to emphasize the correct location and construction of anti-noise measures, as well as their further development.[2]

2 Noise's impact on the environment

Epecially effects on humans and animalsare analysed. The whole chapter draws mainly from the information of the State Health Institute and the information provided by the doctors and assistants of the Occupational Disease Ambulance of the University Hospital in Ostrava-Poruba.

For simplicity, these negative effects of noise on humans can be divided into organ effects, activity disruption (insomnia), or effects on subjective human feelings (harassment). Also, noise can cause deterioration of existing illnesses that have multifactorial causes.

The organ effects can be further divided into specific and non-specific. By non-specific we mean an effect on multiple organs, influencing their activity on the basis of stress or nervous irritation. Specific include disorders of hearing and hearing organs.

Generally, we most commonly experience hearing damage due to noise (increased hearing threshold, poor speech comprehension, tinnitus, etc.), cardiovascular system (increased blood pressure, pulse and vasoconstriction), sleep disturbances or even noise harassment. [3]

3 Legislative framework of noise impact on human health

The legislative regulation of this issue in the Czech Republic contains a number of legal norms, some of which are based on EU law and international standards. Below we choose the two most important ones.

3.1 Decree No. 107/2013 Coll.

For the purpose of assessing the impact of physically or mentally demanding work on humans, the Government of the Czech Republic issued on 22 April 2013 Decree No. 107/2013 Coll. for categorizing positions where workers are exposed to increased dust, chemical toxic substances, noise, vibration, non-ionizing radiation, cold, heat, physical or mental stress, or work in unfavorable or elevated air pressure.

In the second category, workers are exposed to steady or variable noise with an equivalent sound

* Corresponding author: michal.petr.hranicky@fpedas.uniza.sk

https://doi.org/10.1051/matecconf/201823500006

pressure level in the range of 80 to 84.9 dB (A) and a peak sound pressure level in the range of 130 to 139.9 dB (A). For the third category, 85 to 105 dB (A) at steady or variable noise and 140 to 150 dB (A) for peak sound pressure apply. At higher values, these employees belong to the fourth category. High-frequency noise, ultrasound, infrasound and low-frequency noise are not considered for these purposes. [4]

3.2 Government Regulation No. 272/2011 Coll.

Pursuant to Directive 2002/44 / EC of the European Parliament and of the Council of 25 June 2002 on the minimum health and safety requirements regarding the exposure of workers to the risks arising from physical agents, on 24 August 2011 the Government of the Czech Republic issued Regulation No. 272/2011 Coll. on the protection of health from the adverse effects of noise and vibrations. This regulation lays down, inter alia, hygiene limits for noise and vibrations at workplaces, the way they are identified and evaluated, and the minimum scope of protection measures for the employee. [5]

Under Section 3, the permissible exposure limit for steady and variable noise at work is equivalent to an 85 dB (A) sound level or 50 dB (A) for workplaces where work is demanding for concentration, attention or creative work. For other workplaces where production or storage activities are carried out, and the noise is not produced by the works itself, but a 70 dB (A) limit is applied to the ventilation or heating equipment. Section 4 deals with impulse noise limits. These are also determined by this Regulation at 85 dB (A), at a peak sound pressure level of 140 dB (A).

Section 9 contains recommendations and regulations for employers. Employees who carry out noise-related work above the permissible values must be adequately trained, especially for the use of protective equipment at work (hearing protectors). These workers are also entitled to a safety break (beyond the exposure to noise exceeding the permissible exposure limit) for at least 15 minutes after two hours of work, and then at least 10 additional minutes after every two hours. The last break must be placed no later than one hour before the end of the shift. The employer must ensure regular and proper maintenance of the means of production, equipment and work tools so that their wear rate is not the cause of noise increase.

If the measurement evaluation reveals that noise during the eight-hour shift exceeds 80 dB (A), the employer shall provide hearing protection devices effective in the area of the frequencies of the noise. If the 85 dB (A) limit is exceeded, the employer must ensure that employees use the aids provided. [6]

4 Possibilities of removing noise in railway transport

The most important source of noise in the Czech Republic is the transport infrastructure. Rail transport is generally ranked among modes of transport that are environmentally friendly. Compared with other transports, this is indeed the case, especially in the area of CO2 emissions and other harmful substances discharged into the air by combustion engines. The component that causes the greatest problems for railways in relation to the environment is noise. In order to address noise abatement options in rail transport, it is first necessary to analyse its sources. [7]

Noise emissions from rail transport cannot be generalized as they come from many sources whose intensity varies depending on the speed and design of the train as well as on the surface of the railway superstructure. When comparing more studies, we can say that at speeds of up to approx. 40-60 km / h the most significant noise is the one of the traction engine. From approx. 40-60 to approx. 160-200 km / h the rolling noise prevails and at higher speeds the dominant aerodynamic noise (see graph in figure 1). The noise of the collector and the local traffic noise are among the less significant but not negligible.

An important role in the spread of noise from rail transport is also played by the current weather situation. This is especially true in distances over 100 meters, but sometimes even in smaller ones. Snow may also absorb some of the noise or there might be a reflection of sound from different layers of air. Higher air humidity usually also leads to better propagation, so that noise can go even further than dry air.

With the current state of the Czech railway infrastructure, we can easily find that the biggest problem in the Czech Republic (and indeed in Slovakia) is just the rolling noise. That is why most anti-noise measures focus on this source. We can basically divide noise-reducing measures into active ones that restrict the sound of noise (noise emissions) and passive ones, which reduce noise and its impact on the environment (noise pollution).

We can then divide the active noise measures into four groups: urban-architectural, urban-transport, traffic-organizational and technical.

Urbanistic and architectural anti-noise measures relate to construction itself. The process of spatial planning and the comprehensive solution of residential units in terms of their functional layout is quite essential. In particular, it is good to locate buildings depending on the location of other sources of noise. These include concentrating these units into blocks that are perpendicular to the noise source, or building terraced objects near sources. As far as the reflection is concerned, it has been found that the reflection takes place at the highest intensity at a height of about ten meters above ground, so we should place high-rise buildings and communications away from noisy roads.

Urban-traffic noise pollution measures are also closely related to spatial planning in conjunction with transport. Long-term transport infrastructure planning needs to be adequately organized at all levels (national, regional and local). As a concept, these measures are covered by a set of many urban and transport road and rail transport tools that can be combined, including the management of new roadways away from protected buildings, tracing motorways of both classes and first class roads outside residential areas and other areas with higher demands for noise protection (nature reserve, etc.), optimizing transport rights and improving transport relations, minimization of transit traffic in city centers and residential areas, minimizing heavy freight traffic in the vicinity of residential areas, creating conditions for preference of mass transport and limiting individual transport within cities and agglomerations, not allowing planning of new acoustically sensitive construction near to the busy roads, the location of parking areas and other transport areas at a sufficient distance from protected buildings, the limited time of entry of supply vehicles in city centers and other sensitive locations and the concentration of different modes of transport main routes and corridors with the possibility of creating uniform noise measures (a successful example is road and railway track jointly led by several kilometers near Hrádek stop, see Figure 1). [2]

Fig. 1. The common noise barriers of the road and railway in the Jablunkov region

Traffic-based anti-noise measures are such as to limit or modify traffic in order to minimize either noise alone or its negative impact. In railway practice, this includes in particular a reduction in line speed; eliminating where you need to reduce the train speed; more efficient traffic management to reduce stopping for transport; relocation of noisy transport processes (for example, changing the place for shunting in a hilly hill or HVV exchange); or even the reorganization of the circulation of the railway vehicles (so that the lesser noisy ride on the affected section).

In particular, technical noise suppression measures have the advantage that it is often possible to establish them retrospectively, even after many years of construction, based on noise reduction requirements. The essence of the technical basis of rail transport means that these measures can be carried out either on railroad tracks or other stationary tracks or directly on the vehicles themselves.

Regular maintenance, grinding and milling of rails, complete overhaul or application of rail absorbers are essential and necessary measures for the track. The disadvantage of these measures, however, is the fact that they are only local in the area of vehicle modifications.

For the rolling stock, a complete upgrading or modification of the chassis and wheel silencers is considered. Another option is to replace the classical cast iron brake jaws with more modern composite materials, reducing not only the braking sound itself, but also the fact that the cast iron blocks rough the surface of the wheels and thus increase the rolling noise (up to 50%). The advantage of vehicle measures is that they reduce noise across the rail network, but for sufficient efficiency they need to be applied to a considerable number of vehicles, which entails considerable investment costs.

Other measures in the runway environment can be broad stripes of greenery, which can be combined with earth valleys. Their advantage is that they capture not only noise, but also dust. They also often have a higher aesthetic level, thus enriching the surroundings of the noise source. The most significant disadvantage is that in the winter when deciduous trees fall, the effectiveness of this measure will be significantly reduced. [8]

5 Analysis of the effectiveness of noise reduction measures

For the first measurement, we chose the locality near the Ostrava-Svinov railway station in the direction of Hranice na Moravě (261st km of the railway line SŽDC number 270). The area on the main track was deliberately chosen on the one hand because of the high frequency of trains and because of the fact that many different types of trains are being driven here. The concrete sites were also meeting the requirements for the presence of the absorption noise barrier, even without it, but the advantage was that recently, in the immediate vicinity of the noise barrier, a new road road linking the D1 motorway exit and the Ostrava-Svinov station was completed and built on a raised embankment, which made it possible to partially shade ambient noise not coming from rail transport (see Figure 2).

Fig. 2. The place of the first measurement

For refinements, we have repeated measurements in other locations. Specifically, at the head of the station Suchdol nad Odrou towards Ostrava (233rd km of the railway line SŽDC number 270). We chose this place on the one hand because of the favorable conditions for noise measurements outside and beyond the wall, but also because of the wider range of vehicle types (for example motor coaches in the local lines which we could not measure in Svinov).[9] Another advantage was that some types of trains (express train and passenger trains) stop at this station, but others (EuroCity, SuperCity) drive through it; this helped us in the subsequent analysis of starting, braking and passing noise. Further measurements took place in Žilina, the local part of Brodno (4th km of ŽSR line 127). The air temperature during measurements was approximately 4 ° C and the air humidity was approx. 50%. In some cases, the noise of the train was exceeded significantly by the noise of the sound signals. However, elimination of these signals can lead to several risks to the carrier as well as the manager of infrastructure and therefore these results are not considered relevant. [7] [8]

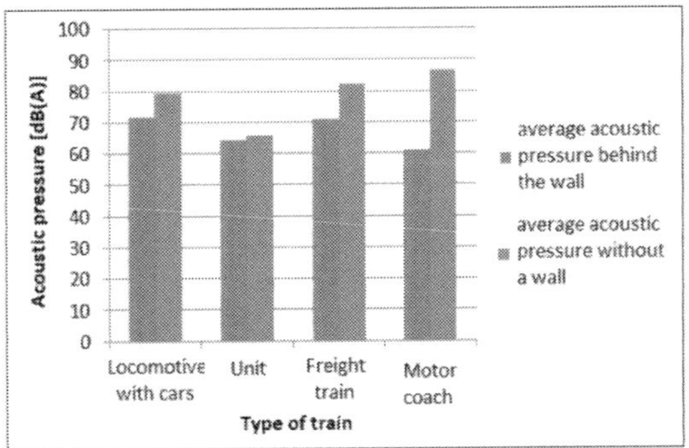

Fig. 3. Comparison of the noise of individual train types in the second measurement

Figure 3 presents the aggregate measurement results and the asbestos differences in acoustic pressures. Percentage of noise reduction is shown in Figure 4.

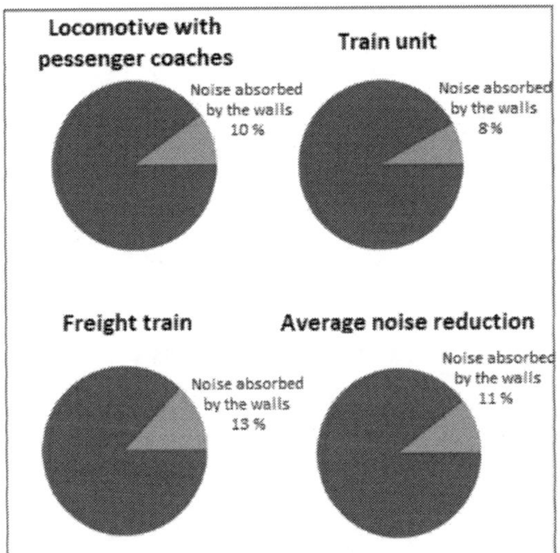

Fig. 4. Total absorbed noise absorption depending on the type of trains

The most effective was the noise barrier for freight trains, the least for units. By considering the fact that freight trains are the loudest and units seem to be the quietest, we can easily infer the logical conclusion that the volume of sound pressure that the wall is able to absorb is not absolute, but directly proportional to the magnitude of the sound pressure coming from the source. On average, however, these walls reduce the noise by more than one tenth, which may seem like a relatively small number, but when compared to the noise thresholds causing permanent health effects, we can judge that these walls can be crucial for human health. In addition, the fact that the decibel scale is logarithmic and therefore a small drop or increase in the measured sound pressure may result in a jump difference in human perception.

From the measurement results, we have also concluded that even though the noise value decreases relatively slowly over longer distances, it decreases considerably at a smaller distance. Therefore, the dependence of the distance on the resulting measured value is not directly proportional, and thus even a small distance change of the exposed subject or building can be key to negative effects of noise.

This research was financially supported by the project VEGA No. 1/0019/17 "Evaluation of regional rail transport in the context of the region's economic potential focusing on the efficient use of public resources and social costs of transport".

References

1. M. Bernard, P. Doucha, *Legal protection against noise*, 199, (2008)
2. J. Havránek et al., *Noise and health*, (1990)
3. Decree No. 395/1992 Coll., Which implements certain provisions of Act No. 114/1992 Coll., On Nature and Landscape Protection, as amended. Prague: Ministry of the Environment of the Czech Republic
4. Regulation of the Government of the Czech Republic No. 272/2011 Coll. on the protection of health from the adverse effects of noise and vibrations. Praha: Office of the Government of the Czech Republic
5. D. Thompson et al., *Assessment of measurement-based methods for separating wheel and track contributions to railway rolling noise*, vol. **140**, 48-62, (2018)
6. R. Kampf, O. Stopka, I. Kubasakova, V. Zitricky. *World Multidisciplinary Civil Engineering-Architecture-Urban Planning Symposium*. (2016) DOI: 10.1016/j.proeng.2016.08.623.
7. E. Nedeliaková, D. Harmanová, M. Panák, Š. Kudláč, *Selected Problems of Risk Management in the Company of Railway Infrastructure Manager*, 877-880, (2016)
8. E. Nedeliaková, M. Hudáková, V. Štefancová, *Application of risk management methods in transport companies*, 237-241, (2017)
9. A. Galierikova, J. Sosedova, NASE MORE. **65**, **3**. 8 – 11. 2018

 https://doi.org/10.1051/matecconf/201823500007

Determining lateral resistance of sleeper in railway ballast

Zdeněk Hřebíček[1], Vladimír Ľupták[1], * Mária Stopková[1]

[1]Institute of Technology and Business in České Budějovice, Okružní 517/10, 370 01 České Budějovice, Czech Republic

Abstract. Within a project of Technology Agency of the Czech Republic No. TA 04030889 "Operational diagnostic of qualitative indicators of railway tracks by means of gauging its spatial deformation." Empirical cross-resistance tests of sleepers in railway ballast were carried out in geotechnical laboratory test field (GLTF) of the Transport research center in Brno (TRC). The tests were done in collaboration with Czech Technical University in Prague, Department of Railway Structures (ČVUT).

1 Description of the testing equipment

Geotechnical laboratory test field in the picture 1 is a unique technical device registered by Office for Patents and Inventions as a Utility Model on which tests for laboratory structural body layers of rail bottoms and for rail superstructures regarding the current simulation effects of a different operating load and climatic conditions can be done. The laboratory equipment allows performing destructive and non-destructive tests of building materials and constructions. The workplace is also equipped with modern information technology supplied with appropriate software. The structural organization of GLTF is depicted in Figure 2. [1]

Fig. 1 Geotechnical laboratory testing field

For laboratory testing purposes within the problematics, a railway track model in the scale 1:1 was implemented. Modeling procedure: Establishment of the ground plane, determination of the bearing capacity of the railway underbody body by static loading test, construction of the track bed plan below the sleeping area of the sleeping area and laying of the rail grid. [2, 3]

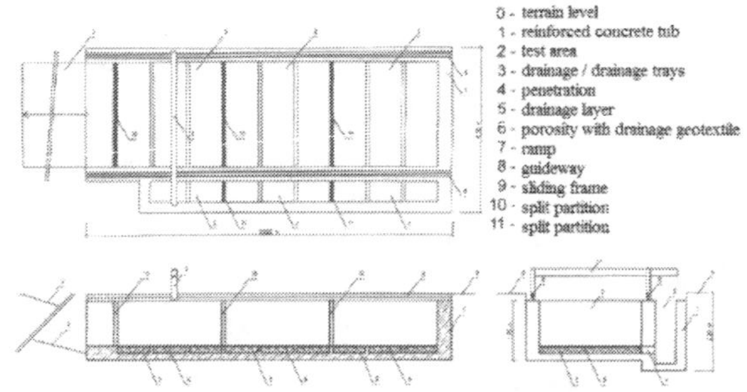

Fig. 2 Structural scheme of geotechnical laboratory test field

Fig. 3 Geotechnical laboratory testing field

2 Calculating the thickness of structural body layers of the rail bottom of the track model on GLTF - input data

Track bed ballast is made of sandy clay, layer thickness 300 mm, capacity of the track bed ballast is calculated to

* Corresponding author: luptak@mail.vstecb.cz

E0= 20 MPa. Track bed of the body of the rail bottom, required load capacity Epl= 40 MPa (for main tracks of the current national and other railways; see Tab. 1 and 4, attachment 6, regulation S4). The estimated thickness of the structural layer is 400 mm. Ballast compositions - type 2, construction material: gravel, fraction 0/32, material consistency of the track bed ballast - solid (sandy clay). Suggested thickness of the structural layer - 400 mm. Reduction factor z = 0.8. Reduced module of deformation of track bed ballast E0r= 0.8. 20 = 16 [MPa]. Module of deformation of the structural layer E1= 60 MPa. [4]

Calculation:

$$k_1 = E_{0r}/E_1 = 16/60 = 0.27$$

$$k_2 = h_1/D = 0.40/0.30 = 1.33$$

Where they mark:

E_{0r} - reduced module of deformation of track bed ballast [MPa],
E_1 - module of deformation of the structural layer [MPa],
h_1 - thickness of the structural layer [m],
D - diameter of the load plate [m].
K_3 - is determined from the diagram in Pict. 8, attachment 6, regulation S4
K_3 - 0.71

The equivalent module of deformation of a two-layer construction of the rail bottom body

$$Eel = k3.\ El = 0.71.\ 60 = 42.60 > 40\ [MPa]$$

Qualitative parameters of the material of the ballast body and structural layers.

Track bed – Ballast from the vicinity of Křivá Borovice, Olomoučany
Locality:
Geomorphological classification:

- System - Hercynský
- Subsystem - Hercynská pohoří
- Province - Česká Vysočina
- Subprovince- Česko-moravská
- Region - Brněnská vrchovina

Geological region: Moravia-Silesia region - borderline of Moravian-Silesia Paleozoic (Moravský Kras) and brunovistulic (Brněnský masív). [5,6]]

Table 1. Further classification

Further classification	
Sandy clay	F4 CS
gravel fraction ratio	g = 8.1 %
sand fraction ratio	s = 46.6 %
ratio s/g	f = 45,3 %
Natural soil humidity	w = 20%
CBR (when w = 14.7%)	CBR = 17%
CBR (when w = 20.2 %)	CBR = 5 %
W_{opt}	18.0 %
Max. density Proctor Standard $\rho_{d,max\ PS}$	1720 kg/m^3
Liquid limit	$w_L = 37.2$
Plasticity limit	$w_P = 15.2$
Plasticity number	$I_P = 22.0$

Structural layer
Locality: Dolní Kounice Quarry – gravel ŠD 0-32 mm

Table 2. Structural layer

Structural layer	
Coarse-grained gravel	CGr (G3 G-F)
gravel fraction ratio	g = 70% - 67%
sand fraction ratio	s = 27%
ratio s/g	f =3% - 6%
heterogeneity number	Cu = 27
curvature number	Cc= 1.65
Relative compaction (min)	$\rho_{d,min}$ = 1669 kg/m^3
Relative compaction (max)	$\rho_{d,max}$ = 2208 kg/m^3

Under the classified sub-grade (F4 - CS) 350 mm of thickness, there is a 600-mm layer of loess soil (LS) of similar properties. [7]

3 Ballast cushion

The new, natural, crushed aggregate of density min. 2 000 kg.m-3, fraction 31.5/63, class BI. Requirements for the aggregate classification are laid down by GTR (General Technical Requirements). The thickness of the ballast cushion from the sleeper is 350 mm. [8]

Load capacity of the ballast cushion

The data for the model construction of the railway were acquired from values of the empirical research at Department of Railway Structures ČVUT in Prague under the following conditions:

- National railway line in operation
- the ballast cushion of the age 2.5 - 3.5 years from the ballast laying; alternatively an older ballast cushion visually regarded as partially polluted (unless defined otherwise)
- The data do not include tests carried out on a heavily polluted ballast cushion and entirely new ballast cushion.

Load capacity on the level of the surface of the ballast cushion (sleeper bed area) - impact test:

Minimum value: 27.8 MPa
Maximum value: 72.0 MPa
Average value: 51.3 MPa
Standard deviation: 9.7 MPa

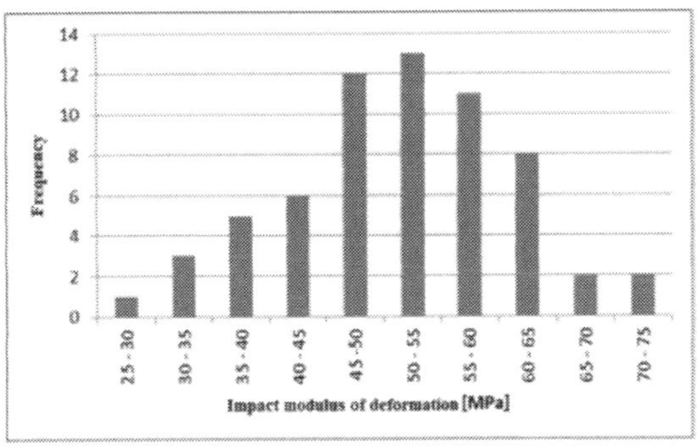

Fig. 4 Frequency chart

Load capacity on the level of the sleeper bed area – impact test:
Minimum value: 40.4 MPa
Maximum value: 75.8 MPa
Average value: 59.0 MPa
Standard deviation: 10.1 MPa. [9]

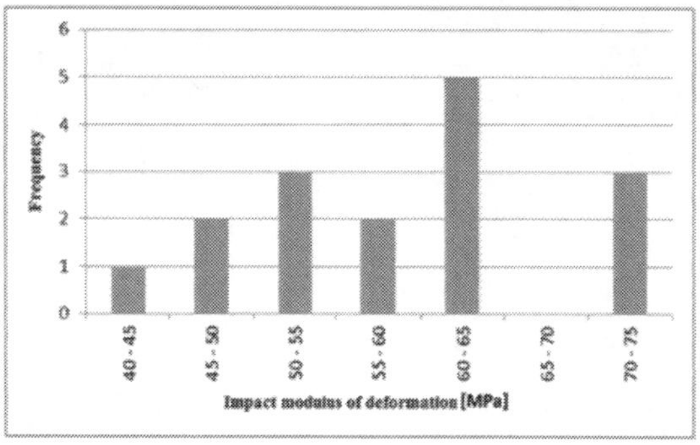

Fig. 5 Frequency chart

The average value of the static module of deformation was, in this case, 71.8 MPa including tests carried out on rails just after a reconstruction. [3]

4 Rail grid

Concrete sleepers SB 8 P (see Pict. 6); consolidation of type K rails; rails S 49, used.

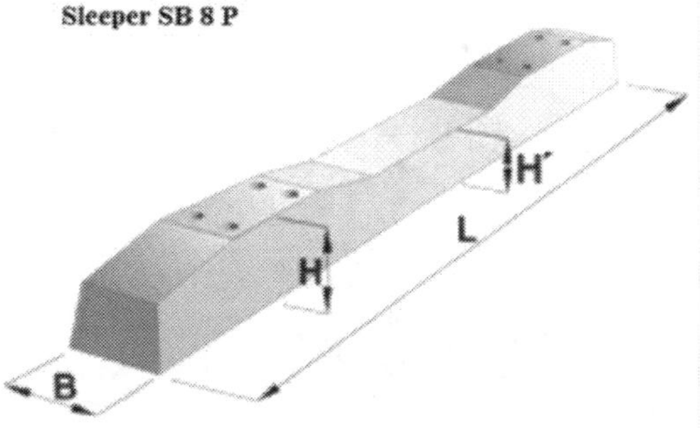

Fig. 6 Concrete sleeper SB 8 P

Table 3. Properties of sleeper

trademark	concrete class	cubic capacity (m^3)	weight
APP 34-21	C 45/55-XF1	0,1040	270
Size (cm)			
L	B	H	H´
242	28,4	21	15,5

Simulation of the operating load
Simulation equipment used is pulsator, axle weight 22.5 t = 225 kN affects the middle of the sleeper through cross-section I profile and spreads through two longitudinal rail profiles to two adjacent sleepers.

Essential requirements:
- Simulated operating load 120 000 t,
- Train speed 80 km h-1,
- Simulated drive of a two-axle freight wagon lbbhps 25,
- Axle distance 8 000 mm.

Simulated frequency:

$$f = 8x\ 3\ 600/\ 80\ 000 = 0.36\ Hz = 0.36\ s\text{-}1.$$

In case of a pulsator being able to exert a force of 22.5 t, the simulation lasts 120 000 t:

$$t = 0.36\ x\ 120\ 000/\ 22.5 = 1\ 920\ s = 32\ min.$$

Gauging of the rail field
Within the paper, for the purpose of gauging, a rail field with three sleepers SB 8 P, consolidation K, rails S49, axial distance of the sleepers 600 mm, ballast cushion of granularity 31.5/63 and thickness 35 cm under the sleeper bed area was installed. Sleepers were marked by numbers from 1 to 3 (see Pict. 7). When constructing a model structure, a static impact test was carried out with the following result: deformation module on the level of the track bed of the rail bottom 90 MPa; on the level of the sleeper bed area 93 MPa and on the level of the ballast cushion 82 MPa. [10]

Fig. 7 The test field

Sleeper lateral resistance in the ballast cushion is defined by the value of the force acting in the longitudinal sleeper axis when shifted by 2 mm. During the experiment, gauging of the lateral resistance before and after the cycling of the railway construction model; it lasted 2 hours, frequency 1.5 Hz, amplitude 5 - 140 Kn (12 000 cycles). The lateral resistance of sleeper No. 1 was gauged before the cycling while that of sleeper No. 2 was gauged after the cycling without an extra load. When gauging the lateral resistance, the chairs of the examined sleeper were always loosened. The railway construction model was subsequently loaded by a force of Q = 150 kN above the sleeper No. 2 (see Pict. 7) while the lateral resistance of sleeper No. 3 was being gauged. [4-5]

Furthermore, the shift of sleeper No. 1 of loading was Q= 0 kN and that of sleeper No. 3 of loading Q= 40.75 was 150 kN; the load was above sleeper No. 3 and the consolidation on other sleepers was removed. Location of sensors for monitoring sleeper shifts is depicted in Pict. 8. At the same time, a significant decrease of the model in the very place of loading was gauged. The results of the loading are suggested in Pict. 9.

Fig. 8 Location of sensors, when monitoring sleeper shifts

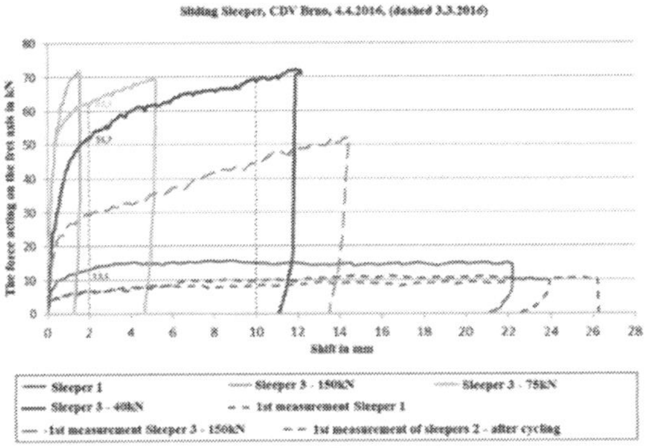

Fig. 9 Graphic view of the results

The research study of gauging the sleeper lateral resistance using variant system of loading shows that its value significantly increases when loading the sleeper; the cyclic loading had only a minimal effect. The experimental results serve as a foundation to calibration of QVW (Querverschiebewiderstand) relationship - the output of the stabilizer DGS90 for assessing the sleeper lateral resistance in the ballast cushion. [6]

Conclusion

The experimental results were applied in other phases of the project, whose outcome was a certified methodological approach for introducing a new method of quality assessment of carriageways by means of exploring spatial deformations of railway line tracks in regard to specific requirements for track speeds. [8]

The essential requirement is the effective use of track mechanization for dynamic stabilization of ballast cushion DGS - dynamic stabilizer. When switching off the stabilization, so called reference height, which demonstrates continual solidity of the rail, is possible to be registered. The calculation of its standard deviation defines qualitative limits of the rail structure. The next goal is to use the registration equipment for determining the lateral resistance of the rail during its stabilization, to connect the results with the value of the sleeper lateral resistance and thus continually assess the structure condition and its safety as required for continuous welded rails. Using dynamic stabilizer for the rail diagnosis, its weak points in the structure are indicated so that a more ingenious economic plan for repair works of railways can be devised. [9]

References

1. UIC Project "USP in track", WP 2: Lateral Track Stability, Results from the SBB test track Kiesen/CH, (Measurements done on 03.- 05. 10. 2010).
2. Roel van den Bosch: Querverschiebewiderstandsmessung mit dem dynamischen Gleisstabilisator, EI – Eisenbahningenieur (58) 6/2007.
3. Zpráva TA 04030889: Provozní diagnostika kvalitativních ukazatelů koleje železničních tratí měřením její prostorové deformace, 2016.
4. Fendrich Lothar: Handbuch Eisenbahninfrastruktur, ISBN-13 978-3-540-29581-5.
5. A. Plakhin, R. Kampf, E. Ogorodnikova, A. Kokovikhin. *18th International Scientific Conference-Logi 2017* (2017). DOI: 10.1051/matecconf/201713400047
6. B. Abramović, J. Gašparík, V. Zitrický. Tehnički Vjesnik. **25**, **4**. 1190-1195. (2018),
7. R. Kampf, L. Lizbetinova, K. Tislerova, Open Engineering, 7, 1 (2017)
8. A. Torok. Transport **24**, 26–29, (2009) DOI: 10.3846/1648-4142.2009.24.26-29
9. J. Gašparík; B. Abramović, M. Halás. Promet Traffic & Transportation. **27**, **4**. 283-290. (2015)
10. V. Zitrický, L. Černá, B.Abramović. Procedia Engineering. **192**. 994-999. (2017)
11. J. Gašparík, V. Zitrický. Transport. **25**,**4**. 387-393. (2010)

Comparing Trains Operated in Western Europe from Passenger Viewpoint

Jan Lizbetin[1*]

[1] Department of Transport and Logistics, Faculty of Technology, Institute of Technology and Business in České Budějovice, České Budějovice , 370 01, Czech Republic

Abstract. This paper provides a comparative analysis of trains used on high-speed railways in Western Europe. The analysis is based on a passenger satisfaction survey of transport service quality on said trains. The survey was carried out en route by interviews with passengers. The second part of the survey consisted in questionnaire where passengers expressed levels of satisfaction with different quality factors in rail transport and in their own opinion, they assessed the importance of the specific quality factors having an influence on user satisfaction with transport services. The survey was carried out in Germany, France, Belgium, Netherlands and Switzerland.

Introduction

The quality of transport service depends on a set of phenomena referred to as quality indicators. These indicators include regularity, reliability, safety, speed, reasonable transportation cost, environmental friendliness, transport comfort, or efficiency and availability. The current state of quality means the specific conditions where the individual aspects may be different and leading in a desired or undesired direction [1].

1 Materials and Methods

The survey of passenger satisfaction with transport service quality was carried out directly on selected routes. It was based on personal interview and questionnaire. [2] The survey was focused on high-speed railways in Germany, France, Belgium, Netherlands and Switzerland. Respondents were asked to fill in a questionnaire based on their experience with rail transport and based on their satisfaction with the actual (ongoing) transport service [3,4].

The questionnaire outputs are divided into a part with questions for passengers such as what is important for them and what is not and when are they satisfied with the provided quality of service and when they expect a higher level of quality [5]. The level of importance, or customer satisfaction scores were expressed on a 1-5 scale where 1 was the least important, or not satisfied at all, and 5 was the most important, or fully satisfied.

In conclusion of the analysis of a specific railway company, pros and cons of the service provided by specific carriers are illustrated [6-8].

2 Results

2.1 Germany

Modern train sets are used across the German network. They are very fast and reliable. As one of just a few, the railway network is not centralized in big cities, but rather connecting the North with the South and the East with the West. Rail line speed reaches the required maximum. The route from Hannover to Nuremberg is made by passengers on a ICE 2 train in three and half hours. This relation that is less than 500 km on road would be covered by a passenger car in a little less than 5 hours. The ICE train only stops at four stations on the route, namely Würzburg, Fulda, Kassel and Göttingen. In terms of commercial speed, this railway is a strong competitor to individual passenger transport [9-11].

Intervals between trains in one direction are 30 – 60 minutes which is a sufficient frequency. Trains keep on the schedule. They leave at a fixed time and arrive on time at the destination.

DB dispatch trains with 6 to 12 cars. The train capacity varies from 360 to 690 seats. However, seat reservations are not is necessary for high-speed ICE trains so the train may be overcrowded and the passenger will not find a seat. It is a nuisance at rush hours which the German carrier is facing on busy routes [12].

The questionnaire results clearly show that high-speed rail transport is more expensive for passengers. Specifically, one-way and round-trip ticket prices without any tariff discount for the above relation Nuremberg - Hannover will be app. EUR 100 and EUR 180 respectively.

During travel, passengers are regularly informed about arrival times at next station directly by the train driver. There is no train personnel on the train. Thus it is

* Corresponding author: lizbetin@mail.vstecb.cz

not possible to contact them en route in case of a question or complaint. Information boards are well-arranged and the DB application offers whatever the customer may expect, including delay announcements and the number of wagons, or the route. [13]

All parts of the train are modern and clean. Toilets are spacy and clean. The interiors are well lit using red and grey colours. Doors are controlled by electronic pushbuttons or by direct contact where the passenger comes up to the door and there is a pressure sensor built in the floor in place of a doormat that will open the door automatically.

ICE trains are designed for high-speed lines and protected by the necessary braking systems. Passengers' safety perception is very positive.

When the train is moving, noise and vibration are more distinct than those produced by other high-speed trains. ICE trains are higher and narrower than e.g. TGV trains, thus having a higher aerodynamic resistance. The front cars ICE 1 and 2 do not have a distinctive nose shape which increases noise and vibrations inside the car. Pressure oscillations are felt strongly by passengers when entering or exiting tunnels [14, 15].

Table 1. Summary of pros and cons resulting from the survey in Germany

Pros	Cons
Modern fleet	High occupancy of trains
Departure and arrival accuracy	High price
Seat reservations not required	Lower seat comfort
Dense transport network	Lack of information
High frequency of services	Noise and vibrations when the train is moving

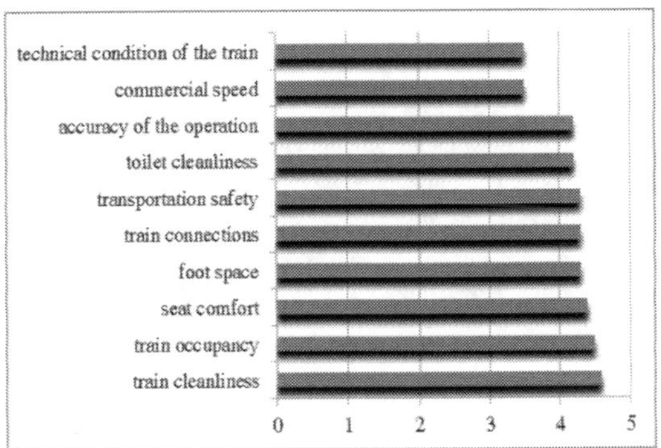

Fig. 1. Important and unimportant aspects for German passengers

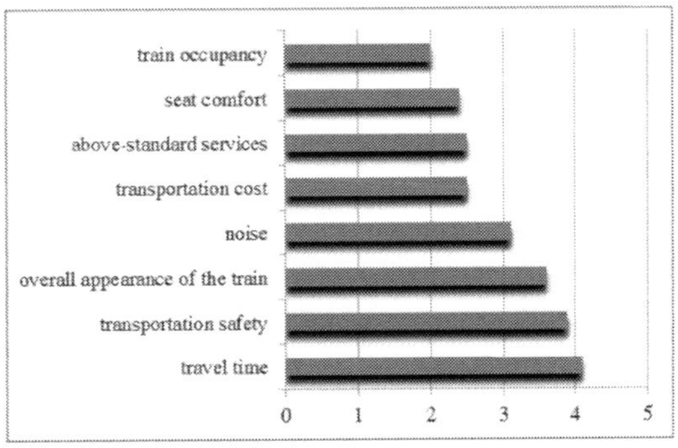

Fig. 2. German passengers' perception of service quality

2.2 France

As a result of co-operation between Belgium and France, the passengers can travel on a high-speed train of the French TGV already from Brussels where they board the train and in less than an hour and half, they reach the downtown of Paris, specifically the main train station Gare du Paris from where they can easily continue to travel as far as Barcelona [16].

The train capacity is always fully used. Seat reservations required from the passenger will not influence the selection of a specific seat. Seats are very wide, but with less space between the rows. The foot space is a little narrow. The on-board temperature is centrally controlled. No individual control by passengers is possible. Trains are equipped with the air-conditioning system. Trains are very long and consisting of two block train sets connected into one.

On the train route from Paris to Barcelona, several respondents noted they were not satisfied with smooth travel. Currently, the route to Barcelona is under construction in the territory of Spain and the running speed was much lower than 300 km/hour in this section. The rest of the route is all in operation with the maximum running speed of 300 km/hour and TGV trains will actually run at this speed for the whole journey [17, 18].

The passengers are used to quickly getting on and off the train. The time spent by the train at a station is extremely short and varies between 3 – 5 minutes. It allows trains to be on time and keep on the schedule. Trains run hourly. The passengers are very satisfied with the density and reliability of rail service.

TGV trains always require seat reservations. The customer demand for transportation is huge because trains are fast. The train capacity is always fully used and the occupancy is 100%. Nevertheless, the passengers were not satisfied with occupancy although every passenger has a booked seat. The current line capacity should be increased and extended because the rail transport demand is really high.

The respondents were highly satisfied with the accuracy of operation and information provided by the personnel. You can easily find your way around train stations and all information you need, can be found on information boards. The French personnel is very helpful and makes all efforts to arrange everything.

Transportation safety and commercial speed are considered by the passengers as a very important factor. The appearance or above-standard services are less important.

TGV trains are designed and constructed to run at high speeds with the minimum air resistance. Trains are lower set which lowers its centre of gravity and quite wider than e.g. ICE trains. Noise or vibrations are not transferred inside the car. TGV trains only reduce the speed to 230 km/hour when entering tunnels due to pressure decrease in the cabin. Pressure is markedly oscillating even in pressure-proof train cars.

Table 2. Summary of pros and cons resulting from the survey in France

Pros	Cons
High commercial speed	Incomplete part of the infrastructure impeding the traffic
Transportation safety	Train occupancy
Operation accuracy	Transportation cost

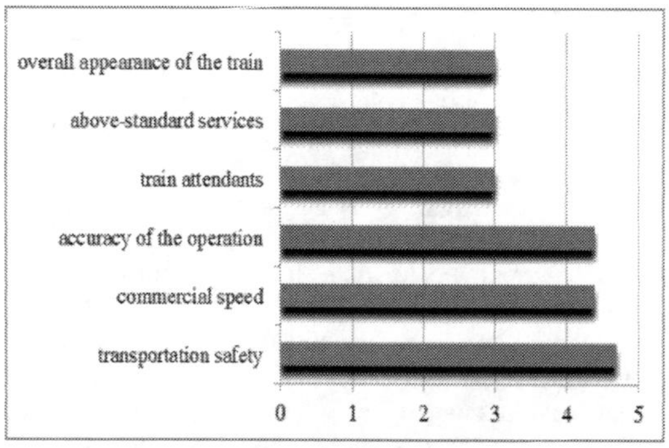

Fig. 3. Important and unimportant aspects for French passengers

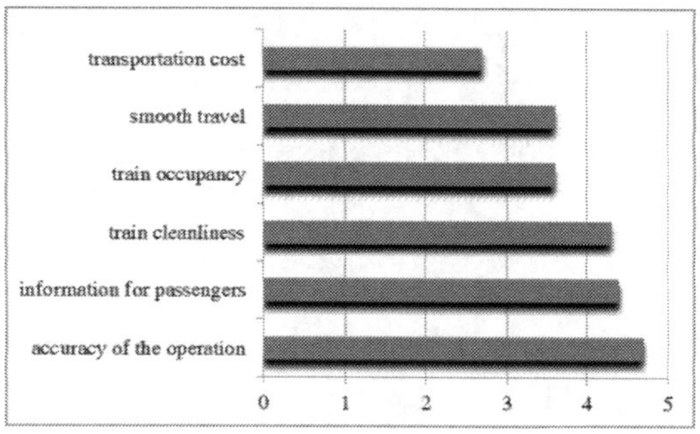

Fig. 4. French passengers' perception of service quality

2.3. Netherlands – Belgium

The countryside in the Netherlands and Belgium is very flat. The construction of corridors is not hindered by any hills or valleys. Railway tracks run directly without the need for tunnels or bridges. By virtue of the flat countryside, trains can reach higher speeds and construction costs are lower than those in hilly parts of Europe. Trains do not need to lower the speed when passing through curve track sections and ascending or descending on a track [16].

Leather seats in trains are very comfortable and provide reasonable foot space. The overall appearance of trains is good, the interiors as well as toilets are clean. In case of double-decker trains, all seats are arranged into groups of four around a side table to allow a good view. Every table is provided with an electric socket 220V and Wi-Fi is free in the whole train.

Train intervals are short. In some sections, passenger trains run ten times per hour. You cannot find a line in the Netherlands with trains running in a less than half-hour interval. This fact was mentioned by one respondent in the interview. Train departure is sharp on time as well as arrival at the destination station.

The passengers were very satisfied with the train occupancy. Trains are extra long with many seats, comprising up to seven wagons. The occupancy is low and every passenger can find a comfortable seat.

The railway routes on the territory of the Netherlands and Belgium are much shorter than routes in other countries. As a result, the train fare is lower. The passengers were satisfied with transportation cost [19, 20].

All personnel can speak English fluently. It is very appreciated by foreign passengers because they can communicate with everyone at the station as well as on the train. Train attendants are well informed and always helpful. The announcement of the next station, arrival time at the next station and information about the alighting side platform is provided by the train driver both in English and French via the passenger information system. Information is continuously provided to passengers throughout the journey. There is no unnecessary delay at stations and the 3-5 minute interval for getting on and off the train is fully sufficient.

Trains are painted yellow and blue and those colours also appear in the interior. Trains are clean and in good technical condition. All cars provide a lowered curb for barrier-free access. Double-decker trains provide a panoramic view from the first deck. Passengers' safety perception is very positive.

Table 3. Summary of pros and cons resulting from the survey in the Netherlands and Belgium

Pros
High train capacity
High speed in the flat countryside
All instructions also given in English
Short interval between trains
Seat reservations not required
Electric power from windmills

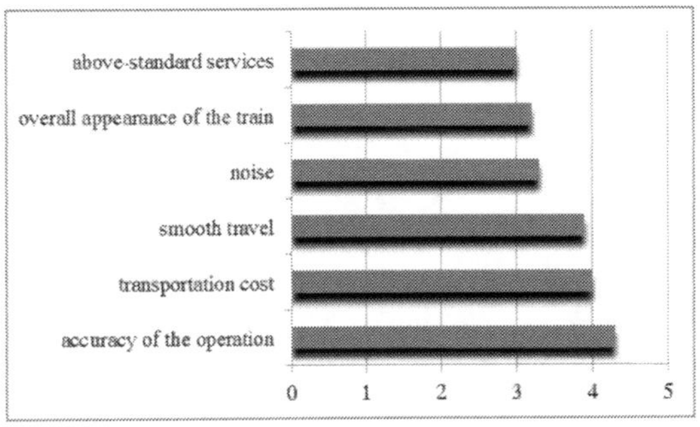

Fig. 5. Important aspects for passengers in the Netherlands and Belgium

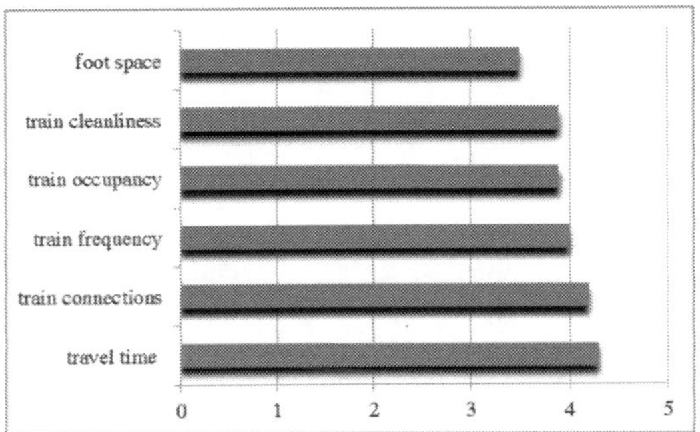

Fig. 6. Passengers' perception of service quality in the Netherlands and Belgium

2.4 Switzerland

Miscellaneous trains of the neighbouring carriers and the local SBB run on the Swiss railways. In general, trains are very comfortable. Rail transport is used by a high percentage of the population which indicates that users are satisfied with the rail travel comfort. The air-conditioning is controlled at a reasonable temperature in respect to outdoor weather conditions [21].

Swiss rail is not the fastest, but far busiest with the shortest train intervals. Short running intervals are allowed by modern signalling installations. Trains reach speeds up to 200 km/hour. There is a high demand for the route from Zurich to Olten by reason of the travel time of 56 minutes. A one-hour travel is considered a kind of limit among passengers who are willing to commute to work on a daily basis. Swiss passengers were fully satisfied with the train frequency and connections. The interval is typically 15 to 30 minutes. [

Across Switzerland, the integrated transport system is used for urban transport and commuter transport as well as long-distance rail transport. Unlike Germany, the Swiss carrier is not struggling with a high occupancy since trains are dispatched in really short intervals with a sufficient capacity [22,23].

As previously mentioned, the integrated transport system in place on the whole territory makes it even more attractive for passengers, mainly in terms of transportation cost. Seat reservations are not required.

Unfortunately due to high diversity of trains dispatched to the neighbouring countries, train personnel is unable to keep up with the information flow as the passenger would expect. They are even unable to get information out to passengers concerning delays.

The passengers were thinking highly about safety and technical condition of trains. Trains are modern and each car provides barrier-free access.

Table 4. Summary of pros and cons resulting from the survey in Switzerland

Pros	Cons
Integrated system by tariff and operations	Lack of information provided by the personnel
Short intervals and travel time	
Seat reservations not required	

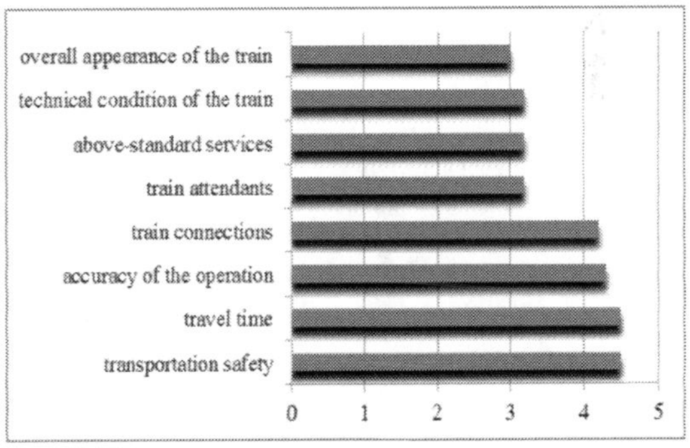

Fig. 7. Important and unimportant aspects for Swiss passengers

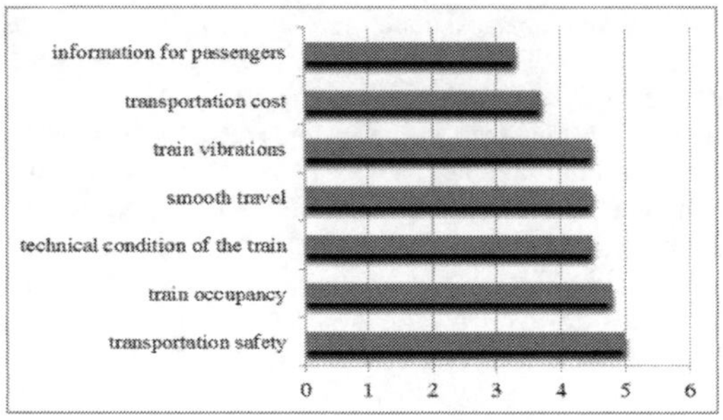

Fig. 8. Swiss passengers' perception of service quality

3 Discussion

The survey summarized in Fig. 1-8 indicates quite a difference in important aspects of the passenger rail transport across the European countries. Germany highly accentuates cleanliness of trains, the overall train occupancy and comfort. French passengers mainly look for transportation safety, travel speed and the accuracy of operation. Passengers in the Netherlands and Belgium consider the accuracy of operation, transportation cost and smooth travel as the most important. Swiss passengers highly insist on transportation safety, travel time and accuracy.

Figures 2, 4, 6 and 8 show that in principle, respondents were satisfied with all examined quality factors. There was no factor assessed as "unsatisfied". The results presented hereinunder show quite efficient operation of the passenger rail transport in given countries from the viewpoint of passengers.

Conclusions

The Czech Republic has the opportunity to use funding from the EU programmes to build a high-speed railway system as well as draw on the experience of other European countries from building good quality high-speed systems. Unfortunately, the financially favourable moment for building railway lines was missed. For building and ordering new trains, the preference must be oriented to the future and not based on the current requirements. The situation experienced during the construction of transit corridors that are currently insufficient in terms of their speed must be avoided.

Technical requirements, standards and the general technical development are the factors that have been ready for a long time. The consequent steps such as the adoption of legislation concerning the construction and the establishment of an executive authority, or the definition of financial and economic responsibility are the priorities that should immediately be tackled by the law makers. Until that time, the imaginary boundaries will be closed to the fast-paced modern times sought after by passengers to use rail transportation as often as passenger cars.

References

1. L. Nemecky. *Aktuální problémy v dopravě 2009*, (2009)
2. R. Kampf, S. Lorincova, M. Hitka, O. Stopka, Sust., **9**, 1561 (2017)
3. K. Lehutova, A. Krizanova, T. Kliestik, *Transport Means - Proceedings of the International Conference*, 258-261 (2013)
4. J. Sekulova, E. Nedeliakova, *Logi*, **6**, 2, 79–89 (2015)
5. A. Nordborg, ACU. **84**, 2, 289-300 (1998)
6. M. Halas, V. Zitricky, P. Blaho, *Logi*, **4**, 1, 53–62 (2013)
7. V. Zitricky, L. Cerna, B. Abramovic, Procedia Engineering, **192**, 994-999 (2017)
8. L. Cerna, V. Zitricky, J. Ponicky, *34th International Conference Mathematical Methods in Economics (MME)*, 126-131 (2016)
9. M. Vojtek, M. Kendra, V. Zitricky, J. Danis, *MATEC Web of Conferences*, **134**, 00061 (2017)
10. G. Fedorko, P. Liptai, V. Molnar, EFA, **83**, 30-46 (2018)
11. J.M. Duart, J.A. Pero-Sanz, J.I. Verdeja, RDM, **41**, 1, 66-72 (2005)
12. H.L Zhao, Y.J. Li, *1st China/Japan/Korea Joint Symposium on Optimization of Structural and Mechanical Systems*, 643-648 (1999)
13. R. Kampf, O. Stopka, I. Kubasakova, V. Zitricky. World Multidisciplinary Civil Engineering-Architecture-Urban Planning Symposium. (2016) DOI: 10.1016/j.proeng.2016.08.623.
14. B. Bukova, E. Brumercikova, P. Kondek, P. Groma, *Procedia Engineering*, **187**, 498-503 (2017)
15. S. Stoilova, T-TG, **25**, 2, 587-595 (2018)
16. S. Yang, Y. Zeng, L.L. Cao, Y.H. Cheng, Y.C. Zhang, *Advances in Education Research*, **100**, 101-106 (2017)
17. G. Fedorko, V. Molnár, J. Strohmandl et al. *Transport Means 2015 – Proceedings of the International Conference*. 466-+ (2015)
18. J. Ponicky, J. Camaj, M. Kendra, *AER-Advances in Engineering Research*, **62**, 191-194 (2016)
19. Y. G. Zhang, Q.Y. Liu, D.Y. Lei, X.P. Xu, J.Z. Zhao, *ICEME 2016 and ABM 2016*, 481-486 (2016)
20. E. Nedeliakova, J. Sekulova, I. Nedeliak, P. Majercak, *Transport Means - Proceedings of the International Conference*, 227-230 (2014)
21. Z. Tomes, *Conference on Regulated and Unregulated Competition on Rails*, 192-203 (2013)
22. W. Jiang, B.H. Mao, T. Chen, L.L. Yang, Y.Z. Yang, *6th Advanced Forum on Transportation of China (AFTC 2010)*, 90-93 (2010)
23. T. Skrucany, M. Kendra, M. Skorupa, J. Grencik, T. Figlus, Procedia Engineering, **192**, 806-811 (2017)

https://doi.org/10.1051/matecconf/201823500009

The Possibility of Capacity Increase on the Modernised and Electrified Railway Line R201 along the Zaprešić – Zabok Section

Ivica Ljubaj[1]*, *Tomislav Josip* Mlinarić[1], *Tomislav* Ležaić[1] and *Martin* Starčević[1]

[1]University of Zagreb Faculty of Traffic and Transport Science, 10000 Zagreb, Croatia

Abstract. This paper is focused on the regional railway track line R201 which begins at the Zaprešić railway station (who is a part of the M101 railway line - part of this Mediterranean corridor) and it separates itself from the international Mediterranean TEN-T corridor. In the forthcoming period is planned electrification and modernization of the R201 railway line, so this paper will be focused on capacity calculation as well on the utilization of the new available capacity. There will be made simulation, with the Opentrack program package for railway simulation, of the possibility to equip this line with the ETCS Level 1 and Level 2 safety systems and then it will be made comparation of the results (available capacity) obtained by simulations with the results without having equipped track with these modern safety systems. As the main idea of electrification and modernization of this track is to include her in the suburban passenger system of the city of Zagreb and to transport freight by railway, there will be given some suggestions how to improve capacity utilization.

1 Introduction

The railway line R201, which branches off the international TEN-T corridor, passes through the counties of Zagreb, Krapina - Zagorje, Varaždin, and terminates at Čakovec train station in the Međimurje County. The line is of major significance for regional passenger transport, since commuters in these areas use the railway as their primary method of transport. The railway line is furthermore important as a bypass connection to Slovenia and Austria via Krapina. As modernisation and electrification works are scheduled for this year on the Zaprešić - Zabok section, this paper will analyse and take into consideration only this section of the railway line. The paper will analyse the capacity of the line fitted with the new signal-safety system that will ensure centralised and remote traffic management, as well as the capacity in case the signal-safety system is upgraded with the European Train Control System (ETCS) Levels 1 and 2. Finally, the paper will propose potential improvements of the utilisation of the capacity on the section of the R201 line.

2 Analysis of the current state and the state after the overhaul

The R201 railway line is used for regional passenger and freight transport and has a length of 100,7 km. The line has always had outstanding importance for the economic development and growth of the region it passes through. [1]This is a single-track, non-electrified line with trains moving in station distance, with stations fitted with very outdated mechanical or electro-mechanical signal relays. There are 13 stations and 13 stops along the line. The railway line is lowland with an average slope of around 4 ‰. The technical condition of the upper construction of the railway line is unsatisfactory because the last capital overhaul the line underwent was in 1978, and it used second-hand material. The line is a B2 category line of relatively low axial weight which would amount to 180 kN/axis per meter and 64 kN/m. The average permitted train speed along the line is 60 km/h with a couple of sections that limit the speed to as low as 20 km/h. The number of trains operating daily on the Zaprešić - Zabok section, the most loaded section, is 27 for passenger and 8 for freight transport. The average journey time on the section is 36 minutes. [2,3,6,7,9]

2.1 Analysis of the Project Status of the Track

This part of the paper will provide an insight into the future project status of the track upon the completion of the planned overhaul.

As earlier mentioned, the track is of great economic importance to the population of the region it passes through. However, there is a more intensive usage of the track for passenger and freight transport on the Zaprešić - Zabok section, so that part has been selected for the modernisation and electrification, in order to include the

* Corresponding author: ivica.ljubaj@fpz.hr

section into the suburban passenger transport system of the City of Zagreb. Namely, most of the population using the railway on the section use it to commute to Zagreb and back. Therefore, this will be a single-track, electrified line fitted with an Absolute Permissive Block (APB) system, managed remotely from a central location. The track will be projected so as to allow train speeds of up to 120 km/h and basic axial weight of 225 kN per axis. Moreover, platforms with 160 m in length will be constructed in all the stations. The minimum track radii will amount to 700 m. [7]

Fig. 1. Railway track R201 Zaprešić – Zabok section [7]

The same data used in the analysis of the future state of the track in this paper will be used to construct a computer model that simulates and calculates the newly-available capacity of the track.

3 European Railway Transport Management System ERTMS

The modernised and electrified railway line R201 on the Zaprešić - Zabok section is scheduled to be fitted with an INDUSI principle-based safety system and balises AUTOSTOP. These systems have been regulated by enforced Croatian legislation and they prescribe the minimum safety level required for the transport of trains with speeds of up to 160 km/h. However, the European Union insists on an interoperable railway network, i.e. the TEN-T corridors, and this line should participate in it, as a part of the bypass or access corridor to the Mediterranean corridor. For this to happen, the track needs to be fitted with up-to-date signal and safety relays such as ETCS Levels 1 and 2 and a mobile communication system required by the ECTS Level 2 – GSM-R. [4]

3.1 ETCS Level 0

ETCS level 0 provides the basic functions for the train driver to supervise the constant (fixed) signals which act as a replacement for the authority/permission to drive. Using signal boxes, all the tasks related to the supervision and control of the track occupancy and for the supervision of the turnouts and relays. On this level, no ERTMS/ETCS equipment is used apart from the eurobalise which is used to transmit a change of the level and all other necessary instructions. Traction vehicles are already fitted with a system for transmitting an electronic telegram using the balise. [4] Figure 2 illustrates the ETCS Level 0.

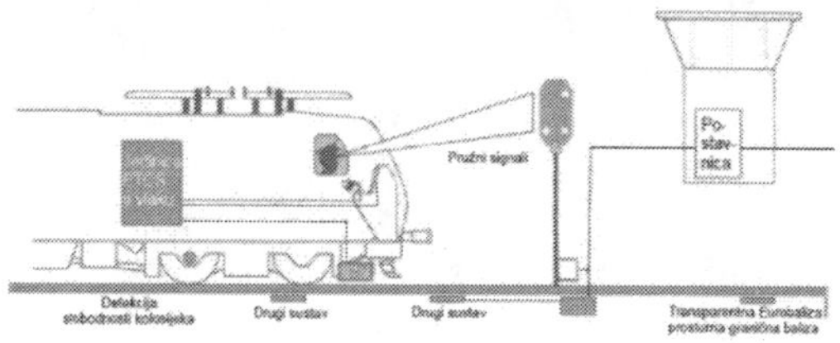

Fig. 2. ETCS Level 0

The main functions provided by the ETCS Level 0 are the supervision of the maximum train speed, maximum train speed when driving in an unfitted area, reading from the eurobalises in order to detect level changes, and other specific instructions. [5]

3.2. ETCS Level NTC (STM)

The NTC level is used for controlling ERTMS-/ETCS-fitted systems of the national system of driving of trains and supervision of speeds. The information related to train driving generated on the track by the national system is transmitted train-to-train via communication channels of the main national systems.

The track devices fitted along the railway line, level NTC, use a system of transmitting track-to-train from the main national system, which is not a part of the ETCS. For the purposes of changing levels, eurobalises are used.

Similarly, traction vehicles are fitted with the devices used to transmit telegrams using the eurbalises, and a part of the vehicle equipment belongs to the national signal and safety system. [4] Figure 3 illustrates the ETCS level NTC.

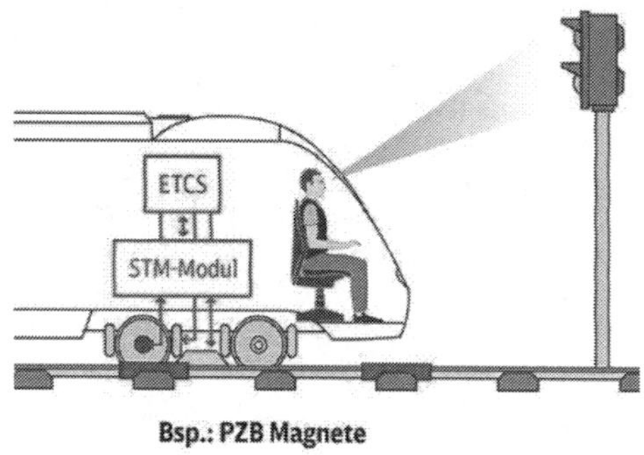

Fig. 3. ETCS Level NTC (STM)

The main functions of the ERTMS/ETCS is that there is no supervision of trains since it has been transferred into the national system, and the balises would be used to detect the changes in levels and other specific instructions, and managing through the national STM system, if the ETCS devices on a train are linked to the national system and the STM.

3.3 ETCS Level 1

The ETCS Level 1 relays data from the track on to the train. Signaling information is obtained via a local Lineside Electronics Unit (LEU) in order to secure the permit to drive. The data is relayed using the balise group, which is electrically connected to the signal or optionally via Euroloop or infill function using the Euroloop or GSM-R. The interlocking remains in charge of the control of the track occupancy, and of the supervision and monitoring of the external equipment. [7]

The track equipment used in this level is eurobalise for the transfer from the track to the train, whereby eurobalises must be able to relay sensitive pieces of information and be used for a semi-continuous infill transmission via Euroloop or Radio infill. Similarly, traction vehicles must be fitted with devices for transmitting the telegram via eurobalise or via radio (if needed).

The main functions of the ETCS on the track are establishing permit to drive based on the underlying signal system for transmitting driving permits and the data on the track features to the train. The signalisation transmits the data on to the DMI (Driver Machine Interface). [4] Figure 4 illustrates the ETCS Level 1.

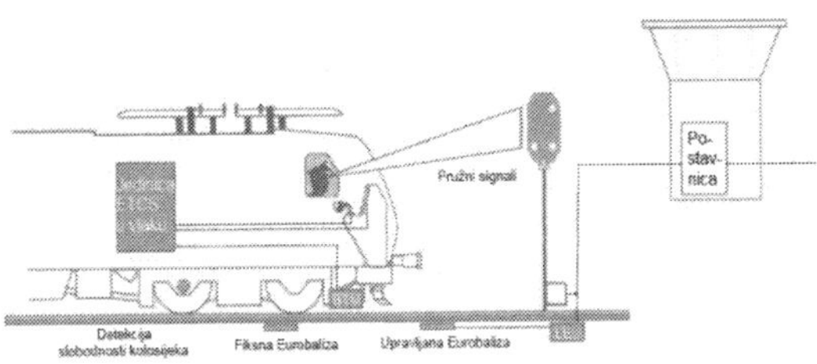

Fig. 4. ETCS Level 1 [4]

ETCS Unit on the train is: Track signals, Detection of the track occupancy, Fixed eurobalise, Operated eurobalise, Interlocking.

3.4 ETCS Level 2

ETCS Level 2 ensures a constant supervision in both directions (track - track, train - track) as well as a constant monitoring by the RBC (Radio Block Centre). Outside signals in this level are not required. The communication is transmitted via GSM-R. The interlocking is responsible for controlling track occupancy, turnouts and signals, and the information on the scheduled routes are transmitted to the RBC.

The required track equipment on this level is a radio block centre (RBC), and EuroRadio for the two-way communication between train and track. Eurobalises are used mainly for transmitting location data. Trains are fitted with devices for transmitting information via eurobalises and euroradio. [4] Figure 5 is a schematic representation of the ETCS Level 2.

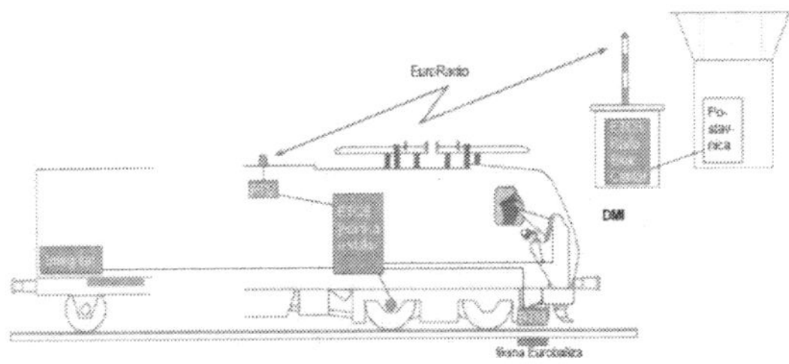

Fig.5. ETCS Level 2 [4]

The main functions of ERTMS/ETCS on the track are familiarity with the equipment of every train with ETMS/ETCS and operating in the system of ERTMS/ETCS within the RBA (Radio Block Area) using its ERTMS/ETCS identity, monitoring the location of every train controlled by ERTMS/ETCS within the RBA, determining permits for driving the train in accordance with the basic signal system for each train, and handing off the supervision of trains, transferring driving permits and track descriptions, and handing off the supervision of trains between different RBCs on the RBC-RBC borders.

The main functions of the ERTMS/ETCS onboard a train are reading of the eurobalises and transmitting the position of the train with regard to the detected balise to the RBC. The train receives permission to drive and track description via Euroradio. Furthermore, the most restrictive value of different train speeds is selected and permitted on every location in front of the train, calculating the dynamic profile of the speed taking into consideration the characteristics of the drive/braking of the train which are available onboard, as well as the data on the track, comparing the speed of the train with the permitted speed and issuing instructions for activating brakes if required, and transmitting signals to the train driver on to the DMI. [4]

This part of the paper will not analyse ETCS Level 3, since the level is still in its experimental phase and not in commercial use.

4 Simulation and the obtained capacity results

This railway line is seen as extremely significant for passenger transport due to the fact that there are plans include it in the suburban transport system of the City of Zagreb. [5] It is expected that, after the liberalisation of the railway passenger transport market, i.e. opening thereof, railway operators will show interest in passenger transport on this section. In order to ensure access to the market, the train fleet must consist of modern and economically efficient trains. In that sense, trains will surely be fitted with modern signal and safety systems (ETCS equipment) which will then ensure interoperability, that is, operating on the railway tracks owned by other railway undertakings. This paper will, therefore, analyse, simulate, and outline potential solutions for the utilisation of the maximum capacity of the track with regard to the signal and safety equipment, i.e. ATP - Automatic Train Protection.

It has earlier been established that the track capacity on the Zaprešić - Zabok section amounts to 57 trains, and that, based on the existing timetable, the utilisation of the capacity is 66%. [5]

After the estimation of the modernisation, maximum utilised capacity was calculated and it amounted to 87 trains, the latter capacity is a 50 percent increase of the original capacity after the modernisation. Figure 6 show this new available capacity. This capacity is available when taking into consideration the newly-installed signal and safety system (APB) and remote train transport management located in Zabok station. [8,9]

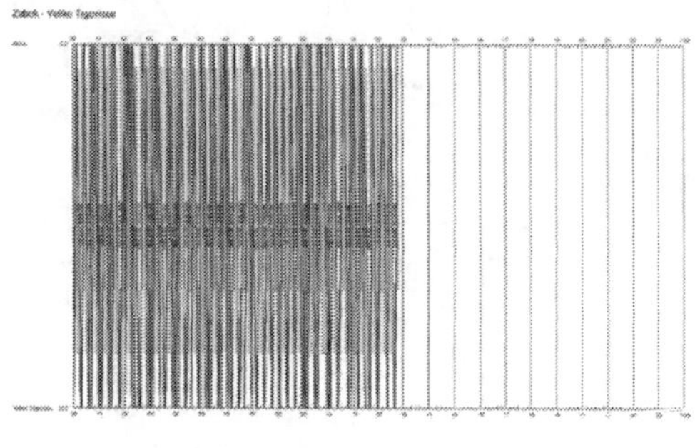

Fig.6. Compressed timetable for maximum capacity calculation

Upon the construction of the model and the change of the signal and safety system by means of modernising the section, the track capacity was determined at 87 trains. Evidently, the latter capacity is a 0 percent increase of the original capacity after the modernisation, which is why the investments in the modernisation of the track become substantially more expensive. [9-11]

ETCS Level 2 was also simulated in the model, whereby the reconstructed section was upgraded with a signal and safety equipment of the corresponding level. The calculation revealed the track capacity of 87 trains. Therefore, the obtained capacity is 0 percent higher than the original one after the modernisation, which further increases the cost of the investment, which is even greater than the investment in ETCS Level 1.

5 Conclusion

The railway line R201 is of great importance for the entire region it passes through, and it is essential that it be modernised. By modernising the Zaprešić - Zabok section, the line can be included into the suburban transport system of the City of Zagreb, thereby making it easier for residents to commute to Zagreb and back. The project documentation reveals that the track is scheduled to be modernised and electrified in order to increase its axial weight load so that the commercial speed may increase from 60 to 120 km/h. This increase would shorten journey time by 5-10 minutes, which may not be a great improvement, but will ensure the provision of a higher quality of service in terms of more frequent departures, more modern, faster and economically efficient motorised trains, as well as the possibility of private operators offering transport service. At the same time, the railway undertaking will benefit from higher profits coming from infrastructure fees, and the fact that the railway line will become more appealing to freight operators, due to the increase in axial load, as trains will be able to load more cargo.

This paper has shown that the upgrade of the modernised and electrified track by means of a signal safety system would not justify the investment since the capacity would largely remain the same. In fact, greater benefit would be ensured by fitting STM systems onboard trains which could then, using modern signal and safety systems, operate using the national ATP system, and ETCS as well.

References

1. D. Badanjak, B. Bogović, V. Jenić, *Organizacija željezničkog prometa* (2006)
2. I. Ljubaj, M. Petrović, T.J. Mlinarić, *Analysis of Public Transport Offer in Krapina- Zagorje County and Proposal of Integration Point // Horizons of Railway Transport* (2017)
3. M. Mikulčić, I. Ljubaj, M. Petrović, *Analysing the Railway Passenger Train Operation Using a Simulation Model // Horizons of Railway Transport (2017)*
4. P. Steinly, ETCS for engineers 1st edition (2011)
5. Fakultet prometnih znanosti & DB Engineering & Consulting GmbH, *Studija uvođenja europskog sustava upravljanja željezničkim prometom (ERTMS)* (2016)
6. J. Klimková, J. Čamaj, J. Gašparík. *Transport means 2016: proceedings of the 20th international scientific conference*. 891-896. (2016)
7. HŽ Infrastruktura d.o.o., *Izvječće o mreži 2018,* (2018)
8. B. Abramović, J. Gašparík, Jozef, V. Zitrický. Tehnički Vjesnik. **25, 4**. 1190-1195. (2018),
9. HŽ Infrastruktura d.o.o., Projektna dokumentacija projekta: *Modernizacija i elektrifikacija pruge Zaprešić – Zabok* (2016)
10. HŽ Infrastruktura d.o.o., *Uputa za utvrđivanje pružnih i kolodvorskih kapaciteta – Uputa HŽI-70* (2014)
11. HŽ Infrastruktura d.o.o., *Materijali voznog reda 2017/2018* (2017)

Wider aspects of deceleration supervision in ERTMS/ETCS

Petr Nachtigall[1,*] and *Jan* Ouředníček[2,**]

[1]University of Pardubice, Faculty of Transport Engineering, Department of Transport Technology and Control, Studentská 95, 532 10 Pardubice 2, Czech Republic
[2]AŽD Praha s.r.o., Technology Division, Research and Development, Zirovnická 3146/2, Zabehlice, 106 00 Praha 10 Czech Republic

Abstract. The paper concerns basic information about implementation of the ETCS system in the Czech Republic especially its influence on the railway operation. Also facts about the principles of the automatic train protection system with the full speed supervision are described as well as the specifics of the speed monitoring experiences. In the next chapter are some principles of the braking curves calculation.

1 Introduction

The ETCS L2 implementation projects currently carried out by the Railway Infrastructure Administration (further SŽDC) of Czech Republic – i.e. in an environment with basically no previous experience with operating an automatic train protection (further ATP) system with full speed supervision – bring either new aspects, or aspects not very well known. A significant part of these aspects share a common denominator: the transfer of responsibility from human to a technical device. Being an ATP system with full speed supervision, it is the responsibility of ETCS as a technical device to ensure that the vehicle doesn't exceed the limit parameters, i.e. the maximum allowed speed in the given position and the maximum position to which a movement of the vehicle is allowed. *(Note: This represents the main safety objective of the implementation of ETCS, with other ETCS features being implemented either to achieve this safety objective or to ensure interoperability).* However, depending on the method of implementation of the ATP system with full speed supervision, the abovementioned transfer of responsibility can lead to operating restrictions compared to the situation before the implementation.

This article presents certain aspects of the introduction of deceleration supervision or stopping supervision as key safety-operational aspects of the ETCS implementation, adding or emphasizing relevant facts to broaden the context of the operational impacts of ETCS and their solutions.

2 Automatic Train Protection systems

In principle, the general purpose of railway signalling systems is to limit the behaviour, or commands of operators to a subset and sequence which in the given operational context are allowed – i.e. they not represent a direct risk to safety. In this regard, it is natural that the introduction of supervision functions of technical devices can limit the operational flexibility until now only achieved thanks to the skilfulness, dexterity, and responsibility of the operator.

Furthermore, as is well known, a defining technical feature of railway signalling systems is limiting the operation (or more precisely limiting the authority degree) in the absence of operational-technical information or where the system loses functionality due to a failure – also referred to as the fail-safe (Fail-safe means safe in a case of failure. In addition to railway signalling technologies, these principles are applied in the chemical industry in processing and manufacturing of dangerous substances, partly in the nuclear industry as well and recently also in industrial automation in general. Related to the last one mentioned applying fail-safe technologies increases both the production quality, and health and safety at work) principle. Even though it is logical and natural, it should be pointed out that a signalling system can only be responsible for such aspects of railway operation and can only perform such supervision functions, for which it has the necessary information and for which unambiguous decision algorithms and calculation algorithms can be determined.

As for ATP systems, in particular, those with full speed supervision, the speed reduction or stopping falls fully within their responsibility. Therefore, they must be able to supervise and ensure a timely deceleration of the vehicle (train, train unit) so that it doesn't exceed the limit speed in the given position, or comes to a standstill in the given position at the latest.

Based on the information available to the ATP system (both "online" – current operational and technical data, and "offline" – configuration parameters), it has to consider the most unfavourable conditions (the worst-case scenario) for the deceleration supervision. Therefore if some information is not available to the ATP at all, or only with limited accuracy, it has to assume that its actual value is neither better, nor average, but worse, meaning that the

* petr.nachtigall@upce.cz
** ourednicek.jan@azd.cz

braking effect is generally assumed to be lower and the deceleration distance longer. In particular, the system is to consider a probability distribution with a mean value (i.e. average value) and other parameters, where the mean value is definitely not the correct representative of a safety parameter. The selection of a "sufficiently safe" mean value deviation depends on the acceptable rate of risk that the deceleration distance will be longer and by how much (In addition to this qualitative approach to safety assessment, it is necessary to apply qualitative approaches as referred to in [1]).

An ideal ATP system with full speed supervision would be a machine having the qualities of an experienced train driver taking into account not only the information on speed and distance, but also other information having an impact on the vehicles' braking, such as typically weather conditions, or adhesion conditions, train set structure and qualities of the material transported (e.g. partly filled tank cars), where this train driver would be concentrated in all circumstances, and basically inerrable. A real ATP system can be implemented in a way that with a necessary safety integrity level (put simply, with an astronomically low probability of failure), it will be concentrated and inerrable. However, the experience can only be algorithmized (Experience, as a phenomenon attached to living creatures, is examined in relation to possible development of artificial intelligence. However, as for the transfer of responsibility for safety (not only of transport), artificial intelligence is a thing of the future; according to authors of this article, a rather distant future) to a very limited extent (for instance the feedback of the current braking effect of a service break to evaluate the need to use an emergency brake), with some information being completely unavailable (e.g. weather conditions leading to lower adhesion).

If a real ATP system with full speed supervision is to take on the responsibility, it has to algorithmize more critical braking curves than those which a vehicle is able to achieve in normal operation in most situations, i.e. in nominal (average) situations. In this way, the ATP system will "act" responsibly also in non-nominal situations, i.e. under critical braking conditions. The difference in braking curves is shown in Fig. 1.

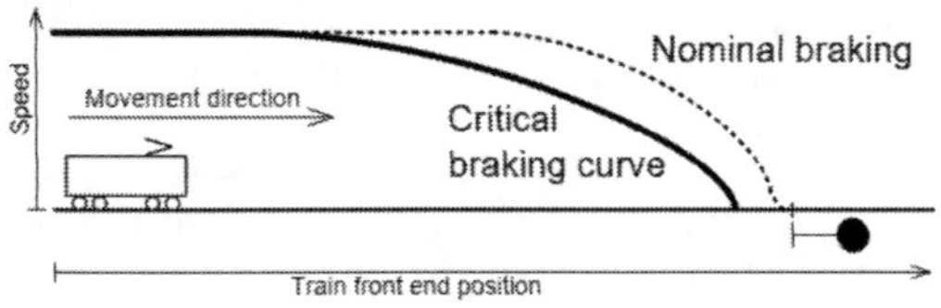

Fig. 1. Deceleration supervision taking into account critical braking conditions, vs. braking under nominal conditions.

Source: Authors

Historically, based on the information available to the authors, it appears that the responsibility has never been assumed by the ATP system without negatively influencing the performance parameters of railway operation compared to a situation where the responsibility remains with the driver, together with the responsibility for distinguishing between nominal and critical situations. However the operational performance is designed for nominal situations!

In implementing ATP systems with full speed supervision, two approaches can be (and logically are) applied:

- The railway operation performance is designed for critical situations – for critical braking curves. It is thus expected that the deceleration supervision will (or can) lead to a longer braking distance to a standstill and longer time necessary for stopping the vehicle. *Note: In practice, this involves creating sufficiently long station tracks (Fig. 2) or overlap sections behind the stopping point (Fig. 3).*
- At a certain phase of the deceleration supervision, the responsibility for speed reduction or stopping is fully or partly transferred to the driver – in a controlled and active manner of the driver – to reduce the negative impact of the absence of information and/or absence of "experience algorithms" in the ATP system on the operational performance. *Note: In practice, this involves introducing a release speed limit, under which the driver can take over responsibility (Fig. 4).*

Both approaches are commonly combined.

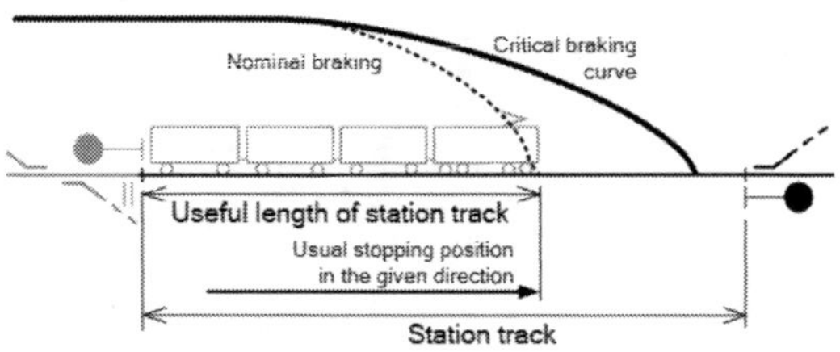

Fig. 2. The length of station tracks is significantly higher than their usable length.

Source: Authors

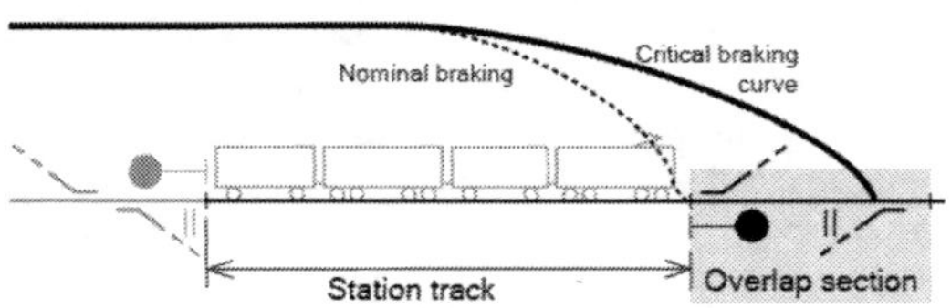

Fig. 3. Overlap sections.

Source: Authors

The possibility of transferring responsibility to the operator is crucial in intermittent ATP systems. In case where deceleration information can only be transferred at one or a few specified positions (equipped relevant transmission device), usually close to the stopping position, the possibility of a release from supervision is the critical from operational point of view.

 https://doi.org/10.1051/matecconf/201823500010

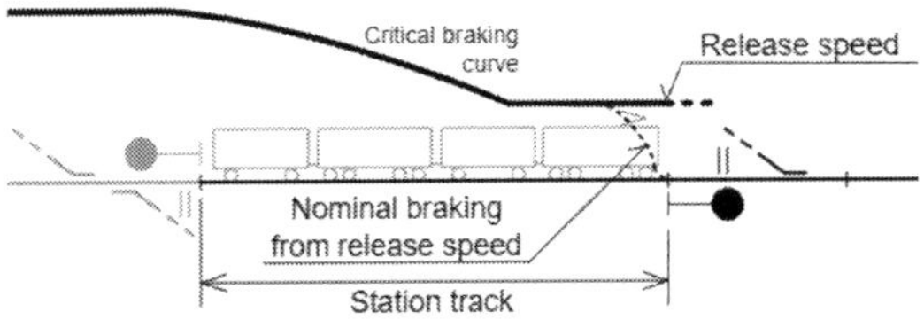

Fig. 4. Release speed.

Source: Authors

3 Current operational conditions and safety restrictions

The current relevant SŽDC operational conditions and safety restrictions are the following:

- The length of the station tracks almost corresponds to the maximum length of operated trains.
- Minimization of operational intervals (a logical and justified requirement of railway undertakings leading to corresponding efforts of the infrastructure manager).
- Stopping before a given position falls fully within the responsibility of the train driver (Through a signal device, the current LS train protection system safely indicates information (on four levels) in front of the vehicle front end , and according to this information (or changes to this information) and the current vehicle speed, it expects / doesn't expect the intervention of the train driver, otherwise intervening (initiating emergency braking) by itself. The system is thus responsible (and performs the respective functions in accordance with the fail-safe principles) for indicating the information to the train driver and intervenes, where the train driver fails to carry out the necessary actions based on this information (or a change to this information). However, it cannot ensure stopping the train before the given location, which remains the responsibility of the train driver).
- The non-existence of overlap sections in front of the end of train routes.

In view of the above, the distance and time within which the train driver has to move the vehicle is very limited. On one hand, it is required to maximize the use of performance parameters of the infrastructure (minimization of time spent train moving through open line, stations and station heads when coming to station tracks with maximally using their length), on the other hand it is not possible (not even in critical, i.e. non-nominal situations) to exceed the position where a movement of the vehicle is allowed. In addition to complying with the principles and rules necessary for ensuring the safety of traffic, the train drivers are under pressure due to the logical and justified efforts to increase the performance of infrastructure.

Safety restrictions inherently requiring certain operational reserves exceeding the physical and performance capabilities of vehicles and infrastructure in fact don't exist. This means that it is not possible, or it is only possible to a limited extent, to use the "free infrastructure capacity" to perform supervision functions of a non-intelligent, but safety (i.e. fail-safe) ATP device to fully assume the responsibility for supervising the limit parameters of a train movement.

4 European Train Control System (ETCS)

General principles, features and application of the ETCS train protection system are described in [2, 3, 4, 5, 6]. The authors of this article present them as possible sources of information as it is not their aim to describe already known information.

4.1 General principles of deceleration supervision in ETCS

In short (more information is provided for in the sources specified above), the ETCS L2 train movement supervision consists in creating and sending to the on board unit a movement authority (further MA) together with static speed profiles (further SSP), a gradient profile and potential other speed limits (i.e. line description) [6]. As for location, this information is related to balise groups (further BG) placed in the railyard. The balise group, to which the current MA and the line description are related, play the role of the last relevant balise group (further LRBG) for the creation of a balise co-ordinate system.
From the various speed limits which can be valid for the given vehicle/train type at the given position, the ETCS on-board equipment creates the most restrictive speed profile (further MRSP).

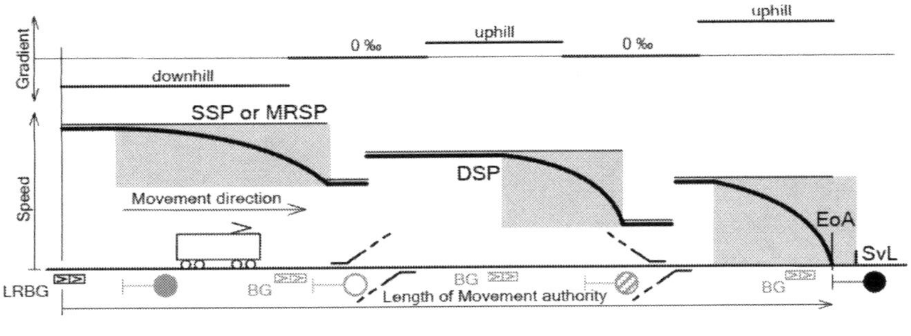

Fig. 5. From the transmission of MA and line description until the calculation of DSP.

Source: Authors

Based on the MRSP, the gradient profile and other vehicle/train parameters, the ETCS on-board equipment calculates a dynamic speed profile (further DSP). According to the DSP, it then supervises the running of the vehicle – with multiple deceleration, i.e. braking curves for one speed change (warning curve and intervention curve of the service brake, warning curve and intervention curve of the emergency brake, etc.), the braking curve area is marked with grey in Fig. 5 (only one curve is shown for illustration purposes). It can be seen that in stopping supervision, some curves can reach behind the EoA – the supervised location (further SvL) is behind the EoA, where allowed by the infrastructure conditions.

4.2 ETCS features related to stopping supervision

As mentioned above, the main and fundamental change brought by the introduction of ETCS L2 on the SŽDC network is the full speed supervision and control through a ATP. Fig. 2 and Fig. 4 show the examples of braking to standstill supervision.

A fundamental prerequisite for the right functioning of this supervision is an exact mathematical-physical description of the vehicles' behaviour during running and braking [7]. The braking model is based on the value of brake weight percentage. Together with the maximum vehicle speed, this variable determines its braking properties. Without knowing other parameters, however, this description leads to the fact that the respective conversion braking model adds extra time to make sure the EoA isn't exceeded. Furthermore, this model is based on historical UIC methodology and measurements, which might no longer correspond to the features of modern interoperable vehicles. This means that the conversion model creates safe curves, which are not practical for the SŽDC operational conditions though (not taking into account adhesion, brake shoe properties, and linear course of the onset of braking performance).

A distinction is drawn between service curves (Service brake intervention - SBI, Service brake deceleration - SBD) and emergency ones (Emergency brake intervention - EBI, Emergency brake deceleration - EBD). The difference in the amount of deceleration for both curves manifests itself in trains with a BWP of more than 135 . In service brakes, for BWP > 135 the value of BWP is always set at BWP = 135. The BWP value is then used for the calculation of v_{lim} according to (1), where in the interval of <0;v_{lim}>, the deceleration is constant.

$$V_{lim} = 16.85 \cdot \lambda_o^{0.428} \qquad [km \cdot h^{-1}] \qquad (1)$$

The amount of deceleration is then calculated according to (2).

$$a_0 = 0.0075 \cdot \lambda_o + 0.076 \qquad [m \cdot s^{-2}] \qquad (2)$$

Other amounts of speed with a change in deceleration are provided for in [7], with the amount of deceleration determined according to the general relation (3).

$$a_n = a_3 \cdot \lambda_o^3 + a_2 \cdot \lambda_o^2 + a_1 \cdot \lambda_o + a_0 \qquad [m \cdot s^{-2}] \qquad (3)$$

The result is the course of deceleration for the emergency or service brakes depending on the speed. This is shown in Fig. 6.

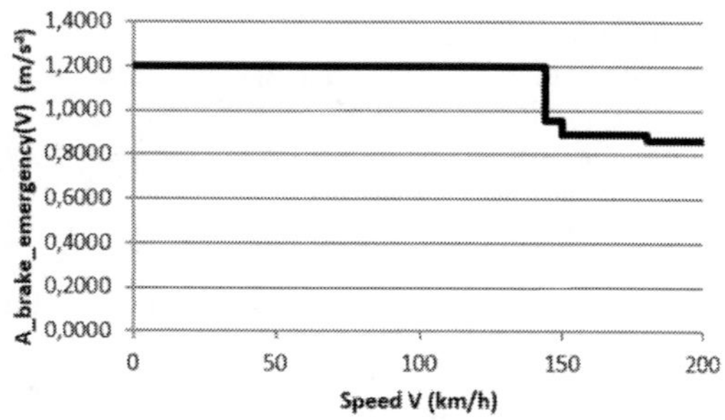

Fig. 6. Emergency brake deceleration.

Source: Authors, [8]

In this way, it is possible to create braking curves for any train. In ETCS, these braking curves are marked with lambda. They have to be functional and safe for any combination of vehicles corresponding to the given BWP. Another option for braking curves is gamma. This curve is usually provided (upon request) by the vehicle supplier. It is used in cases where it is possible to guarantee certain braking properties, i.e. in cases where a train unit has a constant structure for a long time. The advantage of these braking curves is the fact that they are in accordance with potential good achievable braking capabilities of the train unit, and a more critical and thus operationally less advantageous approach to braking properties is thus not necessary. In the context of the Railway Infrastructure Administration, this applies to the Pendolino, railjet, or Interpanter train units. However, there is no limitation to this approach being applied to common train units with a constant structure. These include container trains (under load, empty), Regiojet train units with a specified number of cars or push-pull train units in general.

 https://doi.org/10.1051/matecconf/201823500010

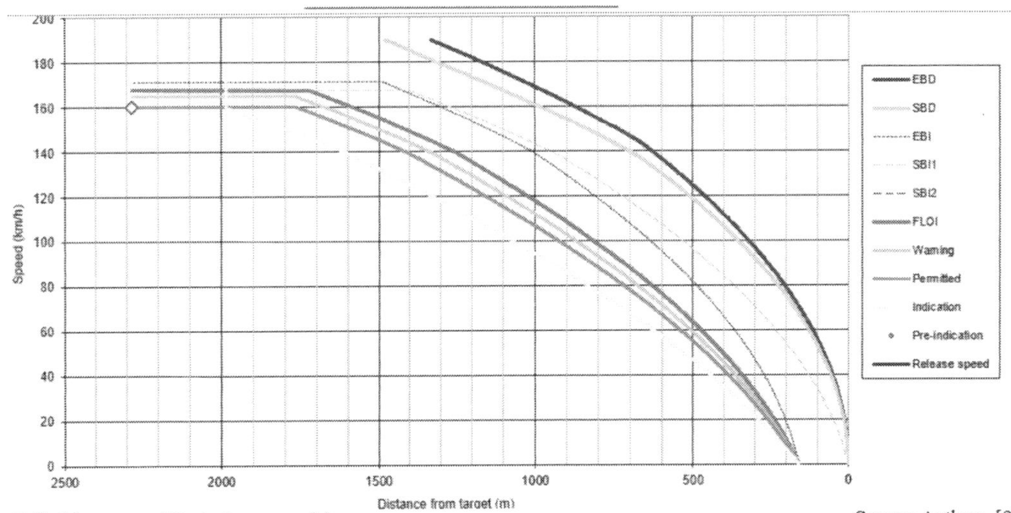

Fig. 7. Braking curves (illustration example). Source: Authors, [8]

In this way, it is possible to calculate and display a whole range of curves, which can be supervised by ETCS. In addition to the above mentioned EBD, EBI, SBD, SBI, also First Level of Intervention - FLOI and Permitted are worth mentioning. All curves are shown in Fig. 7.

It is the Permitted Speed curve that is displayed to the locomotive driver on the Driver Machine Interface - DMI and ideally, it is the responsibility of the locomotive driver to monitor this curve. In Fig. 8, this curve is already a warning one.

Another interesting feature of braking curves, as shown in Figure 6, is the different position of the individual curves for $v = 0$ km·h^{-1}. EBD is supervised at the danger point, SBD is supervised at the end of the MA and other curves are moved forward before the end of the MA by the size of odometry inaccuracy. For the least advantageous case, this inaccuracy is calculated as 5 m + 5% distance from the last balise (Also here, there is a parameter determining the accuracy of balise positioning (Q_NVLOLACC). The national value thereof is 12 metres). Where the balise distance is 500 metres, this value is 30 metres. It is this distance that causes the "shortening" of the useful length of tracks for freight trains and in a combination of the end of a MA and the end of a platform, it can cause problems to passenger trains approaching the platform. A solution already put to use is the installation of relocation balise groups specifying the vehicle's location and decreasing this inaccuracy. In new planned projects (high-speed railway line Praha-Brno [9], high-speed railway line Brno-Ostrava), the engineers and technologists already determine the line topology in a way excluding these problems.

Fig. 8. Braking curve on the DMI.

Source: Authors

Critical ETCS braking curves are calculated in a more critical way as they are to ensure that ETCS fulfils its responsibility to perform stopping supervision in a broad range of operational conditions in different railway infrastructures of different countries. Higher ETCS (from 1.1 onwards) implemented according to the Baseline 3 specification version [7] allow for parametrizing the braking curves of the supervised ETCS on-board equipment (indeed of a version ETCS 1.1 or higher) based on the trackside ETCS values (i.e. according to the infrastructure manager). The question is what added value such parametrization can bring compared to other influences on the critical ETCS braking curves. This must be further examined together with the implementation of ETCS 1.1 on the SŽDC network.

 https://doi.org/10.1051/matecconf/201823500010

5 Conclusion

This article points out the operational aspects of the implementation of ETCS L2 in the Czech Republic. It is the very first automatic train protection system with full speed supervision. Until now, the first ATP systems gave the train driver quite a high degree of freedom in the train speed deceleration. This is flexible from operational point of view (a delayed train could brake within a shorter distance than an early train, which could make an efficient use of drift, recuperation, etc.), but from a safety perspective, the probability of passing through a stop signal at danger was more likely. On the other hand, the introduction of full speed control reduces the flexibility of operation. A partial elimination of the problem – to be further examined, ideally by all ETCS stakeholders in the Czech Republic – is available through parametrizing the curves using national values, which is possible from ETCS system version 1.1 onwards according to Baseline 3. In other countries, the experience with train protection systems with full speed supervision is more extensive, which means that the risk of decreasing the line operational capacity is not so great there. One of the elements with a positive effect is for instance overlap sections.

The work was supported from ERDF/ESF "Project PosiTrans - University of Pardubice cooperation and application sphere in application oriented research of localization, detection and simulation systems for transport processes" (No. CZ.02.1.01/0.0/0.0/17_049/0008394).

References

1. CENELEC, EN 50129, *Railway applications – Communications, signalling and processing systems – Safety related electronic systems for signalling*, (2003), 93.100
2. Railwaysignalling.eu, *The ERTMS/ETCS signalling system*, [Online] (2014) [cit. 09/2018], available on: http://www.railwaysignalling.eu/wp-content/uploads/2016/09/ERTMS_ETCS_signalling_system_revF.pdf
3. SŽDC, s. o., *Zvláštní technické podmínky na zhotovení projektu a stavby ETCS úrovně 2*, (2016)
4. SŽDC, s. o., M. Klega, 17. konference Železniční dopravní cesta 2013, Praha 27. – 29. 3. 2012, *Zavádění systému ETCS na tratích v ČR*, [*.pdf file], (2012), [cit. 09/2018], available on: https://www.szdc.cz/soubory/konference-a-seminare/zdc-2012/a06-klega-szdc.pdf
5. M. Ghazel, *Formalizing a subset of ERTMS/ETCS specifications for verification purposes*, Transportation Research Part C. (2014), 42.
6. UNIFE, The European Rail Traffic Management System, [Online, cit: 06/2018] http://www.ertms.net.
7. European Railway Agency, *ERTMS/ETCS System requirements specifications.*. [Online] 15/06/2016. [cit. 10/2018], available on: https://www.era.europa.eu/node/641/211_en
8. European Railway Agency, *Braking curves simulation tool*, [Online] (2016) [cit. 10/2016.], available on: http://www.era.europa.eu/Core-Activities/ERTMS/Pages/Braking-Curves-Simulation-Tool.aspx
9. Ministerstvo dopravy ČR, *Národní implementační plan ERTMS*, [Online] (2017) [cit. 09/2018.], available on: https://ec.europa.eu/transport/sites/transport/files/rail-nip/nip-ccs-tsi-czech-rep-cz.pdf

Energy consumption and greenhouse gases emissions on relation Brno - Jihlava

Vladimír Ľupták[1*], *Martina* Hlatká[1] and *Rudolf* Kampf[1]

[1]Institute of Technology and Business in České Budějovice, Okružní 517/10, 370 01 České Budějovice, Czech Republic.

Abstract. Currently, the influence of transport on the environment represents a serious problem, especially in terms of energy consumption and generation of gases causing global warming. The contribution on the topic of comparing energy consumption and greenhouse gases emissions on relation Brno – Jihlava thus aims to describe generation of emissions in the current types of transport and their calculation according to EN 16258:2012. The results of the calculations will be expressed in emission intensity units to the number of passengers transported. The EN 16258:2012 calculations will be applied to passenger rail transport. In particular, energy intensity and generation of greenhouse gases will be monitored in relation to travelling the distance between Brno and Jihlava.

1 Introduction

Transport and transportation of either material or passengers currently plays an important role in many aspects of our lives. Due to the increasing number of means of transport, a long-term strategy and the role of the railway sector in a national economy and transnational ecology context are very important. In recent years, there has been a significant increase in oil consumption and exhaust gases emissions. The impact of transport is clearly visible also in the landscape affected by land take and construction activities associated with repairs or construction of railway lines. The most discussed topic, however, is the impact of exhaust gases and the pollutants contained on the environment and human health. The problem of the influence of pollutants from exhaust gases on human health is even more significant in urban agglomerations, where their concentration is up to 100 times higher than in uninhabited areas. The need to starting a political fight against transport emission generation is even more urgent due to the upward global warming trend, which is caused predominantly by greenhouse gases, their main producer being transport. Transport is reported to emit about 23 % of all anthropogenic CO_2 production on Earth. [1-3]

The contribution is focused on this negative aspect of transport and will try to compare two modes of transport on the aforementioned transport relation so that it is possible to better understand all side effects and burden these options bring along. For individual calculations of emission intensity, a method according to EN 16258:2012 standard will be used. [4,5]

An important point in discussion about transport emissions is to realize that the burden on the environment does not result only from the operation of means of transport, but also from their production and especially from the waste arising from their disposal. Emissions in transport are generated mainly during the chemical processes caused by incomplete combustion. Their composition varies according to the type and condition of the engine, type of fuel used, driving mode, intensity or presence of equipment to reduce emissions (e.g. catalysts). It is estimated that the mass emission unit from motor vehicles transport in city centers is up to 10 times higher than emissions from other sources (heating, industry) and up to 100 times higher than in non-urban areas. Motor vehicles exhaust gases are mixtures of chemicals with various concentrations and with different effect on population. People more sensitive to exhaust gases are children, elderly people and people with cardiovascular or respiratory problems. [6]

2 European standard EN 16258:2012

This European standard specifies and describes methodology for calculating and determining energy intensity and volume of GHG emissions in relation to any mode of transport. It can be applied to various modes of transport (road, sea, etc.) as well as for various transported items (passengers, cargo, or both). The method describes in detail general steps, definitions, relations, calculation methods, breakdowns and recommended information as a support of precise and standardized calculation which shows energy intensity and greenhouse gases generation in all modes of transport.

Each calculation of energy intensity and greenhouse gases generation requires necessary volume of information about the specific situation. [7-10]

* Corresponding author: luptak@mail.vstecb.cz

Steps in calculating energy consumption and greenhouse gases emissions of one transport service (the 3 following must be included in the calculation):

- step1: Identification of various sections of this transport service
- step 2: Calculation of energy consumption and greenhouse gases emissions in each section
- step 3: Sum of the results from each section.

A great advantage of the standard is the fact that it does not consider only secondary generation of emissions and energy intensity that occurs during the combustion process, that is, extraction of energy from fuel and its transformation into kinetic energy, but also the primary emissions associated with the process of extraction, production and distribution Depending on the type of emissions, the Standard considers the following unknown variables:

- E_w - well-to-wheels energy factor for the given types of fuel,
- G_w - well-to-wheels emission factor for the given types of fuel,
- E_t - tank-to-wheels energy factor for the given types of fuel,
- G_t - tank-to-wheels wheels emission factor for the given types of fuel. [11]

Well-to-wheels factor includes both primary and secondary emissions including consumption. This factor is sometimes referred to as LCA – "life-cycle analysis" [10]. On the other hand, tank-to-wheels considers only the secondary emissions and consumption. The final calculations of these unknown variables indicate the information on energy consumption and greenhouse gases emissions both in the summing of primary and secondary emissions, and only primary emissions. [12-14]

Partial steps of energy consumption and greenhouse gases emissions calculations from one section of a transport service

Calculations for one section of a transport service must be carried out using the following four main steps:

- Partial step 1 – Determining vehicle operation system (VOS) for the given section. For each section of the transport route, it is necessary to start with choosing VOS for a particular section. As a minimum requirement, VOS must be a consistent set of vehicles operation related to a particular section. When determining VOS, the factors influencing the VOS extent and composition (number of vehicles, operation period of these vehicles). Choosing VOS is an optional variable relating to the criteria chosen, but it is necessary to take into account also unladen journeys.

- Partial step 2 – Calculation of the total energy consumption for this particular VOS. For the calculation, the aforementioned category of values is used. If the vehicle uses more types of energy, it is necessary to carry out the calculations for each energy carrier.

- Partial step 3 – Calculation of the total energy consumption and greenhouse gases emissions for this VOS – conversion from the total fuel consumption for VOS to the volume of energy consumed and greenhouse gases emissions must be carried out using the following formulas:

Table 1 Indicators of total energy consumption and greenhouse gases emissions

Indicator	Formula
for well-to-wheels VOS energy consumption	$E_w\ (VOS) = F\ (VOS)\ x\ e_w$
for well-to-wheels greenhouse gases VOS emissions	$G_w\ (VOS) = F\ (VOS)\ x\ g_w$
for tank-to-wheels VOS energy consumption	$E_t\ (VOS) = F\ (VOS)\ x\ e_t$
for tank-to-wheels VOS greenhouse gases emissions	$G_t\ (VOS) = F\ (VOS)\ x\ g_t$
Where: F (VOS) is a total consumption of fuel used for VOS, ew well-to-wheels energy factor for fuel used, gw well-to-wheels greenhouse gases emissions for fuel used, et tank-to-wheels energy factor for fuel used, gt tank-to-wheels greenhouse gases emissions for fuel used.	

- Partial step 4 – Assigning a share of each of the four results of the partial step to a section – after carrying out steps 1,2,3, the transport service sections must be assigned a share of Ew (VOS), Gw (VOS), Et (VOS), G t (VOS) corresponding to its relative share of transport activity carried out within VOS.

The relevant formulas are thus as follows:

$$S\ (leg) = T\ (leg)\ /\ T\ (VOS) \quad (1)$$
$$Ew\ (leg) = Ew\ (VOS)\ x\ S\ (leg) \quad (2)$$
$$Gw\ (leg) = Gw\ (VOS)\ x\ S\ (leg) \quad (3)$$
$$Et\ (leg) = Et\ (VOS)\ x\ S\ (leg) \quad (4)$$
$$Gt\ (leg) = Gt\ (VOS)\ x\ S\ (leg) \quad (5)$$

where:

- S (leg) is a factor used for calculating the energy and emissions share of the VOS assigned to transport service for a particular section. The share is based on the ratio of transport activity for a section and the related VOS,
- T (leg) is a transport activity service for a particular section,
- T (VOS) is a VOS transport performance in the given field.
- T (leg) and T (VOS) must be assigned the same parameters and units. [15]

3. Application of EN 16258:2012 European standard – Case study

The main objective of the case study is to determine energy intensity and greenhouse gases emissions for passenger rail transport. After obtaining the results, environmental performance will be assessed according to the European standard and the impact on the environment will be evaluated. The case study will be applied in the relation Jihlava – Brno.

The distance between Jihlava and Brno is 104 km by rail. The railway car is fully occupied, which means there are 98 passengers. The fuel consumption to gain enough energy to travel the given distance is 59.43 l / 100 km. The fuel used is diesel.

Due to the fact that several types (814, 841, 842 and 854) of motor railway cars (integrated railway units) are deployed on this particular transport section, thus for objectification within this research study, calculation of the average values of fuel consumption of individual deployed motor units and the average capacity value of individual units are taken into consideration. [16] Consumption of individual units is determined based on the technical parameters and specifications declared by the manufacturer

Step 1. Determining VOS related to section

First section is a direct route between Jihlava and Brno, railway line No. 240. The distance between the FINAL DESTINATIONS is 104 km, the occupancy rate of the means of transport is 100 %.

VOS (1st section) = 104 km

Step 2. Calculation of VOS total fuel consumption

F(VOS) = consumption x km = 59.43 x 1.04 = **61.8072**

Step 3. Calculation of total energy consumption and greenhouse gases emissions.

$$E_w(VOS) = F(VOS) \times e_w = 61.807 \times 42.7 = \mathbf{2639.17\ MJ}$$

$$G_w(VOS) = F(VOS) \times g_w = 61.807 \times 3.24 = \mathbf{200.26\ kgCO_{2e}}$$

$$E_t(VOS) = F(VOS) \times e_t = 61.807 \times 38.9 = \mathbf{2404.3\ MJ}$$

$$G_t(VOS) = F(VOS) \times g_t = 61.807 \times 2.67 = \mathbf{165.03\ kgCO_{2e}}$$

Step 4. Determining share for each section

$$S(leg) = T(leg) / T(VOS) = (1 \times 104) / (98 \times 104) = \mathbf{1.02 \times 10^{-2}}$$

$$E_w(leg) = E_w(VOS) \times S(leg) = 2639.17 \times 1.02 \times 10^{-2} = \mathbf{26.92\ MJ = E_w(TS)}$$

$$G_w(leg) = G_w(VOS) \times S(leg) = 200.26 \times 1.02 \times 10^{-2} = \mathbf{2.04\ kgCO_{2e} = G_w(TS)}$$

$$E_t(leg) = Et(VOS) \times S(leg) = 2404.3 \times 1.02 \times 10^{-2} = \mathbf{24.52\ MJ = E_t(TS)}$$

$$G_t(leg) = Gt(VOS) \times S(leg) = 165.03 \times 1.02 \times 10^{-2} = \mathbf{1.68\ kgCO_{2e} = G_t(TS)}$$

In the same way, the indicators for a 50% occupancy rate of the vehicle on the relation Brno – Jihlava adding the return route. Table 2 shows the final resulting values of energy consumption and greenhouse gases emissions in the given transport relation. [13, 17]

Table 2. Final results and impact of number of passengers

	Resulting values of rail transport 1st section (100% occupancy)	**Resulting values of rail transport 1st section (50% occupancy)**	**F Resulting values of rail transport 1st section (0% occupancy)**
Ew(TS)	26.92 MJ	80.76 MJ	2666.09 MJ
Et(TS)	24.52 MJ	73.57 MJ	2428.82 MJ
Gw(TS)	2.04 kgCO2e	6.13 kgCO2e	202.30 kgCO2e
Gt(TS)	1.68 kgCO2e	5.05 kgCO2e	166.71 kgCO2e

The case study deals with the significance of the number of passengers as an aspect influencing final results of energy intensity and greenhouse gases emissions. Before the application, it was not clear what the influence of the number of passengers transported on the calculations and the final result is. For the calculations, two variants of return route were chosen, that differed only in the number of the passengers transported by the means of transport. Using this simulation, it was possible to show a maximum possible impact of the number of passengers, since both maximum and minimum value of the occupancy rate has been included in the calculation. [18] The Table above shows the difference in values caused only by changing the number of passengers transported.

As seen in Table 2 showing all final results analyzed in the study, the number of passengers has a significant influence on final results. Exaggeratedly it can be said that the extreme differences in the values are caused only by the effectiveness the operation of the vehicle was carried out with. The transport was effective if it was able to attract passengers, which resulted in the values on the left-hand side of Table 2, which are significantly lower. The values on the right-hand side are much higher only due to the fact that there were no passengers for a half of the transport route. In principle, the emissions generation and energy intensity of vehicles do not change much. This methodology shows a new direction in thinking and understanding GHG emissions generation from a completely different point of view.

3. Conclusion

Lastly, it is necessary to focus on developing proposals, taking into account the study and data processed in the paper, where it is necessary to point out the need for gradual electrification of railway lines in the Czech Republic. In the Czech Republic, there is currently about one third of railway network electrified, which is below average in comparison with Western Europe. Unfortunately, the rail transport in the Czech Republic is not at the required level, and this is also one of the reasons why the share of railway transport in the Czech Republic is also below average.

An important tool for the development of quality public transport, which is the main way of reducing global production of emissions from transport, are also integrated transport systems of territorial units. Simplicity, continuity, but also uniformity of carriers, when it is possible to use various modes of transport with one travel pass, this all leads to preference of public transport and reducing congestions and other negatives of heavy traffic in the future.

A possible way to reduce emission gases generation is using alternative power, such as maglev trains or hybrid traction. There are even more options in road transport, ranging from electric engines to hydrogen vehicles. However, it is a fact that many alternative technologies have not been developed enough to represent an equivalent alternative to current modes of transport for acceptable financial investment.

References

1. E. Demir, W. Burgholzer, M. Hrušovský, E. Arıkan, E., W. Jammernegg, T.V. Woensel, Transportation Research Part B: Methodological, **93**, 789-807, (2016)
2. T. Skrucany, S. Semanova, T. Figlus, B. Sarkan, J. Gnap, Communications - Scientific Letters of the University of Zilina, **19**, 3-9, 2 (2017)
3. F. Kellner, Energy Policy, **98**, 565-575. (2016)
4. F. Kellner, Logistics Research, **9**, 1. (2016)
5. T. Skrucany, M. Kendra, M. Skorupa, J. Grencik,T. Figlus, Procedia Engineering, **192.** 806 – 811. (2017)
6. F. Cavallaro, F. Maino, V. Morelli, European Transport - Trasporti Europei, **55**, (2013)
7. G. Dalkic, O. Balaban, H. Tuydes-Yaman, T. Celikkol-Kocak, Journal of Cleaner Production, **165**, 746-761. (2017)
8. G. Fedorko, V. Molnar, J. Strohmandl et al. *Transport Means 2015 – Proceedings of the International Conference.* 466-+ (2015)
9. T. Kirschstein, F. Meisel, Transportation Research Part B: Methodological, **73**, 13-33, (2015)
10. M. Mohsen, M.B. Ahmed, J.L. Zhou, Transportation Research Part D: Transport and Environment, **62**, 112-124, (2018)
11. Y. Ge,.J. Long, F. Xiao, Q. Shi, Transportation Research Part D: Transport and Environment, **60**, 1-6, (2018)
12. F. .Ma, W. Wang, Q. Sun, F. Liu, X. Li, Sustainability (Switzerland), **10**, 2 (2018)
13. A. Galierikova, J. Sosedova. Ekologia:**35**, **3**. 279-288, (2016)
14. A. Torok. Transport **24**, 26–29, (2009) DOI: 10.3846/1648-4142.2009.24.26-29
15. W. Yang, S. Wang, X. Zhao, Sustainability (Switzerland), **10**, 5 (2018)
16. V. Hromadka, E. Vítkova, H. Kovarova, P. Vařbuchta, *International Multidisciplinary Scientific GeoConference Surveying Geology and Mining Ecology Management, SGEM*, (2014)
17. M. Kendra, 1st *International Conference on Engineering and Applied Sciences Optimization, Proceedings*, (2014)
18. MM. Zefreh, F. Meszaros, R. Junevicius, A. Torok. Promet-Traffic&Transportation, 29,1, 77-84, 10.7307/ptt.v29i1.2072
19. M. Vojtek, M. Kendra, V. Zitricky, J. Danis, MATEC Web of Conferences, **134**, 00061, (2017)
20. L. Hansut, A. David, J. Gasparik. *Bussines Logistic In Modern Management*. 29-42. (2017)

Irregularities in Czech Integrated Periodic Timetable

Michal Drábek[1*]

[1]Czech Technical University in Prague, Faculty of Transportation Sciences, Department of Logistics and Management of Transport, Horská 3, 128 03 Praha 2, Czech Republic

Abstract. The aim of this paper is categorization of various irregularities in Integrated Periodic Timetable (IPT), according to their impact on passengers. For instance, minor deviations in departure minutes can cause almost zero impact, provided that all periodic transfer connections are ensured for each service. On the other hand, skipping of a periodic service at off-peak hours can considerably lengthen some passenger transport chains. Chosen categories of irregularities are illustrated by examples from 2017/2018 Czech railway timetable.

1 Integrated Periodic Timetable (IPT)

Contrary to periodic timetable, as used, for instance, in urban public transport, IPT requires unified service period and unified symmetry time among all involved public transport lines.

Unified period means that there is some basic network period $t_{p,b}$ (e.g. 60 min), and any other period $t_{p,i}$ has to fulfil following equation

$$t_{p,i} = 2^k \cdot t_{p,b} \tag{1}$$

where k is an integer.

If a 60 min basic period is chosen, periods of 120, 30 and 15 min can be used as well.

Unified timetable symmetry is necessary for symmetrical transfer connections within both directions. Symmetry time of a public transport line occurs when two services in opposite directions meet each other. In European railway network, symmetry time in (or slightly before) the minute 00 has been established as a rule. It is desirable to design an IPT-node in a station where two or more public transport lines intersect. In an ideal IPT-node, all services arrive before the minute 00, and all services depart after the minute 00. For the sake of maintaining unified zero symmetry, each public transport line should keep following rule. If the service from the station A arrives x minutes before 00, the service in opposite direction (to A) should depart x minutes after 00. Thus, the first arrivial corresponds to the last departure back (within the particular public transport line and direction).

Unified basic period, unified symmetry time and system of IPT-nodes imply some mathematical preconditions – arc and circle equations [1-3].

If two services (with 60 min period) depart from two neighbouring «zero» IPT-nodes, towards each other, they meet each other in another symmetry time – the minute 30. If they stop, another IPT-node in the minute 30 can be designed there. If generalized to any basic period, arc equation comes out, formulated as follows:

$$t_{T,A\leftrightarrow B} = (n/2) \cdot t_p \tag{2}$$

where $t_{T,A\leftrightarrow B}$ is travel time between IPT-nodes A and B, including both half of dwell time in node A and half of dwell time in node B, and n is natural number. Scheduled travel time between two IPT-nodes, rounded up to fulfil the arc equation, is called system travel time.

In any IPT, there are two disjunct groups of IPT-nodes (e.g. for 60 min period and zero symetry, it is group of nodes around the minute 00, and group of nodes around the minute 30). IPT works properly provided that all services arrive into any IPT-node before the same symmetry time and depart thereafter. If the symmetry time is the minute 00, any arrival before the minute 30 leads to unacceptable transfer time between services. Thus, even stronger precondition than arc equation has to be maintained – formulated as cycle equation:

$$\forall(A,m)\colon t_{T,A\leftrightarrow A,m} = n \cdot t_p \tag{3}$$

where $t_{T,A\leftrightarrow A,m}$ is travel time along m-th cycle from node A into node A, including at least 2 another IPT-nodes and dwell time in A, and m, n are natural numbers.

All preconditions for proper working of IPT for passengers are summarized below:

- Exact basic period, or its 2^k-multiples (implies unified travel time for each section).
- Unified symmetry time (implies equal travel time there and back).
- Arc equation (system travel time).
- Circle equation (two disjunct groups of IPT-nodes).

* Corresponding author: xdrabek@fd.cvut.cz

2 Irregularities with network impact

Network impact means that at least one periodic transfer connection is broken due to irregularity in IPT. Either a periodic service is skipped, or the minimum transfer time cannot be kept due to late scheduled arrival of particular service.

Skipping of periodic connections at off-peak times is motivated by (at least variable) cost saving for public service orderer. However, two or more skipped services in «unlucky» time sequence can considerably lengthen public transport chain. For instance, a passenger travelling from Česká Lípa to Náchod has to transfer twice – in Nymburk and Hradec Králové. Fast train lines operate in 120 min period. In the ideal case, the travel time between two cities by train lasts about 3.5 hours. However, at weekends there are some services skipped. As a result, in 2017 Czech timetable, from 7 daily connections, four ones lasted 1 hour longer, due to skipped services on one or two involved public transport lines (R22, R10 [4] and a regional line). All three lines contained some skipped services at weekends. In other words, only three connections daily offered travel time, planned by public service orderers [5].

One regional public service orderer, Liberec Region (Liberecký kraj), via its subsidiary KORID LK and in cooperation with neighbouring Central Bohemia Region (Středočeský kraj), has patched a pair of skipped fast train services of the R22 line [4] around noon by a pair of regional train services 6022 and 6009 between Rumburk and Mladá Boleslav (railway lines 081 and 080). It sounds like irony of fate, since both fast and regional trains on this corridor are operated by state-owned carrier Czech Railways (České dráhy). But one orderer is forced to fix a hole that another orderer was unable to fix himself – both by one railway carrier. Another interesting fact is that this pair of train services serves between Česká Lípa and Bakov nad Jizerou 7 stops more (mostly on request) than the fast train. Nevertheless, interposition into exactly periodic times has been kept – likely thanks to off-peak time and thus low passenger demand. Further, from Mladá Boleslav to Kolín, the pair of fast train services is skipped without any patching [6].

The longer waiting time has considerable impact on public transport quality, as estimated for instance by Gašparík et al. [7].

The officially declared cause of skipped periodic services is lack of available funding [8]. The most likely reason, however, is the fixed amount of compensation per train-kilometre in public service contracts (which comes out from negotiated sum of train-kilometres and total amount of compensation). The vast majority of sum of publicly ordered train-kilometres in the Czech Republic is operated by Czech Railways. Although Czech Railways very often can profit from more efficient use of their rolling stock thanks to network-wide operation, they insist on fixed price of train-kilometre even if this would cost no extra trainset.

An interesting case is shift of periodic service about half period. The reason may be diluting of the offer during off-peak time (especially before noon or during late evening – see Figure 1), or plenty of timetabling constraints. The other variant occurs on international line of express trains between Poland and Vienna. One pair of services terminates in Budapest instead. In the Czech Republic, this line is marked as Ex4 [4]. Its services keep periodic times, but their period varies. Either it is 120, or 180 or even 240 minutes [6].

km	SŽDC, státní organizace / ČD, a.s. Vlak	Rx 729 R	8239	8269	R 731 R D	R 733 R D	8259 R D
	Ze stanice						
0	**Praha**-Holešovice 091				21 48	23 18	
3	**Praha** hl.n.	20 31			22 01	23 31	
6	**Praha**-Vršovice	20 36			22 06	23 36	
52	**Benešov u Prahy** 222	21 11			22 41	0 11	
	Benešov u Prahy 222	21 12		22 21	22 42	0 12	
57	Bystřice u Benešova			×22 26	J		
62	Tomice			×22 31	A		
67	**Olbramovice** 223	21 21		22 36	N 22 51	0 21	
	Olbramovice 223	21 22		22 37	22 52	0 22	
71	Votice			×22 41	H		
76	Heřmaničky			22 48	U		
81	Ješetice			22 55	S		
83	Červený Újezd u Votic			×22 58			
85	Střezimíř			23 01			
89	Mezno						
92	Sudoměřice u Tábora						
97	Chotoviny						
106	**Tábor** 201,202,224	21 56			23 23	0 54	
	Tábor 201,202,224	21 58	22 45		23 24		0 59
108	Tábor-Čápův Dvůr		22 47				
110	Sezimovo Ústí		22 49				1 03
113	Planá nad Lužnicí	22 03	22 53		23 30		1 07
117	Doubí u Tábora		22 56				
120	Roudná		22 59				
126	Soběslav	22 14	23 05		23 41		1 18
129	Řípec-Dráchov		23 08				
133	**Veselí nad Lužnicí** 225,226	22 19	23 11		23 46		1 24
	Veselí nad Lužnicí 225,226	22 21	23 18		23 47		1 25
136	Veselí nad Lužnicí zastávka		23 21				1 29
139	Horusice		23 24				
143	Dynín		23 28				
146	Neplachov		23 32				
150	Ševětín 108		23 37				1 41
154	Chotýčany 108		23 41				
162	Hluboká nad Vltavou-Zámostí 104		23 48				
165	Hosín 104		23 51				
167	Hrdějovice 101		23 53				
170	Č. Budějovice severní zastávka 190 101		23 57				
172	**České Budějovice** 101	22 52	24 00		0 17		2 06

Fig. 1. Cut-out from 2017/2018 timetable Praha – České Budějovice. During late evening, 90 min period is used [6].

Another case is an "almost-transfer" when both services operate, but minimum transfer time cannot be kept. In this category, few minutes do not matter, because arrival time of the service is completely different – mostly intentionally.

The first variant, more common at the morning peak, is a service adjusted to start of school tuition, or shift by large employer. This irregularity is difficult to cope in more sparsely populated areas, since an extra trainset for extra «periodic» connection cannot be justified by sufficient passenger demand.

The second variant occurs if system travel time cannot be fulfilled due to infrastructure constraints – or due to number of stops.

In the Czech Republic, there is another very widespread timetable irregularity – non-systematic differences in days of operation of the train services. This issue is so complex that it cannot be fully described in this paper. One illustrative example is the R26 fast train line Praha – Beroun – Písek – České Budějovice [4], regarded by the orderer as «the least evil» whenever some services have to be reduced for financial reasons [8]. Thus, some services run for instance only on Mondays or on Fridays. On the other hand, South Bohemian Region (Jihočeský kraj) has ordered fast trains between České Budějovice and Písek město when the R26 fast trains do not run. However, this order is not ideal. For instance, the pair of 633 and 634 fast trains serve the station Písek město on workdays, but without Fridays, since fast train to and from Prague runs on Fridays from the station Písek through another line – see timetables of railway lines 200 and 190 [9].

3 Irregularities with no network impact

No network impact means that these irregularities do not break periodic transfer connections off. They can be divided as follows:

- Minor irregularities that enable maintenance of minimum transfer time.
- Local irregularities (periodic arrival and departure times in transfer nodes are maintained).
- Major irregularities.

The first group is represented by few minutes' deviation in periodic arrival and departure times in the node stations. If the scheduled transfer time is shortened, the buffer time can be exhausted this way, so such connection can lower stability of the traffic. In some cases, buffer time is added to absorb the irregularity and keep it limited to a particular section. An interesting example is local shift of departure and arrival times of the R18 long-distance line between Přerov and Staré Město u Uherského Hradiště [4] (railway line 330). This shift is driven by fluctuating period of Ex4 line, mentioned in the chapter above. If the train of Ex4 runs, the R18 train departs from Přerov 3 min earlier to avoid timetabling conflict. The buffer time as long as 3 or 4 min is added in Staré Město where electric locomotive is changed to the diesel one. If there runs no Ex4 train, the R18 train waits in Přerov and arrives in Staré Město to dwell just as long for the locomotive changing. This irregularity occurs in both directions, but the difference in departure times backwards from Staré Město towards Přerov is only 1 min [6].

In extreme cases, hold-up for extra stoppings is absorbed by buffer time and low demand thanks to off-peak time (see the "patching" regional trains mentioned in rhe chapter above). [10]

The second group is represented by terminus that differs from the periodic pattern or by passing through some periodically served stops.

The third group is represented by extra peak services that do not replace periodic services.

Figure 2 displays several types of irregularities that are built in rather periodic timetable.

The first, major irregularity, is termination of 7452 train in Stod. Thus, the rest of line to Domažlice město is not served 100% hourly.

The second, middle irregularity, is infrastructure-driven waiting of regional trains in Staňkov station (in even hours only) to let run the express train in the opposite direction. [11] Thus, precise interposition into hourly service is no more possible between Staňkov and Domažlice město. In the opposite direction, waiting time is shorter, probably because of preference of morning commuting to afternoon travelling back. This irregularity affects timetable symmetry between Staňkov and Domažlice město.

The third, minor irregularity, is scheduled running of 7416 train between Domažlice and Domažlice město at weekends only. The reason may be rolling stock utilization at the afternoon peak (fast turn-around in Domažlice).

The fourth, minor irregularity, is skipping of stop in Dolní Kamenice for several trains (sometimes the train stop there at weekends).

The fifth, minor irregularity is a slight difference in scheduled runtimes of 7410, 7412 and 7416 trains. The reason might be different dynamic traction characteristic of different trainsets [6, 9].

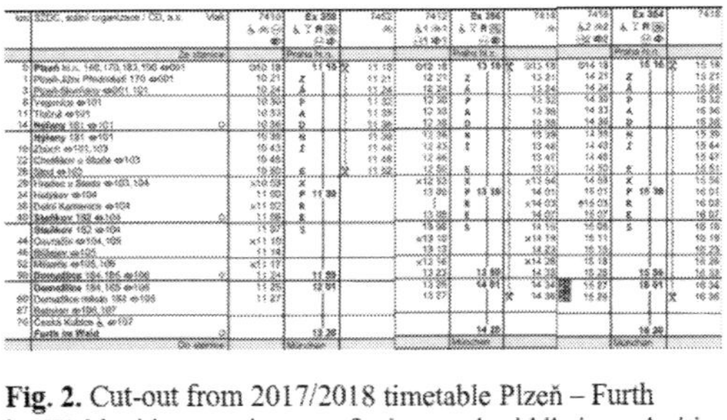

Fig. 2. Cut-out from 2017/2018 timetable Plzeň – Furth im Wald with several types of minor and middle irregularities [6].

4 Recommendations for reduction of network impact

The author proposes three, rather soft, criteria for the decision on seriousness of network impact of irregularity in IPT:

- Daytime.
- Significance within railway network.
- Complexity of transfer connections.

4.1 Daytime

The most serious irregularities in IPT, like major shift of departure and arrival times of the service, are acceptable at the morning peak, preferably in the direction with larger passenger flows, provided that

- There is more passengers profiting from the irregularity than losing due to it (e.g. they may lose the transfer connection).
- Otherwise there would be an extra trainset needed and the amount of passenger flows does not justify it.

Another times with acceptable major irregularities are «edge» times – early morning or late evening. The demand is comparably low and trainsets may depart from depots or arrive towards them. However, as much transfer connections as possible should be kept. So the irregularity should spread outside the transfer node if possible.

4.2 Significance within railway network

The author recommends following criteria for significance:

- Significance of settlement served.
- Temporal proximity to agglomeration centre.
- Centrality within railway network.

The sense of the first criterion is evident.

The second criterion is significant in proportion to size and transport attractiveness of the agglomeration.

For instance, 60 min travel time to Český Těšín is much less significant than 60 min travel time to Brno.

The third criterion is linked to railway capacity utilization. The more central (i.e. more utilized) element of railway network is affected by the irregularity, the larger impact on another periodic railway services. Thus, for instance, irregularity between Kolín and Pardubice has much larger impact that irregularity between Břeclav and Znojmo.

4.3 Complexity of transfer connections

The author recommends following criteria for complexity of transfer connections:

• Number of node stations with transfer connections along the route of public transport line.

• Amount of the highest appeared number of transfer connections in one station.

5 Conclusion

There is clear evidence that IPT has increased quality of passenger railway and demand for it [2]. However, its inconsistent application by many public service orderers, combined with low emphasis on efficient rolling stock utilization and on potential synergistic effects between various orders by Czech Railways has led to plenty of irregularities that undermine the very nature of ITF – to offer periodic, lucid, uncluttered public transport service to passengers.

Of course, the number of passengers equipped with mobile internet is steadily growing. These passengers can find their timetable online anytime. Nevertheless, if they can only memorize simple information, such as «departure in the minute 42, and at peak times in the minute 12 in addition», the public transport would be much more attractive and user-friendly.

The research and publishing of this paper was supported by Zéta Programme of The Technology Agency of the Czech Republic (TAČR), as a part of the research project TJ01000162 "Synergy in Railway Public Transport Line Planning – Improvement of Efficiency of Spatial Public Railway Transport Service" – co-financed by CTU in Prague. Actual (so far unpublished) version of the project research report [10] was cited in this paper. The author hereby acknowledges the funding sponsors.

T A
Č R Program Zéta

References

1. Forschungsgesellschaft für Straßen- und Verkehrswesen, e. V. *Merkblatt zum Integralen Taktfahrplan* (FGSV-Verlag GmbH, Cologne, Germany, 2001, pp. 18-20)
2. V. Borza, V. Janoš, I. Neumann, *R. Roth, H. Jacolin, Eastern European railways in transition: Nineteenth to twenty-first centuries*. 327-342 (2013)
3. V. Janoš et al., *Lecture in the subject Transport Technology and Logistics*. (2017)
4. SŽDC, s.o *Portál provozování dráhy* (2017). Available online: http://provoz.szdc.cz/PORTAL/Show.aspx?path=/Data/Mapy/linky_dalkove_dopravy.pdf
5. R. Papež, *ŽelPage*, Oct 19, (2016) Available online: http://www.zelpage.cz/zpravy/9697?lang=cs&oddil=6#komentare
6. SŽDC, s.o., *Jízdní řády (platné od 10. 6. 2018)*, (2017). Available online: https://www.szdc.cz/provozovani-drahy/knizni-jizdni-rady.html
7. J. Gašparík, V. Ľuptak, P. Mesko, *Proceedings of the 3rd International Conference on Traffic and transportation Engineering (ICTTE)*, 388-392 (2016), WOS 000391016300055)
8. J. Klimková, J. Čamaj, J. Gašparík. *Transport means 2016 : proceedings of the 20th international scientific conference*. 891-896. (2016)
9. Ministerstvo dopravy, *Plán dopravní obsluhy vlaky celostátní dopravy*. (2017). Available online : https://www.mdcr.cz/getattachment/Dokumenty/Verejna-doprava/Financni-ucast-statu/Plan-dopravni-obsluhy-uzemi-vlaky-celostatni-dopra/Plan-dopravni-obsluhy-uzemi-2017-2021.pdf.aspx
10. SŽDC, s.o., *Sbírka služebních pomůcek pro jízdní řád 2017/18* (2017) - Unpublished
11. M. Drábek et al., *Analytická část. Zpráva o dosažených výsledcích projektu TJ0100162 řešeného s finanční podporou TA ČR*. Research report in progress, to be finished in 2019, (2018) – Unpublished

Logistics of information in intermodal transport

Cezary Mańkowski[1,*] and *Dariusz Weiland*[1],

[1]Department of Logistics, Faculty of Economics, University of Gdansk, Poland

Abstract. The contemporary economy is characterised by a high degree of reliance on modern technologies and information systems. Modern technologies have become a source of innovativeness and a tool for building competitive advantage across all sectors of the market, including transportation. Such technologies as telematics, RFID and GPS/GPRS have contributed to the development of transport systems. Information logistics is crucial in order for the modern technologies to be utilised effectively in transportation. This necessity is driven by the fact that information is a type of resource and hence requieres adequate logistics. The same applies to intermodal transport, where it is vital to manage the stores of information appropriately and efficiently through well organised logistics. This article discusses logistics of information as one of the key auxiliaries to intermodal transport. It also outlines an innovative approach to information as viewed from the perspective of domestic railway companies, which may benefit from implementing information logistics in order to utilise such assets as railway infrastructure, supply trains, train stations, sidetracks and ground adjacent to railtracks. This, in turn, will allow them to transition from being traditional transport suppliers to the rank of logistics integrators responsible for handling intermodal transport. The authors elaborate on the possible functional mechanisms of an intermodal transport market whose management and supervision is assumed by a domestic railway service supplier, thereby becoming an integrator of logistics for intermodal transport whose operating draws on information flows between operators working in different sectors of transport.

1 Inntroduction

The contemporary economy forces businesses into searching for new, often innovative and above all economically more viable theories and practices of organizing logistics. These trends would certainly not have emerged had it not been for the development of the theory of logistics as well as the contemporary IT systems. It was precisely the synthesis of logistics and IT tools that allowed companies to further tailor their products and services to customer needs, thus building a competitive advantage [1]. For all these decision processes to occur, however, it is crucial that information resources be utilized adequately. What characterizes the type of information that companies use these days is the necessity for it to be handled by means of information and informational systems, which lie at the core of the system of logistics. Both decision-making processes and the constant evolution of the logistics of material assets have been and will continue to be a direct result of the changes in systems of information delivery [2]. The advancement of computer systems and networks has created a specific environment for the growth of information logistics across all sectors of trade and economy. The expansion of information logistics, in turn, created fertile ground for the emergence of ideas, conceptions and utilizations of logistics on a heretofore unprecedented, global scale, a framework for handling economy-related processes within the 5 primary purposes of logistics (5R) – to supply the adequate resource, to an adequate place, at an adequate time, in an adequate amount and at an adequate cost [3].

The transport market has also used the implications of logistics for the development for IT tools. Transportation businesses are now increasingly taking an advantage of such systems as telematics, RFID, GPS and GPRS. Despite the importance of these, it is information logistics that is most crucial for the effective utilization of modern IT tools. The ideas and conceptions created by logisticians have been a major contributor in implementing IT tools in transport and within logistics in general. While these tools do play a vital role in transportation by car and by air, the theories founded within these sectors are being increasingly employed in maritime and railway transport.

It is important, however, to create new implementations for IT tools and logistics especially in the domain of the development of intermodal transport. The purpose of this article is to delineate an outline of how the already existing IT tools may be used for the growth of the branches of intermodal transport.

2 Information as a resource in logistics

Regardless of which sector of the economy a given business is operating in, it necessarily produces, stores

* Corresponding author's: Cezary Mańkowski – ekocm@ug.edu.pl ; Dariusz Weiland – d.weiland@ug.edu.pl

and utilizes information. In general terms, information can be not only produced, stored and utilized (for example in the decision making-processes), but also sold, which qualifies it as a full-fledged resource. Information is of particular importance when multiple businesses cooperate with one another in order to provide satisfaction to the target customer. In the process, information becomes a substrate for providing added value for the customer and building a competitive advantage. However, the idiosyncratic properties of information set it apart from other resources in logistics. These include [4]:

- the inexhaustibility of information and the possibility of processing it without causing any depletion whatsoever,
- substitutionality,
- complementarity,
- the objectivism of information,
- the virtual nature of information,
- the synergistic nature of information,
- diversity,
- its unlimited capacity to be multiplied and moved in time and space,
- the subjectivity of assessments,

One of the most prominent features differentiating information from other resources is that it does not get depleted during its processing or utilization in the chain of the production processes, one of which being the decision-making procedures. The implication of this fact is that multiple agents may use information simultaneously without any necessity to restore it. The inexhaustibility, however, ought not to be confused with obsolescence and outdatedness of information, since in today's world the lifespan of information is reduced to a minimum.

What also sets information apart from other resources is its virtual nature, which enables it not to be permanently attached to any data carrier. The same piece of information may be stored on multiple drives without any detrimental effects on its value to the recipient. This virtuality is closely related to information's capacity to be multiplied ad infinitum across both space and time. It can be transferred from one data carrier to another and between users no matter how far apart they are.

Aside from the above-mentioned features, information is also characterized by a complex of four properties designated as 4V, which properties have been heretofore primarily applied only to big data [5]:
volume – great amounts of information,
velocity – great variability of information,
variety – great variety of information,
value – validation of information,

Regardless of their intentions, all companies are increasingly encountering big data in reference to large stores of data and information, which compels them to seek out and create new IT systems that would allow them to purify information off its many irrational elements in accordance with the goals of logistics. These systems are being designed to also support the decision-making processes, discover new interconnections and relations, and to optimize processes. At the core of logistics lie the functions that it performs: servicing, processing and integrating. It must be observed, however, that logistics is never the only process in a system, rather, it serves as an auxiliary to the primary process.

Providing logistical support for this process is realized via a series of sequentially arranged activities, whose purpose is to realize the great 5 objectives of logistics (5R). Each of these activities, in turn, is accompanied by an appropriate logistical process. Moreover, in accordance with the integrational function of logistics, all the individual sub-processes that the logistical process consists of need to come together into an integrated unit of subcomponents in relation to the primary process. Successful realization of any trade-related process depends on a well-coordinated functioning of the components of the logistical support for said process [3].

3 The logistics of information as a component of the system of logistical support for intermodal transport

Intermodal transport may be defined as an unobstructed movement of payloads via more than one means of transport, with a caveat that said payload is seen a single unit of transportation; during trans-shipment, what is handled is a loading unit (such as a container) rather than goods themselves (the contents of the container) [6]. What characterizes and differentiates intermodal transport from other means of transport is its relative level of safety, its capacity to meet the deadlines and optimization of costs [7]. Other definitions [8,9,10] emphasize the fact that there is only transit agreement – meaning that the responsibility for the entire length of transportation falls on one transport service supplier – or that there is no need to use the services of more than one operator, although in practice it appears more reasonable to arrange a close cooperation and integration between several operators rather than to employ of a cohort of means of transport managed by a single company.

The development of this branch of transport is unfortunately severely restricted mainly due to a lack of road, tool and information infrastructure. There still are not enough pre-loading terminals in Europe, and those that do exist have a low functioning capacity [11]. Another important problem is a lack of unified standards for the exchange of information between transport providers and the customers, which should be uniform on an international level, while the software used by companies ought to be made compatible with information and informatics systems and in the networks that land and sea-land transport chains use. One faces similar obstacles in the relations between transport operators and customers, whereby the latter often receives an insufficient amount of information from the former.

Intermodal transport is characterized by an immense flow and importance of information. Information is crucial if this type of transportation is to operate correctly, which is a consequence of two factors: the necessity to acquire and process information for the purposes of a particular process, and its importance in communication. Said communication obtains not only within a company responsible for given transportation, but also between companies and their customers or other participants of cooperation. Hence, providing logistical support of intermodal transport consists primarily in acquiring, processing and transferring the information that is relevant at every stage of the process; further, it consists in updating said information along the way and organizing an undisturbed flow of it both with a company and between the company and its surroundings [12].

Information is delivered to businesses via IT systems, which makes said systems a key component of the system of logistical support. It is an information system that is responsible for acquiring and processing data for user-defined purposes. These data are obtained from within the company and its surroundings. Once data is introduced into the system, it becomes information. Another function of IT systems is that they mediate communication between different company departments, between customers and contractors. If well thought-out, thoroughly planned, built in line with the objectives of logistics and equipped with the indispensable infrastructure, an IT system should be capable of delivering information to all workplaces along every level of organizational hierarchy without neglecting any of the purposes of logistics (5R) in reference to the information stores. Unfortunately, intermodal transport poses a big challenge for IT and logistics sectors, as it is necessary to define and construct a system that would offer multiple points of data acquisition, as well as service multiple information recipients simultaneously. The inevitable consequence of this is reduced output of the traditional data bases, which has been incentivizing attempts at creating NoSQL data bases and data bases with blockchain technology as their foundation. Furthermore, the unique properties of intermodal transport make it necessary to integrate at least two different means of transport, thus necessitating implementation of such systems enable multiple businesses to establish integrated and coherent channels of communication and means of exchanging data and information. In intermodal transport, data are downloaded primarily through RFID devices, telematics tools, GPS in transport chains, via transport-monitoring devices, and other GSM equipment of short- and long-ranges. These data are produced *en masse* and go out of date rapidly due to incessant movement, which is why contemporary information was compared to big data in the previous part of this article. Another challenge of equal importance is how to provide sufficient securing of the information system from unauthorized use of its data via encryption techniques based on complex algorithms, which are necessary because these data constitute private business information. The threats that these systems most frequently encounter can be subdivided into the following categories [13]:

- obtaining access to the date being transmitted via the internet or stored on servers by unauthorized agents;
- obtaining unauthorized access to other resources, such the computational power;
- loss of data due to malware or as a result of cyber-attack,
- falsification of data,
- attacks that make a machine or network resource unavailable to its intended, authorized users (Denial of Service).

Whether intermodal transport is successful within the coming years will hinge on the rate at which these obstacles and limitations are successfully diffused through implementation of innovative IT solutions that would integrate various systems of communication between different links along the chain of intermodal transport. The information regarding documents, storage units, trans-shipment equipment, transport chains, infrastructure, prices, etc., is expected to be stored in one place.

4 Railway transport companies as integrators in intermodal transport

Railway transport companies play a dominant role in the majority of intermodal transport systems. These service providers have at their disposal their own fleets and a fully-operational infrastructure, and they are often owners or managers of trans-shipment terminals [14]. What also distinguishes railway transport companies is possession of multiple other assets, which often remain underutilized such as inactive sidetracks and the land adjacent to the rail tracks, which, in the future, may be used to build a competitive advantage within the intermodal transport sector and facilitate development of terminal infrastructure. What follows is a two-section outline of a model of organizing intermodal transport. The first section discusses how railway transport providers may start serving a integration role in intermodal transport, while the second section elaborates on possible routes of implementation and utilization of innovative IT systems that would enable unobstructed flow of information and acquisition of new customers.

As previously discussed, one of the key functions of logistics is integration, and for integration to be appropriately implement, the role of the rail transport provider needs to be changed from that of a transport company to one of a 5PL logistics operator (5 logistics partner). A logistics operator (5PL) integrates multiple components within a single information system, including recipients, users, contractors and the existing infrastructure in order to maximize target customer satisfaction and optimize the costs [15]. At the same time, the information system of the operator creates a digital equivalent of the logistical processes that occur in intermodal transport [16] (Figure 1).

https://doi.org/10.1051/matecconf/201823500013

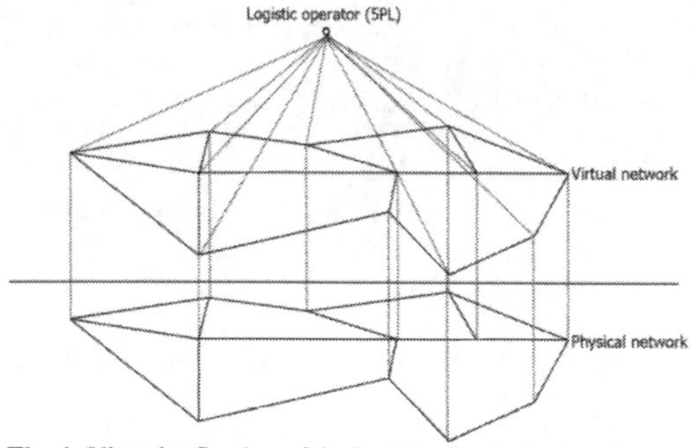

Fig. 1. Virtual reflection of the logistic physical network
Source : Own elaboration

Creating these digital records consists in defining a virtual network that would mirror the on-the-ground network, thus allowing for an effective management of all resources along the entire system and exerting adequate control of individual operations, including those relating to trans-shipment and intermodal terminals. This makes the role of a logistics operator equivalent to that of an integrator of information flow along the entire chain of transport. Supply Chain Control Tower is a concept that translates this idea into the realm of intermodal transport most accurately. However, the tasks of the Control Tower do not end here, as it should also serve as a means of monitoring the centers of costs in a company and organizing the information flow and the entire process of transport from the point at which the arises a need of transportation up until the moment of its conclusion at the doorstep of the target customer.

Supply Chain Control Tower is defined in a variety of ways across the literature. According to one of the broadest and most multi-faceted definitions, SCCT is a centralized system of the required technological level that aims at organizing component processes and acquiring and analyzing data flowing from the chain of delivery in order to provide greater transparency [17, 18]. Such a holistic, real-time perspective on all of the subcomponents of intermodal transport may allow operators of railway logistics sufficient space to make appropriate short- and long-term decisions that would be in accordance with the objectives of the operator as well as the goals of the entire process (Figure 2).

The Control Tower is responsible for monitoring, controlling and support intermodal transport in real time on three levels [19]:

- strategic – it mediates control over the entire on-the-ground network and its virtual counterpart,
- tactical – it is a tool used for planning routes and adjusting business decision to market demands, all the time taking into account the costs of the whole process,
- operational – it comprises a variety of activities and decisions connected with transport management and monitoring, as well as the area of reacting to the arising obstacles.

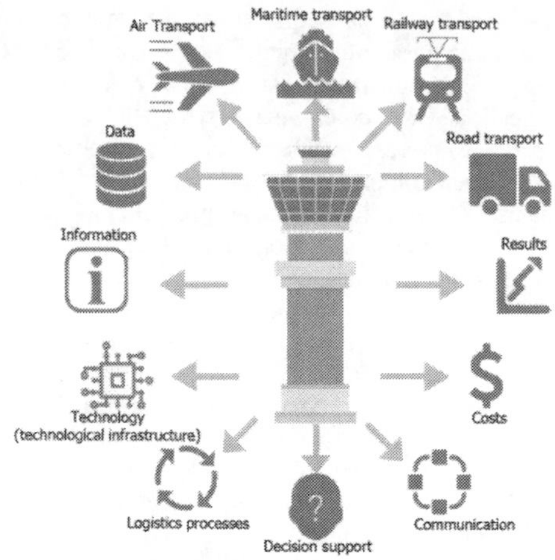

Fig. 2. Control tower for intermodal transport
Source : Own elaboration

Considering the unique characteristics of intermodal transport, it is impossible to directly incorporate the already existing SCCT solutions; what is necessary instead is to create a system based on this framework from ground up and make all the adjustments that will enable it to meet current market demands.

Creating SSCT and position 5 logistics operator by the railway transport provider ought to happen alongside making an information system designed specifically for intermodal transport (the clients and the cooperators within the network of delivery service providers). The other component of the idea presented herein is building a market of intermodal transport that would operate in a similar fashion to other sectors of transport pertaining to transportation via car. The implementation of these solutions would, however, require investments connected with the already owned assets, expanding intermodal terminals, creating new terminals in the vicinity of railway stations, sidetracks and areas of land owned by railway transport providers, and all this with the use of such trans-shipment and transportation technologies as Cargobeamer, Modalohr, Flexiwaggon, and Rollende landstrasse.

An appropriate synthesis of these two segments will allow for a harmonious and uninterrupted organization of a multi-branch transport. A single company would be capable of monitoring the entire process of intermodal transport, thus increasing the overall importance of railway transportation. By investigation the transport market, the clients would be able to localize vacant train cars waiting to transport delivery units and then move their products to that destination. The entire flow of delivery units would be managed by a single company, which would contribute a greater synchronization between different means of transport.

5 Summary and directions for future research

Creating position 5 logistics partner based on the SCCS framework makes it possible to react swiftly to the ever-changing management conditions, market demands and the difficulties experienced by deliverers operating within a single system. The development of data transfer networks aids and facilitates the process of professionalizing activities within logistics with the use of the already existing logistics operators. The idea discussed in this article suggests creating and utilizing a virtual network of data and information flow between all of the participants and agents of intermodal transport along with creating a platform in the form of an electronic market for intermodal transport. For the degree of control that such a system would guarantee to be fully realized, appropriate applications and support by information acquired from the market would be necessary. As a consequence, the process of managing the logistics would be faster even in case of intermodal transport.

The future directions for research in this area should focus on a model-based discussion of the leading idea and on creating a realizability study together with a SWOT analysis.

References

1. D. Weiland, *Studia Ekonomiczne : zeszyty naukowe Uniwersytetu Ekonomicznego w Katowicach.* **306**, 97-110 (2016)
2. M. Chaberek, Informatyczne narzędzia procesów logistycznych, Eds. M. Chaberek, A. Jezierski, pp. 13-23, (2010)
3. M. Chaberek, *Makro- i mikroekonomiczne aspekty wsparcia logistycznego*, p. 15. (2002)
4. E. Skrzypek, G. *Annales Universitatis Mariae Curie-Skłodowska*, Sectio H, Oeconomia 39 (2005)
5. D. Weiland, Research Journal of the University of Gdańsk. Transport Economics and Logistics (Modelling of Logistics Processes and Systems*)* **71**. 179-190 (2017)
6. J. Neider, D. Marciniak-Neider., *Transport intermodalny*, (1997)
7. H. Zielaskiewicz, *Transport intermodalny na rynku usług przewozowych*. (2010)
8. H. Adelsberger, *Intermodal terminals in an integrated transport system of the Baltic Sea region*, (2012)
9. K. Liljestrand, Intermodal transportation from a haulier's perspective. An analysis on how to increase the usage of intermodal road-rail transportation for hauliers in Sweden. *Chalmers University of Technology*, No. 087 (2010)
10. M. Ruesch, A. Heiko, R. Karrer, Integrated Services in the Intermodal Chain (ISIC). *Final Report Task D: "Improving Quality of Intermodal Terminals"* (2005)
11. M. Zalewska-Turzyńska, Technologie informacyjno - komunikacyjne w transporcie intermodalnym, *Infrastruktura transportu dla rodzowju regionów. Zagadnienia teoretyczne i pragmatyczne*, (2014)
12. G. Karwacka, Monitoring informatycznych komponentów systemu logistycznego przedsiębiorstwa. *Informatyczne narzędzia procesów logistycznych*, Eds. M. Chaberek, A. Jezierski, pp. 77-85 (2010)
13. S. Sokół, Zagrożenia w sieci. Available from http://home.agh.edu.pl/~szymon/artykuly/legionowo.html [Accessed 2 September 2018]
14. J. Stokołosa, Š. Liščak, M. Jaśkiewicz, K. Ludwinek, Systemy transportu intermodalnego – kierunki rozwoju w świetle europejskich doświadczeń. Logistyka No. 6 (2014)
15. Fifth party logistics model (5PL), Logistics Glossary. Available from http://www.logisticsglossary.com/term/5pl/
16. Z. Bentyn, E-logistyka narzędziem rozwoju globalnych sieci dostaw. *Gospodarka materiałowa i logistyka* No.5 (2015)
17. B. Benz, Supply Chain Control Tower help organizations respond to new pressures, *Supply Chain Management Review* 18(4), pp. 34-39 (2014)
18. A. Trzuskawska-Grzesińska, Control towers in supply chain management – past and future, *Journal of Economics & Management*, 27, pp. 114-133, (2017)
19. M. Makar, M. Karkula, Monitorowanie i kontrola łańcucha dostaw: koncepcja Supply Chain Control Tower i jej zastosowanie w przedsiębiorstwie handlu detalicznego. *Prace Naukowe Politechniki Warszawskiej, Transport*, No. 120 (2018)

NEW TRAINING SCHEMES FOR THE FUTURE EDUCATION IN RAILWAY SECTOR

Dana Sitányiová[1]* and *Jaroslav* Mašek[2]

[1]University of Žilina, Faculty of Civil Engineering, Univerzitná 8215/1, 010 26 Žilina, Slovakia
[2] University of Žilina, Faculty of Operation and Economics of Transport and Communications, Univerzitná 8215/1, 010 26 Žilina, Slovakia

Abstract. Railway transport is a sector that is rapidly developing, changing and being influenced to the maximum extent by the technological development and innovation, among others, thus facing problems in staffing its several domains with appropriate and qualified personnel. This fact, makes the need for changes in training and education of future transport professionals. SKILLFUL project vision is to identify the skills and competences needed by the transport workforce of the future and define the training methods and tools to meet them. Paper focuses on mid-term results of the project, with special focus on railway transport.

1 Introduction

The transportation sector employs over 10 million persons in the EU today. At the same time, transport is a social sector that is rapidly developing, changing and being influenced to the maximum extent by the development of automation, electrification and greening of transport, among others, thus facing problems in staffing its several domains with appropriate and qualified personnel. This fact, makes the need for changes in training and education content, curricula, tools and methodologies absolutely imperative, incorporating lifelong learning aspects for the professionals in all transports areas. Additionally, the demographic trends (such as population age) are also going to play a key role over the next years, since large groups of professionals taking retirement should be replaced by younger generation of employees. So, an additional challenge is whether enough professionals having the right skills could be attracted to the transportation sector workforce [1].

SKILLFUL project vision is to identify the skills and competences needed by the transport workforce of the future and define the training methods and tools to meet them. Employability is strongly connected by the project to future transport job requirements for all transportation modes and multimodal chains for all levels/types of workers, while all training modes are included and integrated in a balanced way.

SKILLFUL is financed under Horizon 2020, reviewing the existing, emerging and future knowledge and skills requirements of workers at all levels in the transportation sector, to structure the key specifications and components of the curricula and training courses that will be needed to meet these competence requirements optimally in order to achieve European wide competence development. Project results are verified through wide range of pilot courses from all transportation modes Europe wide [2].

Rail is a modern, technologically advanced industry that has an important and growing role. Demand for rail services has increased through significant infrastructure investment, increased demand for passenger and freight services, technological advancement, development in the resources sector and increased customer expectations including customer service standards and reliability. [3]

The rail sector in the next decade will face simultaneously the challenges of an ageing working population, probably leading to workforce shortages. Rail industry needs skilled and capable staff to continue to grow and meet future demand. Increased productivity and attractiveness of rail transport will lead to demand increases and investments such as in new rolling stock from which working conditions are likely to benefit. However, the availability of a skilled and highly motivated labour force in the transport sector is essential for the supply of efficient and competitive transport services. It is vital to continuously improve the skilling and professional development of people across all sectors of the rail industry, at all levels. SKILLFUL project ultimate goal is to establish a suite of training solutions for all transport modes including rail to take the qualification to any desired level.

2 Future challenges for education

The aforementioned changes and developments in the transportation sector of Europe are expected to bring changes to the employability of the sector. They are expected to cause the alteration of many jobs (sometimes even their disappearance), as well as the emergence of various new occupations. In the light of this, the education and training sector will need to adapt rapidly and

* Corresponding author: dasit@fstav.uniza.sk

effectively in order to insure training/education provision efficiency and to fulfil new jobs skills and competences requirements effectively. Education will constitute a pivotal role in order to properly drive new potential workers on covering the future and each emerging demand. As a result, new education schemes will become increasingly focused, not just on knowledge, but also on strategic skills that students need to get a job. [4]

In this respect, SKILLFUL project provides a general overview of new and emerging training tools, methodologies, and schemes, which will be the most promising for the transport education provision in the near and distant future. In doing so, new and emerging training facilities were identified and analysed using the broadest possible spectrum at several levels:

- Including a wide knowledge base and relying on multiple knowledge communication channels;
- Including all user groups and areas, such as blue collar workers, white collar workers, engineers, management teams or relevant competent authorities and political boards;
- Considering all forms and levels of education;
- Exploring potential initiatives and innovation proposed by or successfully applied, not necessarily from the transport sector but with emphasis on applicability in the transport sector;
- Exploring potential initiatives and innovation in the whole European region and beyond;
- Considering potential applications to transport workforce education provision for all areas including all transport modes (maritime, rail, road and air), as well as intermodality/cross modality.

To ensure the best coverage of the topic under investigation, the methodology relied on multiple sources of knowledge, respectively:

- A wide consultation of experts – either in the field of transport or education/training or both – form all over the European continent;
- An extensive review of the recent scientific literature;
- A complementary review of literature on local and regional initiative and innovation (grey literature).

Experts were first asked to rate skills and competences importance for the different transport job categories. From this question, it appears that hard and technical skills are of particular importance for low to middle-skilled worked (i.e. blue collar and professional drivers). Of less importance but nonetheless important are language (especially mother tongue mastery), ICT skills and several soft skills such as flexibility/adaptability, proactivity and engagement and interpersonal skills. For the white-collar job profiles, mother tongue mastery, team working/management and proactivity and engagement were rated as very to extremely important while all other skills were seen as fairly important. Noticeable is the fact that technical skills specific to the job were not seen as the most important. [5] Considering the relative high importance dedicated to soft skills, this suggest however that expert attach higher importance to the skills necessary to an effective adaptation to a rapidly evolving workforce than to technical skills by themselves.

For researcher, engineer and manager all skills categories were rated as very important, with the exception of interpersonal skills which were rated as fairly important (but not very to extremely important) for researchers and engineers. For the decisions making job profiles (i.e. authorities – politics), almost all skills were rated as very important with the exception of the skills that are specific to a particular job, ICT skills and innovation/creativity. This is quite consistent with the fact that while decision makers have to acquire a broad and general knowledge.

As for the improvements for specific education type/level, the improvements 'Adding special themes modules', 'Wider use of new training methods', 'Adequate and sufficient practice training', 'Faster adaptation to new legislations, technologies and labour market' and 'Better adequation between education, job requirements and industry' were quite consistently identified for all education type/level. The improvement 'higher versatility, multiskilling, flexibility and knowledge transfer' was proposed for the 'college-university' category but also – to a lesser extent - for VET education. Finally, the improvement 'Bigger and more specialized offer for in-house training in company' was proposed for in-house training [6].

3 Most promising training methodologies

The last aspect considered in research concerned the identification of the most promising training/education tools and methodologies. In total, 12 promising existing, new and emerging tools and methodologies identified by the expert panel in diminishing importance order are the following:

- E-learning (Distance/mobile/connected learning and online learning environment);
- Virtual/augmented reality;
- Gaming environment;
- Human led individualized training (e.g. executive coaching);
- Blended learning;
- Peer led/mentoring learning programme;
- Traditional lectures;
- Networked learning (e.g. social media networking);
- Smart learning technologies: personalized learning processes;
- Scenario/story based learning;
- Training on the job/experiential learning;
- Informal learning.

In the context of the review of the literature, 150 papers and reports were reviewed and analysed (112 from the scientific literature and 38 from the grey one) covering all the training tools, methodologies and schemes thought to be relevant for the SKILLFUL objectives.

Overall, the new and emerging tools and methodologies identified through the literature review were quite similar to those pointed out by the expert panel with some very few exception. Indeed, from the expert consultation, it appears that some standard educational tools/methods – that are not necessarily new (e.g. traditional lectures and/or human led individualized training) – are still of great importance and may still usefully be adopted in combination with more recent technologies. [7]

The two major lessons learned from this review are on the one side the necessity to base learning on multiple resources, medias and techniques - as known as the blended learning approach - to insure learning performance and engagement into it and on the other side the necessity to place the learner as a central and active agent of his/her learning – process implicitly understood, for example, under the heutagogical, the authentic/active and/or the discovery-based learning approaches.

Blended learning may be conceived as a new way of education that depends on the use of multiple education resources (typically blended face-to-face learning with other information technology based features) which makes it a powerful approach for creating educational programs that can take into account the individual differences between students and bring different learning methods [8].

Among the recognized and expected benefits of a successful blended learning approach are: increased active and authentic learning experience and performance, enhanced engagement and motivation, broader knowledge and skills (e.g. soft skills) acquisition, etc. [9]. Finally, it is important to note that the benefits of a particular blended learning scheme will primarily depend on the components that are included into it (collaborative/peer-led components will, for example, enhances soft skills such as negotiation of communication skills).

Learner as a central and active agent of his/her learning: more and more, the pedagogical models suggest that the learnings should be conceived as active and proactive processes with a central place offered to the learner constructing his/her own learning experiences. In doing so, the learning experience is thought to be more authentic and leading to an increased learning engagement, motivation and autonomy from the learner as well as to increased performance and broader knowledge acquisition – as, for example, critical thinking, self-reflection, metacognition and soft skills [5].

Another major benefit is that it involves the learner in realistic requirements and scenario work practices that are aligned to authentic situation and hence better prepared them for personal life accomplishment and for the workplace and to become a lifelong autonomous learner. As compared to traditional educational models – with the learner considered as a passive recipient of learning content – this implies a major revolution and an essential mental shift in the way teachers and learners roles have to be considered.

All 12 tools, methods and supportive technologies presented above are thought to serve the purpose of these two components (i.e. blended learning and placing the learner at the center of his./her learning) and to facilitate their effective implementation. However each of these categories present benefits and drawback and cannot be seen as a panacea for meeting all learning requirements. Instead, it is the well-designed and well-thought combination of several methods – adapted to the learning content as well as to the learner needs – that ensured effectiveness. It is important to note that the rapid developing technology will increasingly impact future workforce skills and competences development in two ways: firstly, as it exerts pressure for change because the new technologies demand a new set of skills and, secondly, as it provides opportunities for transforming pedagogy because it provides access to information, networks for communication, and new means of presenting learning [6].

4 Critical review of the educational and training systems and mechanisms in the transportation sector

The one of the aims of the project was to critically review the current practice of skills development for the needs of all transport modes in EU and beyond in order to gain a better understanding of existing training schemes and programmes/courses, their content and targeted audience, as well as the training provider and/or programme owner.

As a result, SKILLFUL is able to identify potential gaps in the availability of training schemes and programmes when looking at job categories in the whole transport sector including rail. Templates were designed and surveys organized to collect information on existing training schemes and programmes/courses in targeted countries for all transport modes (road, air, maritime and rail) [2].

The templates follow the three management levels for transport planning and operations [7], namely: strategic, tactical and operational. This was originally developed for the rail sector but is applicable to all sectors.

Strategic includes long term planning of company development. Decisions made at this level set the strategic goals of the company, which include assessment resources, strategic changes in the company structure, redesign and reconstruction of infrastructure, relocation of railway facilities, construction of new rail lines, acquisition of new resources and technologies, etc. This is the highest level of management in railway organizations. Although the decisions made at this level are capital intensive they should provide the minimum amount of required resources for "normal operation" [8].

Tactical includes medium term planning. At this level all the plans, timetables and schedules are developed. As stated by [9] tactical planning is "to ensure, over a medium term horizon, an efficient and rational allocation of existing resources in order to improve the performance of the whole system". At this level capacity research analysis of congestion and performance assessment are conducted.

Operational includes short term planning, which might be executed over the same day of service delivery. At this management level the plans, timetables and schedules are implemented on a "day-to-day" basis in order for the system to provide the service effectively [10, 11].

A career matrix was designed for each transport mode, specifying job categories under each management level.

To complete the template, different types of training schemes and programmes were identified. They include: College; University; Professional Training; Apprenticeships; VET & CVET; Others. Space has been provided within the template for examples of training providers and/or programme owners to be listed, as well as for examples of existing training schemes and programmes to be given. These templates were developed by the partners involved in the project and external contributors, using their expertise in each of the transport modes, in order to identify the existing job categories for each transport mode. The templates were circulated to all SKILLFUL partners, external authorities and stakeholders. Templates were distributed to each country: for each mode and each level, a number of job categories were identified. A comprehensive set of information has been gathered and analysed for 16 countries: Australia, Belgium, Bulgaria, China, Germany, Greece, Ireland, Italy, The Netherlands, Poland, Portugal, Romania, Slovakia, Spain, Sweden and the UK.

A summary report has been produced for every country as part of the analysis (Deliverable D2.1) to provide a brief insight on the content of training schemes and programmes identified. The text below attempt to summarise the gaps found during the analysis. However, it must be noted that it is very difficult to draw any conclusions on the gaps existing in countries as information was difficult to find. Training may be available in some countries for which we have identified gaps but we were unable to find the information on the training. In addition, training on specific course content was very difficult to obtain: in many cases only the title of courses was available and detailed information was not available. Therefore, any conclusions drawn must be treated with considerable caution [2].

High response to demand (United Kingdom)

High supply of rail oriented university degree programmes.

Comprehensive training system for rail in place.

Apprenticeship scheme in place.

Many certified companies provide courses to meet the needs for daily operation.

Easy access to information.

Information available in a living world language.

Average response to demand (Australia, China, Portugal, Romania, Spain, Italy, Germany, Bulgaria, Belgium)

Average supply of rail oriented university degree programmes.

A comprehensive training system for rail in place.

Apprenticeship scheme not in place. [11]

There are a reasonable number of certified companies providing courses to meet the needs for daily operation.

Challenging to access information.

Information unavailable in a living world language.

Low response to demand (Ireland, Poland, Slovakia, Greece)

A small amount of rail oriented university degree programmes.

Apprenticeship scheme not in place.

Monopoly by a single rail company providing training for jobs in rail.

In-company training – information not disclosed to external parties.

None or a small number of certified companies providing courses to meet the needs for daily operation.

Uneasy access to information.

Information unavailable in a living world language.

Others (Sweden, the Netherlands)

Not enough information available.

Difficult to analyse and conclude.

Generally it can be said that here is a sufficient amount of university degree courses (for example: BSc or MSc in railway engineering) including, courses for the maintenance of rail vehicles, locomotive engineers, signalling inspectors, sale assistants or administrative assistants.

Significant gaps are demonstrated in college and apprenticeships, where for all work categories these courses are not present. There is also a deficiency in professional training and VET & CVET. With regards to professional training, there are only courses for the job categories of track inspector, maintenance of rail vehicles, locomotive engineer and train crew. Only those training to be track inspectors, timetable planners, train crew members, signaling inspectors, sales assistants and administration managers have the possibility of taking CET & CVET.

At the tactical level, it is revealed that there are small amount of courses provided for the infrastructure planner, vehicle design engineer or timetable planner.At the operational level, usually the only degree courses available are designed for training track inspectors and research/teaching assistants, as a result, more courses should be provided.

There are not have enough courses specializing in the rail industry in general. The courses are either very general or specialized in non-rail areas, for example, the civil engineering or mechanical engineering course. Students take degrees in other sector, with a focus on transportation. At the operational level many times one rail company have a monopoly in terms of rail service provision, train their drivers and crew members. However, no information is available online as this training is provided in-house.

5 Pilot courses

The one of objectives of SKILLFUL project is to design appropriate training/education modules for the Transport sector to fulfil emerging and foresighted required competences and skills. In the context of previous steps, five types of training/educational schemes for various types of workers in all transport modes were designed to be tested as pilot training by partners as follows:

- **Transport infrastructure operators' training schemes.** Operators need new skills and competences to catch up with emerging technologies. For them life-long learning, with emphasis on new technologies, constitutes a necessity. But they typically have very little available time as they work under pressure and in tight shifts. Thus, for them webinars and other ICT-based training (and certification through web proctoring) schemes might be more appropriate;

- **Young scientists' seminars** as a joint initiative of ECTRI, FERSI and FEHRL from all Transportation sectors;
- **Lifelong training schemes** for low to middle-skilled segments of transport professionals;
- **Interdisciplinary thematic courses** on key technologies, services and trends with a cross modal dimension and/ or where important know-how can be transferred from one mode training to another one;
- **Pan-European Transport master curriculum**. A first core syllabus will be realised towards a Pan-European Transport Engineering master. Trials of trans-national specialty courses will be tested to allow students from one country to attend courses from another university within their master degree.

Several courses were developed for rail sectors (or as multimodal too) as follows:

- The management of stress and fatigue in Air and Rail Traffic Management
- Authorised Economic Operator (AEO) and safe supply chains
- Security in transport: influences of new media & digitalisation
- Advanced intermodal operations and logistics
- The role of Big Data in understanding travel behaviour, traffic and developing policy
- Towards global freight forwarders: New training scheme based on future scenarios
- Greening and electrification of transport module
- Rail operations, signals and model operations through simulation, offered by TUB, IST, UNEW for middle and high-skilled professionals.

5 Conclusions

1. All developments, relate to and affecting transportation, whether they relate to technological discoveries and developments, or to social and economic factors, directly affect also the employability, as well as the education and training needs of professionals in the transportation sector;
2. Some existing occupations are expected to change, fall dramatically or even disappear, some new professions are going to make their appearance to meet any emerging needs;
3. A number of gaps have been identified in relation to training. At the same time, a number of new training tools and technologies have also been identified;
4. The availability of technology in the education sector means that approaches using Virtual Learning Environments are now more possible: opening the potential to allow workers to retrain and up-skill. In addition, there have been a number of developments in relation to gamification and simulation which could be adopted in training for automation or in relation to cybersecurity;
5. It is expected that a number of these methodologies will be tested in the pilot training courses. The impact of these methods has not generally been quantified so pilots will offer an opportunity to test the methodologies;
6. More information on project can be found at http://skillfulproject.eu/

This paper is published within the implementation of the project SKILLFUL – Skills and competences development of future transportation professionals at all levels, funded under Horizon 2020 – Research and Innovation programme. H2020-MG-2016-SingleStage-INEA – GA no. 723989.

References

1. SKILLFUL consortium. 2017 Deliverable D1.1: Retrieved from: http://skillfulproject.eu/library?id=7603
2. SKILLFUL consortium. 2017 Deliverable D2.1 - Best practice on current and emerging training schemes, methodologies and tools in the transport sector and mapping to future training requirements and scenarios.
3. M. Azhar, S. Mustapa, M. Ibrahim, A. Yusoff. Procedia – Social and Behavioral Sciences, **Vol. 204**. Pages 127–135. (2015).
4. N. Canning, Journal of Further and Higher Education, **Vol. 34** (No.1), 59–71. Routledge. Retrieved December 13, 2016, from http://www.tandfonline.com/doi/abs/10.1080/03098770903477102. (2010).
5. L. M. Blaschke, S. Hase. Heutagogy. Springer Berlin Heidelberg. Retrieved December 13, 2016, from http://link.springer.com/10.1007/978-3-662-47724-3 Pages 25–40. (2016).
6. V. Hannon, A. Patton, J. Temperley, M. Stevenson. Developing an Innovation Ecosystem for Education. White Paper. Cisco, Innovation Unit. (2011).
7. M. Marinov, I. Sahin, S. Ricci, G. Vasic-Franklin, Research in Transportation Economics, Vol. 41(No. 1). Pages 59–75. (2013)
8. J. Pachl, T. White. Implementation of Heavy Haul Technology for Network Efficiency. Pages 6–69. (2003).
9. T. Crainic, G. Laporte. (1997). European journal of operational research, **Vol. 97**(No.3), pages 409–438. (1997)
10. J. Camaj, M. Kendra, J. Lalinska. *International Educational Technology Conference,* Chicago, IL, USA, IETC 2014. **176**. 659-666. (2014).
11. J. Camaj, A. Dolinayova, J. Danis, M. Kendra. *ICERI: 9th International Conference of Education, Research and Innovation*. 5822-5831. (2016).

Methodology for calculation of minimum transfer time in the transport hub

Martin Vojtek[1*], *Tomas* Skrucany[1], *Martin* Kendra[1] and *Jan* Ponicky[2]

[1] Universtiy of Zilina, Faculty of Operation and Economics of Transport and Communications, Slovakia
[2] Railway of Slovak republic, Slovakia

Abstract. The most important element in railway passenger transport is a customer – traveller, who requires the transport from one place to another. A basic precondition for accomplishing the main requirement – transport, is making the complete offer which provides not only transport, but also other associated services. Practically, there are many associated criteria of transport, for example safety, duration, price, reliability, comfort and complementary services. Passenger transport is generally considered as an activity, which arises as the consequence of spatial division of places, where people are in exact time and their need to move. Motivators for moving could be commuting – job or education, dealing with personal or working matters, travelling for vacation – hiking, sport, health, cultural and social facilities, visiting relatives and friends. Requirements for transport of passengers originate in the need to move, while the passenger transport is dependent on the willingness of travelling. In passenger transport, there are mostly individual passengers, so it is difficult to determine all transport requirements. The paper is focused on one of the key factors of passenger transportation - connectivity of trains. Connectivity of passenger trains and other means of transport can be distinguished also from temporal and spatial point of view. Temporal connectivity is such sequence of arrivals and departures of different passenger trains and other means of transport, which allows changing the different passenger vehicles easily in regard to necessary time. Spatial connectivity means the distance between two passenger vehicles, among which the passenger is moving. In the paper, there is described the general methodology for calculation of minimum transfer time in the railway station. Railway passenger station is some kind of transport hub – a starting and finishing point for flows of passengers. Passengers have the opportunity to change the train type from long-haul train to regional train or contrariwise or simply enter or leave the system of railway transport. In the methodology, all necessary aspects are taken into account.

1 Introduction

Transport network consists of transport routes and transport hubs [1, 2]. The transport hub is important part of passenger transport process. In the transport hub, a vehicle could be changed in the same or different type of transport [3]. It is necessary to ensure suitable constructional layout, which would provide safe, fast and comfortable transfer of passengers among vehicles of the same or different type of transport [4, 5]. Additional services for passengers are also one of the significant factors [6, 7] because passengers could use them while they are waiting. In the transport process, it is very important to minimize waiting time from temporal point of view, but all conditions for safe transfer and sufficient time for it must be kept [8, 9]. The paper is focused on a methodology, which can exactly stipulate the minimum transfer time in the transport hub among various vehicles and types of passenger transport. Currently, transfer times are set empirically or by estimation. When some lack is identified, transfer times are operationally adapted to passengers' requirements [10]. General methodology for solving transfer problems in transport hubs is described in the paper.

2 Characteristics of passenger transport chain

In the passenger transport system, the element is an object, which cannot be divided into smaller parts during transport process. Transport requirements represent the relocation of passengers therefore the object of transport is the passenger. All passenger transport systems could be characterized by [11]:

- Places of getting on, getting off and getting among vehicles,
- Transport hubs,
- Sections between pairs of transport hubs, which are performed always with one vehicle,
- Mutual aim of fast, safe, comfortable and reliable transportation.

Appropriate methodology of temporal elements quantification could detect some bottle necks of transportation process, which must be subsequently modified for purpose of improving quality and efficiency of public passenger transport.

* Corresponding author: martin.vojtek@fpedas.uniza.sk

https://doi.org/10.1051/matecconf/201823500015

3 The methodology for setting the minimum transfer time in transport hub

The proposed methodology is based on exactly established approaches, which must be abided for reaching the aim – determination of necessary minimum transfer time in transport hub for various groups of passengers.

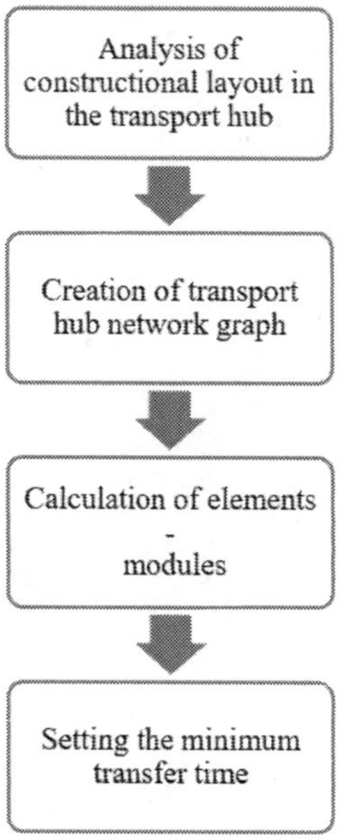

Fig. 1. Process of proposed methodology

3.1 Analysis of constructional layout in the transport hub

Constructional layout in the transport hub significantly influences the minimum transfer time, which are necessary for passengers in case of changing the vehicle or mode of transport in the transport hub [12].

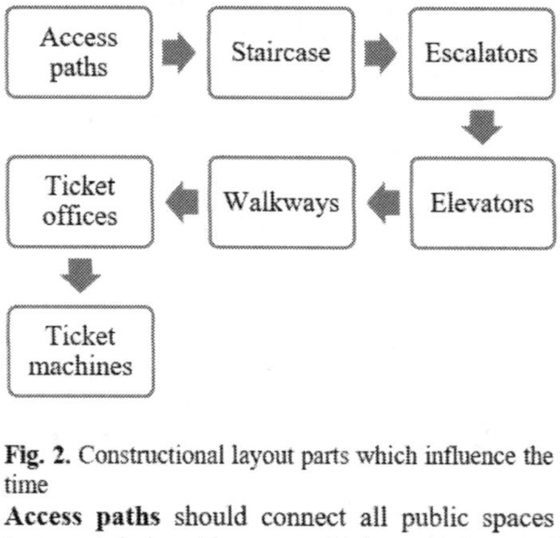

Fig. 2. Constructional layout parts which influence the transfer time

Access paths should connect all public spaces in the transport hub with way, which must be as short as possible. They should be signed by visual tools for better orientation and their surface should be anti-skid. In the analysis, it is necessary to identify all access paths, which could be used by passengers. Also it is necessary to identify access paths for PRM passengers. Identification of access paths length is necessary for calculation of elements (modules) in the transport hub. Another part, what impacts the transfer time, is a **staircase**. [13,14] In the analysis, it is important to determine their location in the transport hub (different passenger frequency) and number of stairs. The **elevator** is a transport device for vertical transport of passengers. In the analysis, it is necessary to determine the location of elevators in the transport hub, because the frequency of passengers could be different. Then, it is important to determine the time of opening doors, transport velocity of elevator and floor size, because these indicators directly influence the transport capacity of passengers. **Escalators and walkways** are devices for horizontal and vertical transport of passengers in the building. Their advantage is higher transport capacity than elevators. It is necessary to analyse their location in the transport hub, transport velocity and constructional length. **Ticket offices and machines** situated in the transport hub ensure travel tickets selling for passengers. In the analysis, it is important to determine the number of them (number of desks, machines etc.) and their opening hours.

3.2 Creation of transport hub network graph

The network graph presents a unique way of mathematical expression for temporal or technological sequence of partial operations. In the proposed methodology, the network graph of transport hub is evaluated by vertices and edges. **Edges of network graph** represent oriented lines between vertices. Edges in the graph show access paths in the transport hub. The evaluation of edges is time t which is necessary for getting across the access paths by passengers. Vertices of network graph represent activities, which must be accomplished by passenger during transfer in the transport hub. [15] For instance: getting off the vehicle, buying travel tickets, etc. Vertices in the graph are carriers of information called modules.

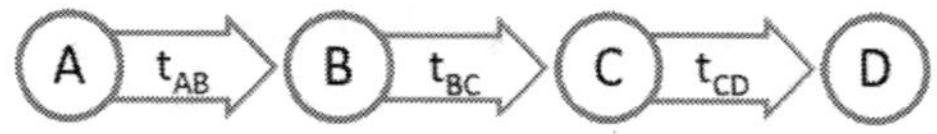

Fig. 3. Sample of network graph in the transport hub

In the figure 3, A means getting off the vehicle, B means going on stairs, C means buying travel tickets and D means getting on the vehicle.

3.3 Calculation of elements (modules) – evaluation of vertices in the network graph

Vertices in the network graph are carriers of information about duration of each activity while transferring in the transport hub – modules. Partial indicators are:

- Getting off the vehicle time,
- Going on stairs time,
- Transport on escalators time,
- Transport in elevators time,
- Transport on walkways time,

- Buying travel ticket time.

3.3.1 Getting off the vehicle time

Getting off the vehicle time is first phase, which is calculated in transfer time. This time starts after the vehicle stopped in the transport hub on the place for getting off the vehicle. Factors with significant impact to getting off the vehicle time are:

- ***Opening door of the vehicle time*** t_{od} – set on measurement according to constructional layout of the vehicle and way of door opening mechanism,
- ***Vehicle door constructional layout coefficient*** C_{vdc} - vehicle door constructional layout (constructional width) has got significant impact to number of passengers and duration of boarding,
- ***Vehicle door load*** L_{vd} – vehicle door load shows proportion on number of getting off passengers and number of vehicle doors. If there are many doors on the vehicle, less time for getting on and off the vehicle is necessary,
- ***Platform length coefficient*** C_{pl} – this coefficient is set on constructional layout of the platform in the transport hub. Constructional layout of the platform could be in one grade when platform is in the same height as the floor in the vehicle, or it could be grade-separated when the platform is not in the same height as the floor in the vehicle.
- ***Luggage handling coefficient*** C_{lh} – it is dependent on size, weight and capacity of the luggage.
- Resultant formula for calculation of getting off the vehicle time is:

$$T_{offv} = t_{od} + \sum_{i=1}^{n} P_{offp} * \sum_{i=1}^{L_{vd}} T^{i}_{offp} * C_{vdc} * C_{pl} * C_{lh} \quad (1)$$

where:
T_{offv} getting off the vehicle time [s],
t_{od} opening door of the vehicle time [s],
P_{offp} number of getting off passengers [passengers],
T^{i}_{offp} unit time of getting off passengers according to type of passenger [s],
L_{vd} vehicle door load [-],
C_{vdc} vehicle door constructional layout coefficient [-],
C_{pl} platform length coefficient [-],
C_{lh} luggage handling coefficient [-].

3.3.2 Going on stairs time

It is necessary to divide going on stairs time into:
- Going upstairs,
- Going downstairs.

The reason of passengers' going on stairs time division into going upstairs and going downstairs is that velocity of going upstairs is lower than velocity of going downstairs. For verification of this hypothesis, the measurement in the transport hub Zilina was done. Passengers were divided into three groups:

- **Group 1** – students,
- **Group 2** – productive ages,
- **Group 3** – pensioners.

The reason of this division is that passengers in each group are going very similar time. Entire number of measurements, which were done in each group, is 100 measurements. Number of stairs was set to 23 stairs.

The formula for calculation of going on stairs time is:

$$T_{st} = \frac{N_{st}}{Øv^{i}_{sw}} * C_{lh} \quad (2)$$

where:
T_{st} going on stairs time [s],
N_{st} number of stairs [stairs],
$Øv^{i}_{sw}$ average speed of walking [stair.s^{-1}],
C_{lh} luggage handling coefficient [-].

3.3.3 Transport on escalators time

Entire transport time depends on individual constructional layout of the escalator in the transport hub. It means that there are these dependent parameters: transport velocity and constructional length of the escalator. For calculation of entire transport on escalators time is necessary to count all passengers, who are going on the escalator in the same time.

The calculation of transport on escalators time should be:
1. passenger: t_1
2. passenger: $t_1 + t_i$
3. passenger: $t_1 + 2 * t_i$
n. passenger: $t_1 + (n-1) * t_i$

Transport on escalators time of the first passenger is set by formula:

$$t_1 = \frac{L}{v_e} \quad (3)$$

where:
t_1 transport on escalator time of the first passenger [s],
L constructional length of the escalator [m],
v_e escalator velocity [m.s^{-1}].

For every other passenger, it is necessary to count with transport on escalator time of the first passenger and increased time on escalator which depends on other passenger order from the first passenger on escalator. Increased time of another passenger on escalator is set by formula:

$$t_i = \frac{l_p}{v_e} \quad (4)$$

where:
lp distance between the passengers on the escalator [m].

If there is possible to stand of more passengers next to each other on the escalator, order of the passengers (1. passenger, 2. passender, 3. passenger,..., n. passenger) means order of the m-tuble of the passengers on the escalator, where m is number of passengers who can stay next to each other on the escalator.

3.3.4 Transport in elevators time

The entire transport in elevators time depends on individual constructional layout of the transport hub. Getting off the elevator time starts after the elevator stops on the floor. It is set by formula:

$$T_{off}^{e} = t_{oed} + \frac{\sum_{i=1}^{n} T_{off_i}^{e} * N_p}{C_{ed}} \quad (5)$$

where:

t_{oed} opening doors of elevator time [s],
T^e_{offi} getting off the elevator time per passenger [s],
N_p number of getting off passengers [passengers],
C_{ed} constructional layout of elevator doors [-].

Getting on the elevator time starts after all getting off the elevator passengers are out of the elevator. It is set by formula:

$$T_{on}^{e} = \frac{\sum_{i=1}^{n} T_{on_i}^{e} * N_p}{C_{ed}} + t_{ced} \quad (6)$$

where:

T^e_{oni} getting off the elevator time per passenger [s],
t_{ced} closing doors of elevator time [s].

Entire transport in elevator time is set by formula:

$$T_e = T_{off}^{e} + T_{on}^{e} + T_{in}^{e} \quad (7)$$

where:

T_e entire transport in elevator time [s],
T^e_{in} transport in elevator time [s].

3.3.5 Walkways module

Walkways are suitable for mass transport of people in shopping centres and transport hubs. In general, transport on walkways time is set by the same formula as for transport on escalator.

3.3.6 Buying travel ticket time

A place, where passengers can buy ticket is a typical example of Queueing theory systems. Basic model is system M/M/n/∞ with n possibilities for buying the travel ticket. Arrivals of passengers are described with Poisson's distribution with λ frequency and duration of service has got exponential distribution with average duration 1/μ. All buying possibilities are equally productive. Passengers are served in the same order as they came to the system. The entire time in the system – average duration in the system is a sum of average time of waiting in the queue and time of buying the travel ticket. It is set by formula:

$$E(R) = \frac{\pi}{n*\mu-\lambda} + \frac{1}{\mu} = \frac{\frac{\rho^n}{n!} * \frac{1}{1-\frac{\lambda}{n*\mu}} * p_0}{n*\mu-\lambda} + \frac{1}{\mu} \quad (8)$$

3.4 Calculation of elements (modules) – valuation of edges in the network graph

Edges in the network graph represent oriented lines between vertices (passengers' activities in transport hub during the transfer). Vertices are access paths in the transport hub.

3.5 Walking on access paths time

Access paths would connect all public spaces in the transport hub by way which is as short as possible. Another module for valuation of transport connectivity in public transport system is setting necessary time for reach these distances. In this measurement, passengers were divided into three groups again, similarly as it was in the measurement of going on stairs time. Entire number of measurements was 100 and the distance was 45,5 meters. According results of these measurements, average time and velocity of walking was set for each group of passengers. The formula for calculation of walking on access paths:

$$T_{wap} = \frac{L}{\O v_{wap}^{i}} * C_{lh} \quad (9)$$

where:

T_{wap} walking on access paths time [s],
L distance [m],
$\O v^i_{wap}$ average velocity of walking according to group of passengers [$m.s^{-1}$],
C_{lh} luggage handling coefficient [-].

In case, when the passenger is getting of the vehicle, it is important to calculate with the distance of the furthest doors of the vehicle.

3.6 Setting the minimum transfer time

The result of the network graph application in the transport hub and by counting the modules is setting the minimum transfer time in all combinations of edges and vertices, which could be possible while passengers are transferring in the transport hub.

Conclusions

1. Proposal of the network graph in the transport hub with valuation of edges and vertices
2. Calculation of transferring passenger partial activities
3. Possibility of constructional layout integration in the transport hub
4. Possibility of constructional layout integration in vehicles
5. Simplification of timetables in integrated transport systems
6. Optimization of transfer times according to passengers' needs

This research was financially supported by the project VEGA No. 1/0019/17 Evaluation of regional rail transport in the context of the region's economic potential focusing on the efficient use of public resources and social costs of transport.

References

1. B. Abramović, PPTE, **45**, 42-47 (2017)
2. J. Camaj, J. Masek, A. Dolinayova, J. Danis, *ICTTE 2016*, 955-963 (2016)

3. R. Kampf, S. Lorincova, M. Hitka, O. Stopka. Sustainability, **9**. (2017) DOI: 10.3390/su9091561
4. J. Gašparík, V. Luptak, P. Meško, *ICTTE 2016*, 388-392 (2016)
5. R. Berezný, V. Konecný, Procedia Engineering, **192**, 40-45 (2017)
6. J. Gasparik, V. Luptak, P. Mesko, P. V. Kurenkov. Communications, **19**, 2, 61-67 (2016)
7. V. Zitricky, J. Gasparik, L. Peceny. *Transport problems*. **10, SI**. 59-72. (2015)
8. J. Gnap, V. Konecny, Kom., **10**, 55-59 (2008)
9. E. Brumercíkova, B. Bukova, P. Kondek, *LOGI 2017 (MATEC Web of Conferences)*, **134**, 00002 (2017)
10. C. Csiszar, D. Foldes, *CogInfoCom 2015 – Proceedings*, 7390578 (2016)
11. L. Bartuska, V. Biba, R. Kampf. *Proceedings of the Third International Conference on Traffic and Transport Engineering.* (2016) WOS:000391016300128
12. L. Hansut, A. David, J. Gasparik. *Bussines Logistic In Modern Management*. 29-42. (2017)
13. D. Foldes, C. Csiszar, *SCSP 2016*, 7501022 (2016)
14. S. Stoilova, TV-TG, **25**, 587-595 (2018)
15. M. Gogola, *Transport Means - Proceedings of the International Conference*, **2015**, 359-362 (2015)

 https://doi.org/10.1051/matecconf/201823500016

Analysis of the system of pricing of railway capacity allocation in the Czech Republic

Jaromir Siroky[1,*] and *Hana* Salakova[1]

[1] University of Pardubice, Faculty of Transport Engineering, Department of Transport Technology and Control, Studentská 95, 53210 Pardubice, Czech Republic

Abstract. The introducing of legal background for capacity allocation in Czech Republic. The analysis of price for capacity allocation in Czech Republic and in Europe. The proposal of the capacity allocation price which should lead the Railway undertaking's path request to the suitable product. Possible impact of the price of capacity allocation to Railway undertaking's decision about the railway transport.

1 Introduction

Railway capacity is allocated using two basic modes, i.e. within processing the annual timetable and its changes, and in an ad hoc request mode. As for timetable 2017, the following (4) products are offered to applicants for railway capacity:

1. Regular request for railway capacity allocation to be taken into account when creating the annual timetable – "RJ",
2. Late request for railway capacity allocation to be taken into account when creating the annual timetable – "PJ",
3. Request for railway capacity allocation until the next change of the annual timetable – "ZJ",
4. Individual ad hoc request for railway capacity allocation.

As for individual ad hoc requests for railway capacity allocation, the applicants can use the following products:

a) Ad hoc request for railway capacity allocation submitted 3 and more working days before the first requested date of train departure (including the request date) – "N3",
b) Ad hoc request for railway capacity allocation submitted later than 3 working days before the first requested date of train departure (including the request date) – "P3" (residual railway capacity),
c) Ad hoc request for railway capacity allocation for technical and safety tests – "TB",
d) Ad hoc request for railway capacity allocation for test rides of vehicles of unapproved type or rides at a speed exceeding the track speed – "ZK",
e) Ad hoc request for railway capacity allocation for Railway Infrastructure Administration infrastructure maintenance – "UI",
f) Ad hoc request for railway capacity allocation due to Railway Infrastructure Administration infrastructure restrictions – "OM",
g) Ad hoc request for railway capacity allocation for other reasons attributable to the Railway Infrastructure Administration – "JD".

2 Railway Infrastructure Administration products related to capacity allocation

Train route and train timetable are determined in assessing the request for railway capacity allocation. For requests for products RJ, PJ, ZJ, N3, TB and ZK, the timetable is created with conflict resolution.

As for requests for products P3, UI, OM and JD, it is up to the allocator to decide whether to allocate ad hoc routes with conflict resolution, or routes within the residual railway capacity without resolving the conflicts. For routes within the residual railway capacity, conflicts are later resolved by operational employees of the infrastructure manager. Table 1 clearly shows that railway undertakings are increasingly using ad hoc requests compared to requests to be taken into account when creating the annual timetable.

This has been a trend for railway undertakings EU-wide due to the increasing demand for greater flexibility, mainly in freight transport. The authors thus tried to summarise the advantages and disadvantages of both railway capacity allocation modes for railway undertakings.

Table 1. Evolution of the number of data timetables for the train traffic diagram 2005–2016.

Year	Number of routes requested for the annual timetable	Number of ad hoc allocation requests
2005	15.468	62.970
2006	14.836	78.595
2007	15.213	127.547
2008	14.366	159.021
2009	14.403	161.223
2010	13.381	180.602

* Corresponding author: jaromir.siroky@upce.cz

2011	13.107	185.100
2012	12.706	181.647
2013	12.023	205.993
2014	11.704	240.460
2015	13.361	252.797

Source: Authors according to (16)

Mode of allocation within the annual timetable – products RJ, PJ and ZJ

Each of the railway capacity allocation modes has its advantages and disadvantages.

Advantages of requests for railway capacity allocation to be taken into account when creating the annual timetable:

- higher quality of timetable processing,
- priority processing,
- determination of technologies at originating, intermediary, destination and frontier stations,
- with a higher number of route-days, the price of railway capacity per route-day is lower.

Disadvantages of requests for railway capacity allocation to be taken into account when creating the annual timetable:

- greater risk of sanctions for unused railway capacity due to a large time gap between the time of booking the route and the actual train ride,
- low flexibility of changes for railway undertakings,
- greater limitation by the booked route parameters (where the business case changes, the railway undertaking has to submit an individual ad hoc request for railway capacity allocation – e.g. upon extension of the original route).

Ad hoc allocation mode – products N3, P3, TB, ZK, UI, OM and JD

Individual ad hoc requests for railway capacity allocation have their advantages and disadvantages as well.

Advantages of individual ad hoc requests for railway capacity allocation:

- greater flexibility for railway undertakings,
- greater probability of using the route, and thus a lower risk of sanctions for unused railway capacity,
- lower price of railway capacity allocation with a lower number of route-days, more flexible adaptation to changing market requirements.

Disadvantages of individual ad hoc requests for railway capacity allocation:

- lower quality of timetable processing (mainly in residual railway capacity routes) without conflict resolution,
- lower priority in timetable processing,
- need for the railway undertaking to negotiate technologies at stations (originating, intermediary, destination and frontier) on a case-by-case basis,
- impossibility to use certain tender prices, e.g. tender price J.

With the increasing number of requests for railway capacity allocation, mainly in the ad hoc mode, it is necessary to manage the railway undertakings to submit their requests for railway capacity allocation in time to ensure a timely and high quality processing of timetables, and to prevent them from submitting requests in the annual timetable allocation mode, which are ad hoc requests in nature. One of the ways to motivate the railway undertakings to use appropriate products is pricing the products in a way that the railway undertakings get the best value for money, and the price covers the costs reasonably incurred in railway capacity allocation at the same time.

Railway Infrastructure Administration products related to capacity allocation and their prices (for timetable 2017) are shown in Table 2. Coefficients K1, K2, and K3 are rates used in formula (1) to calculate the price of railway capacity allocation.

Table 2. Railway Infrastructure Administration products related to capacity allocation for timetable 2017.

Product		K1	K2	K3
RJ	regular request for railway capacity allocation to be taken into account when creating the annual timetable	1.700	8	10
PJ	late request for railway capacity allocation to be taken into account when creating the annual timetable	1.700	10	20
ZJ	request for railway capacity allocation until the regular change of the annual timetable	1.700	10	20
N3	ad hoc request for railway capacity allocation submitted 3 and more working days in advance	100	0	70
P3	ad hoc request for railway capacity allocation submitted later than 3 working days in advance	100	0	160
TB	ad hoc request for railway capacity allocation for technical and safety tests of railway vehicles	480	0	70
ZK	ad hoc request for railway capacity allocation for test rides of vehicles of unapproved type or rides at a speed exceeding the track speed	960	0	70

Source: [4]

The payment for railway capacity allocation was introduced in 2004 (before 2004, railway capacity allocation was not subject to a charge) to regulate the behaviour of railway undertakings in railway transport planning.

The system was devised as a rate times the number of route-days (days the train spends on the route). Originally, 3 rates had been determined as follows: CZK 15 per route-day for allocation within the annual timetable, CZK 25 per route-day for requests submitted 3 and more working days before the first day of the train ride, and CZK 120 per route-day for requests submitted later than 3 working days before the first day of the train ride (residual railway capacity). These rates had applied until 2010.

With timetable 2013, rates K1, K2 and K3 were introduced, and in addition to the rate per route and route-day, rates per kilometre of the planned route started to be charged as well. K1 is the price of processing and determining of the timetable and railway capacity allocation, K2 is the price of train route creation, and K3 is the price of train route allocation per day.

K1 amounting to CZK 1,700 applies to all products within the annual timetable (regular request, late request and request until the next change of the annual timetable – products RJ, PJ and ZJ), and K2 in the amount of CZK 100 applies to ad hoc requests for railway capacity submitted 3 and more working days in advance (product N3) and requests for railway capacity submitted later than 3 working days in advance (product P3). K1 for ad hoc requests for railway capacity allocation for technical and safety tests of railway vehicles remained at CZK 480 (product TB), and for ad hoc requests for railway capacity allocation for test rides of vehicles of unapproved type or rides at a speed exceeding the track speed remained the same as well (CZK 960) (product ZK).

K2 for RJ was fixed at CZK 8 per 1 route kilometre, and for products PJ and ZJ at CZK 10 per 1 route kilometre. For the other products N3, P3, TB and ZK, K2 was fixed at CZK 0 per 1 route kilometre.

K3 for RJ is CZK 10 per day of train route allocation, for PJ and ZJ it's CZK 20 per day of train route allocation, for N3 it's CZK 70 per day of train route allocation. For P3, K3 was fixed at CZK 160 per day of train route allocation and for TB and ZK, K3 amounts to CZK 70 per day of train route allocation.

Revenue from railway capacity allocation should cover the total direct costs of railway capacity allocation, but cannot exceed them. Using MS Excel, prices of railway capacity allocation will be modelled that would take into account all economically justifiable costs of railway capacity allocation, and would motivate railway undertakings to book the individual products in an optimal way at the same time.

2 Requests for railway capacity allocation

To properly regulate the demand of railway undertakings for individual products, it is first necessary to define the requirements on target groups of these respective products.

2.1 Requests for railway capacity allocation to be taken into account when creating the annual timetable and its changes – products RJ, PJ and ZJ

These products are meant for requests for high-frequency train rides (ideally with daily frequency) requiring the creation of a fixed train route to be strictly adhered to and containing other requirements:

- creation of transfer connections,
- creation of passenger service train cycles,
- creation of offered departure or arrival times,
- determination of technologies at originating, intermediary, destination and frontier stations.

The requests are submitted at the times agreed on European level, and timetables proposed by capacity allocator are submitted at the times defined by national legislation and European agreements as well.

a) Regular request for railway capacity allocation to be taken into account when creating the annual timetable – product RJ

For RJ, requests are processed with highest quality (in a "blank" train traffic diagram sheet). Subsequently, detailed information is provided on the timetable implementation in a train diagram, working timetable and booking timetable. A RJ request should be more cost effective than PJ or ZJ ones for the applicants to be motivated to submit requests earlier, which is advantageous for creating a timetable as there is more time for the route creation. With a higher number of route-days, the price of railway capacity per route-day should be lower. The proposed price of RJ, PJ and ZJ should also be lower for railway undertakings than the price of N3 for a number of route-days greater than the limit number.

b) Late request for railway capacity allocation to be taken into account when creating the annual timetable – product PJ

PJ is meant for requests submitted after the regular request deadline. For these requests, it is not always possible to guarantee the requested time frame, the route is created within an already created timetable, and the necessary effort is similar, or slightly higher than for regular requests.

The proposed price should be higher than for RJ requests.

c) Request for railway capacity allocation until the next change of the annual timetable – product ZJ

ZJ is meant for requests which compared to ad hoc requests require higher quality and the processing of the necessary instruments is more demanding. The route creation (within an already created timetable) is similar and requires similar effort as in late requests. The proposed price should be higher or the same as in late requests.

2.2 Individual ad hoc requests for railway capacity allocation – products N3, P3, TB, ZK, UI, OM and JD

This category includes individual and less frequent requests within the period between timetable changes, and sometimes also longer-term requests with a departure date within the period between timetable changes (with limited requirements on links to other requests or on defining arrival and departure time frames for selected stations).

a) Individual ad hoc request for railway capacity allocation submitted 3 and more working days in advance – product N3

For this request, a timetable is created within a period between timetable changes, which is then included in the next timetable change. These are often requests in response to planned closures, and the route creation requires similar effort as in requests until the next change of the annual timetable. What is different is the lower quality of timetable information, requiring railway undertakings to use more information sources.

In some cases, there is a need for more consultations with railway undertakings, technologists, etc., and the processing effort is thus higher. For a number of route-days greater than the limit number, the proposed price of N3 should be higher than for ZJ, but at the same time lower than for P3.

b) Individual ad hoc request for railway capacity allocation submitted later than 3 working days in advance – product P3

P3 includes one-time, casual or irregular transport requests (the applicant doesn't know the exact ride date in advance) submitted very shortly before the actual ride. There is less time for request processing and time frame allocation (in extreme cases only in the order of minutes), and conflicts between the new route and existing ones are not resolved at the route creation stage (they are only resolved operationally later). The requested and actually allocated route parameters differ more than in above mentioned requests, timetable quality is lower, the most important criteria being the speed of processing whilst respecting a certain level of route safety (sticking to absolute route parameters), and giving all necessary information on the route to all operational employees. Technologies at originating, intermediary, destination and frontier stations are negotiated by the railway undertaking on a case-by-case basis. The proposed price should be higher than for N3 requests.

c) Individual ad hoc request for railway capacity allocation for technical and safety tests of railway vehicles, for test rides of vehicles of unapproved type or rides at a speed exceeding the track speed – products TB and ZK

Categories TB and ZK include requests for railway capacity allocation for technical and safety tests of railway vehicles, for test rides of vehicles of unapproved type or rides at a speed exceeding the track speed, submitted by railway undertakings in the individual ad hoc mode as these types of train rides are not made regularly, but only as needed, and the exact date of these rides cannot be predicted long time in advance.

The proposed price of railway capacity allocation (for TB and ZK) must reflect the highly demanding nature of request processing, as finding a suitable route and creating a timetable with the necessary measures is time consuming due to a greater need to consult several aspects of route allocation, and the costs of the necessary instruments are higher due to several technical and organisational measures required for the train ride. The request must be submitted for processing more than 3 working days before the intended departure date. The proposed price is based on the price of N3, taking into account route creation specifications.

d) Individual ad hoc request for railway capacity allocation for maintenance, restoration and increasing of usable capacity – products UI, OM and JD

For train rides made in connection with diagnostics and maintenance of infrastructure of the Railway Infrastructure Administration (UI, OM and JD), i.e. for the needs of the Railway Infrastructure Administration, 100% discount is provided on railway capacity allocation. For train rides made due to infrastructure restrictions attributable to the Railway Infrastructure Administration (i.e. planned and unexpected closures and irregularities), capacity is provided for free, these being rides capacity has already been allocated for (train redirections), or railway undertakings' rides due to infrastructure restrictions which otherwise wouldn't be necessary (bringing and carrying away necessary items).

The aim of the regulatory function of the price of railway capacity is to regulate requests so that long-term requests are submitted in the mode of the annual timetable and its changes, and short-term requests in the ad hoc mode. At the same time, it is necessary to take into account the effort and processing quality, and the following:

a) Quality of route creation and the effort required,
   - dependence on route length and the necessary negotiations with other entities,
   - passenger trains – negotiation with state administration – with ordering railway undertakings,
   - adopting other measures related to the route use,
   - dependence on the length of the period of route use, as the probability of finding a suitable route decreases and its creation becomes more complicated, and the probability of free slots being occupied by requests submitted earlier increases,
b) Quality of the necessary instruments and the effort required,
   - creation of a train diagram and working timetable,
   - creation of a booking timetable for passengers,
   - creation of a data timetable and its distribution to subsequent information systems,
c) Time period of train route use – route booking.

The maximum number of route-days for regular and late requests and requests until the next change of the annual timetable established by the authors is based on the deadlines for the creation of the annual timetable and its planned changes, the maximum number of route-days for individual ad hoc requests submitted 3 and more working days in advance is based on the defined requirements on target groups of requests for individual and less frequent requests within the period between timetable changes, and sometimes also longer-term requests with a departure date within the period between timetable changes. The maximum number of route-days for individual ad hoc requests submitted later than 3 working days in advance starts from the assumption of the most inconvenient accumulation of Christmas bank holidays (3 working days, 3 bank holidays on working days and twice 2 weekend days).

Table 3. Allocated data timetables in 2016 by products in the annual timetable

Allocated data timetables in 2016 – by products in the annual timetable				
	RJ	**PJ**	**ZJ**	**Total**
Number of data timetables	13.686	788	485	14.959
Number of route-days	2.899.051	132.265	32.170	3.063.486
Total length [km]	731.593	37.661	33.727	802.981
Total price [CZK]	58.109.454	4.361.513	1.805.166	64.276.133
Average number of route-days per data timetable	211.8	167.8	66.3	204.8
Average length [km per data timetable]	53.5	47.8	69.5	53.7
Average price [CZK per data timetable]	4.245.9	5.534.9	3.722.0	4.296.8

Source: Authors according to (23)

The difference between the allocated data timetables and allocated route-days, as shown in Table 3, consists in the fact that one data timetable can be allocated for more route-days. The Table shows the calculated average number of allocated route-days, route length and price of railway capacity allocation by products in the year concerned in the annual timetable mode.

Table 4 shows the difference between allocated data timetables and allocated route-days in the ad hoc mode. The Table also shows the calculated average number of allocated route-days, route length and price of railway capacity allocation by products in the year concerned in the ad hoc mode.

Table 4. Allocated data timetables in 2016 by ad hoc products

Allocated data timetables in 2016 – by products in the annual timetable					
	N3	**P3**	**TB**	**ZK**	**Total**
Number of data timetables	13.534	159.968	196	285	173.983
Number of route-days	26.400	160.943	196	367	187.906
Total length [km]	1.357.650	11.491.204	9.742	11.492	12.870.088
Total price [CZK]	3.201.400	41.747.680	107.800	299.290	45.356.170
Average number of route-days per data timetable	2.0	1.0	1.0	1.3	1.1
Average length [km per data timetable]	100.3	71.8	49.7	40.3	74.0
Average price [CZK per data timetable]	236.5	261.0	550.0	1.050.1	260.7

Source: Authors according to (24)

Table 4 does not contain products UI, OM and JD (ad hoc products); even though these represent a considerable number of allocated data timetables, route-days and kilometres of route length, they do not factor in the price of railway capacity allocation as their price hasn't been set.

The price of railway capacity allocation is calculated using formula (1):

$$CK = K1 + K2 \cdot Dtr + K3 \cdot Pdj \text{ [CZK]} \qquad (1)$$

where:

- CK is the total price of railway capacity allocation [CZK],
- $K1$ is the price of processing and determining of the timetable and railway capacity allocation [CZK],
- $K2$ is the price of train route creation [CZK per km],
- $K3$ is the price of train route allocation per day [CZK per day],
- Dtr is the route length; the distance between point of origin and point of destination on railway network, where the Railway Infrastructure Administration functions as infrastructure manager, or capacity allocator [km],
- Pdj is the number of route-days the route is allocated for [day].

Based on formula (1), the authors calculated limit values for the number of allocated route-days of the individual products subject to charge in the annual timetable, i.e. RJ, PJ, and ZJ. For requests matching the product definition, a condition was set stipulating that products in the annual timetable mode must be more cost effective than ad hoc products for a number of days greater than the limit number. A one-time ride in the annual timetable mode will always be more expensive than in the ad hoc mode. As a result, the following limiting conditions: RJ<N3, PJ<N3, ZJ < N3 and motivational conditions: RJ < PJ and N3 < P3 can be formulated. Limiting conditions apply to a number of days greater than the limit number and motivational conditions apply to an arbitrary number of route-days. The current limit values were calculated using average values (Table 5) of individual products from the 2016 dataset.

Table 5. Allocated data timetables in 2016 by ad hoc products

Allocated data timetables in 2016 – by products in the annual timetable					
	RJ	**PJ**	**ZJ**	**N3**	**P3**
Average number of route-days per data timetable	211.8	167.8	66.3	2.0	1.0
Average route length [km per data timetable]	53.5	47.8	69.5	100.3	71.8

Source: Authors

The authors modified formula (1) to fit inequations to calculate limit values for allocated route-days for products RJ, PJ and ZJ (annual timetable) compared to product N3 (ad hoc).

The limit values were calculated to establish the break-even point in the number of route-days, based on which requests should be moved from the ad hoc mode to the respective products in the annual timetable mode. The author marked the calculated break-even point in the number of route-days as the unknown x. For the system to work, the price of railway capacity allocation for RJ, PJ or ZJ for x-1 and fewer route-days should be higher than the price of railway capacity allocation for N3 for x-1 and fewer route-days. For limit number of route-days, the following inequalities (2), (3), (4), and (5) apply.

$$(RJ,PJ,ZJ) \leq C(N3) \quad (2)$$

$$K1 + K2 \cdot DtrRJ + K3 \cdot PdjRJxN3 \leq K1 + K2 \cdot DtrRJ + K3 \cdot PdjRJxN3 \quad (3)$$

$$K1 + K2 \cdot DtrPJ + K3 \cdot PdjPJxN3 \leq K1 + K2 \cdot DtrPJ + K3 \cdot PdjPJxN3 \quad (4)$$

$$K1 + K2 \cdot DtrZJ + K3 \cdot PdjZJxN3 \leq K1 + K2 \cdot DtrZJ + K3 \cdot PdjZJxN3 \quad (5)$$

where:

(RJ) is the price of railway capacity allocation for RJ [CZK],
(PJ) is the price of railway capacity allocation for PJ [CZK],
(ZJ) is the price of railway capacity allocation for ZJ [CZK],
$(N3)$ is the price of railway capacity allocation for N3 [CZK],
$K1$ is the price of processing and determining of the timetable and railway capacity allocation [CZK],
$K2$ is the price of train route creation [CZK per km],
$K3$ is the price of route allocation per day [CZK per day],
$DtrRJ$ is the length of the average RJ route; the distance between point of origin and point of destination on railway network, where the Railway Infrastructure Administration functions as infrastructure manager, or capacity allocator [km],
$DtrPJ$ is the length of the average PJ route; the distance between point of origin and point of destination on railway network, where the Railway Infrastructure Administration functions as infrastructure manager, or capacity allocator [km],
$DtrZJ$ is the length of the average PJ route; the distance between point of origin and point of destination on railway network, where the Railway Infrastructure Administration functions as infrastructure manager, or capacity allocator [km],
$PdjRJxN3$ limit number of route-days for RJ compared with N3 the respective route is allocated for [day],
$PdjPJxN3$ limit number of route-days for PJ compared with N3 the respective route is allocated for [day],
$PdjZJxN3$ limit number of route-days for ZJ compared with N3 the respective route is allocated for [day].

The calculation was made according to formulas (3), (4) and (5), where the number of route-days Pdj was used as the unknown x.

After substitution, the inequation for RJ is as follows:

$$K1RJ + K2RJ \cdot DtrRJ + K3RJ \cdot PdjRJxN3 \leq K1N3 + K2N3 \cdot DtrRJxN3 + K3N3 \cdot PdjRJxN3) \quad (6)$$

$$1\,700 + 8 \cdot 53.5 + 10 \cdot x \leq 100 + 0 \cdot 53.5 + 70 \cdot x$$
$$33.8 \leq x$$

The limit value x for the number of route-days for RJ was rounded to 34. The same substitution was performed for products PJ and ZJ. The results are interpreted in Table 12.

After substitution, the inequation for RJ is as follows:

$$K1PJ + K2PJ \cdot DtrPJ + K3PJ \cdot PdjPJxN3 \leq K1N3 + K2N3 \cdot DtrPJxN3 + K3N3 \cdot PdjPJxN3 \quad (7)$$

$$1\,700 + 10 \cdot 47.8 + 20 \cdot x \leq 100 + 0 \cdot 47.8 + 70 \cdot x$$
$$41.6 \leq x$$

The limit value x for the number of route-days for PJ was rounded to 42.

After substitution, the inequation for ZJ is as follows:

$$K1ZJ + K2ZJ \cdot DtrZJ + K3ZJ \cdot PdjZJ \leq K1N3 + K2N3 \cdot DtrZJxN3 + K3N3 \cdot PdjZJxN3 \quad (8)$$

$$1\,700 + 10 \cdot 69.5 + 20 \cdot x \leq 100 + 0 \cdot 69.5 + 70 \cdot x$$
$$45.9 \leq x$$

The limit value x for the number of route-days for ZJ was rounded to 46. The calculated limit values are summarised in Table 6.

Table 6. Calculated limit values for allocated route-days and comparison of prices for selected products

Product	RJ	PJ	ZJ
Number of route-days x-1	**33**	**41**	**45**
Price of product RJ, PJ, ZJ [CZK]	2.458	2.998	3.295
Price of product N3 [CZK]	2.410	2.970	3.250
Limit number of route-days x	**34**	**42**	**46**
Price of product RJ, PJ, ZJ [CZK]	2.468	3.018	3.315
Price of product N3 [CZK]	2.480	3.040	3.320

Source: Authors

The Table shows that with 34 and more allocated route-days for RJ, the price of allocated railway capacity is lower than the price of railway capacity allocation for N3 with the current coefficients K1, K2 and K3. With 42 and more allocated route-days for PJ, the price of allocated railway capacity is lower than the price of railway capacity allocation for N3.

With 46 and more allocated route-days for ZJ, the price of allocated railway capacity (for the average route length of RJ, PJ and ZJ) is lower than the price of railway capacity allocation for N3.

The aim of the proposed change to the price of railway capacity allocation was for the price to have a regulatory function and at the same time, to cover the direct costs incurred by the Railway Infrastructure Administration in connection with railway capacity allocation, with the revenue from the railway capacity allocation not exceeding these direct costs. Economic assessment was conducted on the timetable 2016 data. The direct costs of railway capacity allocation stated in the annual report of the Railway Infrastructure

Administration (25) were established by the Railway Infrastructure Administration economic department to be about CZK 106,780,000.

3 Creating a model for calculation

The authors start from the consideration that the total amount collected for railway capacity allocation (revenue from railway capacity allocation) must be lower or equal to the direct costs of railway capacity allocation. The following relation thus applies:

$$\sum_{c=1}^{c=10} v_c \leq \sum_{c=1}^{c=10} n_c \quad (9)$$

where:
v_c is the total revenue from railway capacity allocation [CZK],
n_c are the total direct costs of rail capacity allocation [CZK].

The individual vc products are marked as follows: v1 is the price of capacity allocation for individual requests for RJ, v2 is the price of capacity allocation for individual requests for PJ, v3 is the price of capacity allocation for individual requests for ZJ, v4 is the price of capacity allocation for individual requests for N3, v5 is the price of capacity allocation for individual requests for P3, v6 is the price of capacity allocation for individual requests for TB, v7 is the price of capacity allocation for individual requests for ZK, v8 is the price of capacity allocation for individual requests for UI, v9 is the price of capacity allocation for individual requests for OM, and v10 is the price of capacity allocation for individual requests for JD. For these vc the following relations apply:

$$v_1 = \sum_{n=1}^{n=c_1}(K_{1RJ} + K_{2RJ} \cdot D_{trRJn} + K_{3RJ} \cdot P_{djRJn}) \quad [\text{CZK}] \quad (10)$$

$$v_2 = \sum_{n=1}^{n=c_2}(K_{1PJ} + K_{2PJ} \cdot D_{trPJn} + K_{3PJ} \cdot P_{djPJn}) \quad [\text{CZK}] \quad (11)$$

$$v_3 = \sum_{n=1}^{n=c_3}(K_{1ZJ} + K_{2ZJ} \cdot D_{trZJn} + K_{3ZJ} \cdot P_{djZJn}) \quad [\text{CZK}] \quad (12)$$

$$v_4 = \sum_{n=1}^{n=c_4}(K_{1N3} + K_{2N3} \cdot D_{trN3n} + K_{3N3} \cdot P_{djN3n}) \quad [\text{CZK}] \quad (13)$$

$$v_5 = \sum_{n=1}^{n=c_5}(K_{1P3} + K_{2P3} \cdot D_{trP3n} + K_{3P3} \cdot P_{djP3n}) \quad [\text{CZK}] \quad (14)$$

$$v_6 = \sum_{n=1}^{n=c_6}(K_{1TB} + K_{2TB} \cdot D_{trTBn} + K_{3TB} \cdot P_{djTBn}) \quad [\text{CZK}] \quad (15)$$

$$v_7 = \sum_{n=1}^{n=c_7}(K_{1ZK} + K_{2ZK} \cdot D_{trZKn} + K_{3ZK} \cdot P_{djZKn}) \quad [\text{CZK}] \quad (16)$$

$$v_8 = \sum_{n=1}^{n=c_8}(K_{1UI} + K_{2UI} \cdot D_{trUIn} + K_{3UI} \cdot P_{djUIn}) \quad [\text{CZK}] \quad (17)$$

$$v_9 = \sum_{n=1}^{n=c_9}(K_{1OM} + K_{2OM} \cdot D_{trOMn} + K_{3OM} \cdot P_{djOMn}) \quad [\text{CZK}] \quad (18)$$

$$v_{10} = \sum_{n=1}^{n=c_{10}}(K_{1JD} + K_{2JD} \cdot D_{trJDn} + K_{3JD} \cdot P_{djJDn}) \quad [\text{CZK}] \quad (19)$$

Under the following conditions:

$$(K_{1RJ} + K_{2RJ} \cdot D_{trRJm} + K_{3RJ} \cdot P_{djRJm}) < (K_{1N3} + K_{2N3} \cdot D_{trN3m} + K_{3N3} \cdot P_{djN3m}) \quad (20)$$

$$(K_{1RJ} + K_{2RJ} \cdot D_{trRJm} + K_{3RJ} \cdot P_{djRJm}) < (K_{1PJ} + K_{2PJ} \cdot D_{trPJm} + K_{3PJ} \cdot P_{djPJm}) \quad (21)$$

$$(K_{1RJ} + K_{2RJ} \cdot D_{trRJm} + K_{3RJ} \cdot P_{djRJm}) < (K_{1ZJ} + K_{2ZJ} \cdot D_{trZJm} + K_{3ZJ} \cdot P_{djZJm}) \quad (22)$$

$$(K_{1N3} + K_{2N3} \cdot D_{trN3m} + K_{3N3} \cdot P_{djN3m}) < (K_{1P3} + K_{2P3} \cdot D_{trP3m} + K_{3P3} \cdot P_{djP3m}) \quad (23)$$

$$(K_{1PJ} + K_{2PJ} \cdot D_{trPJ} + K_{3PJ} \cdot P_{djPJ}) < (K_{1N3} + K_{2N3} \cdot D_{trN3} + K_{3N3} \cdot P_{djN3}) \quad (24)$$

$$(K_{1ZJ} + K_{2ZJ} \cdot D_{trZJ} + K_{3ZJ} \cdot P_{djZJ}) < (K_{1N3} + K_{2N3} \cdot D_{trN3} + K_{3N3} \cdot P_{djN3}) \quad (25)$$

$$K_{1RJ} + K_{2RJ} \cdot D_{trRJ} + K_{3RJ} \cdot (P_{dj} - 1)_{RJ} > K_{1N3} + K_{2N3} \cdot D_{trN3} + K_{3N3} \cdot (P_{dj} - 1)_{N3} \quad (26)$$

$$K_{1PJ} + K_{2PJ} \cdot D_{trPJ} + K_{3PJ} \cdot (P_{dj} - 1)_{PJ} > K_{1N3} + K_{2N3} \cdot D_{trN3} + K_{3N3} \cdot (P_{dj} - 1)_{N3} \quad (27)$$

Where:
n is the number of kilometres, route-days of the nth request for railway capacity for product c_n, n = 1, 2, 3, …, cn,
m is the number of limit kilometres, route-days for product c_n, m = 1, 2, 3, …, m,
n = cn is the total number of requests for railway capacity for product cn,
vc_1 , vc_2, vc_3, … , $vc1$ is the revenue from railway capacity allocation for product $c1$, $c2$, $c3$, … , $c10$ [CZK],
c_1, c_2, c_3, … , $c10$ is the total number of requests for railway capacity for product (RJ, PJ, ZJ, N3, P3, TB, ZK, UI, OM and JD) [CZK],
$K1$ is the price of processing and determining of the timetable and railway capacity allocation [CZK per route],
$K2$ is the price of train route creation [CZK per km],
$K3$ is the price of train route allocation per day [CZK per day],
$DtrRJ$, $DtrPJ$, $DtrZJ$, $DtrN3$, $DtrP3$, $DtrTB$, $DtrZK$, $DtrUI$, $DtrOM$, $DtrJD$ is the route length for products (RJ, PJ, ZJ, N3, P3, TB, ZK, UI, OM, JD), the distance between point of origin and point of destination on railway network, where the Railway Infrastructure Administration functions as infrastructure manager, or capacity allocator [km],
$PdjRJ$, $PdjPJ$, $PdjZJ$, $PdjN3$, $PdjP3$, $PdjTB$, $PdjZK$, $PdjUI$, $PdjOM$, $PdjJD$ is the number of route-days for RJ the respective route is allocated for [day],

The authors didn't include UI, OM a JD in the calculations as they do not factor in the price of railway capacity allocation as in accordance with applicable legal provisions, their price hasn't been set.

The authors simplified inequations (20), (21), (22) and (23) in the way described below, creating several invariables. Since coefficient K1 is the price of request processing, all requests processed in IS KANGO16 (products RJ, PJ and ZJ) have the same K1, and requests processed in IS KADR (products N3 and P3) have the same coefficient K1. The ratio between K1 value for requests processed in IS KANGO and requests processed in IS KADR is based on the current ratio and was established as 0.06. For coefficient K1 for individual products, the author established the following relations:
$K1RJ = K1RJ$
$K1PJ = K1RJ$
$K1ZJ = K1RJ$

The authors first modelled Option 0 reflecting the current situation and the current allocation charging scheme using the current coefficients K1, K2 and K3.

Table 7. Calculation for all requests based on coefficients (for total values)

Product	K1	Number of data timetables	K2	Total number of kilometres	K3	Number of route-days	Price of product [CZK]	Ø price of 1 request [CZK]
RJ	1.700	13.686	8	731,593	10	2,899,051	58,109,454	4,246
PJ	1.700	788	10	37,661	20	132,265	4,361,513	5,534
ZJ	1.700	485	10	33,727	20	32,170	1,805,166	3,721
ZK	960	285	0	11,492	70	367	299,290	1,051
TB	480	196	0	9,742	70	196	107,800	550
N3	100	13.534	0	1,357,650	70	26,400	3,201,400	240
P3	100	159.968	0	11,491,204	160	160,943	41,747,680	260
Total price [CZK]							**109.632.303**	**15.602**

Source: Authors

The total price in Table 7 is the current revenue from railway capacity allocation. The current direct costs of railway capacity allocation were established by the Railway Infrastructure Administration economic department to be about CZK 106,780,000. This amount shouldn't be exceeded. The difference between revenue and costs is approximately 2.67% (CZK 2,852,303). It is approximately this amount the revenue from railway capacity allocation needs to be reduced by while complying with the conditions of formulas (20), (21) ~ (22), (23), (24) ~ (25), (26) and (27). The break-even point of price between products is shown in Figure 1.

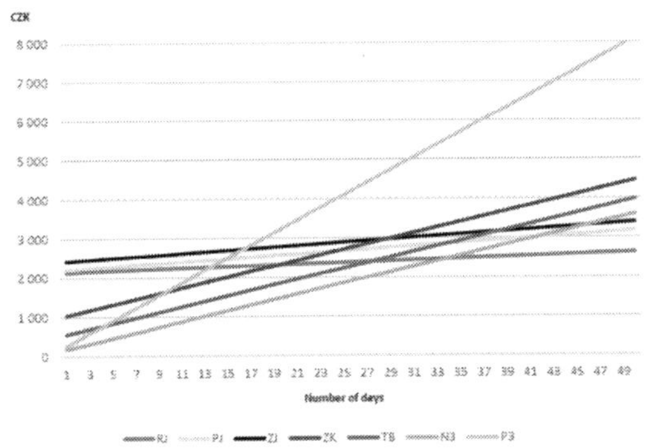

Fig. 1. The break-even point of price between products.

Table 8 Comparison of revenue from railway capacity allocation with total direct costs of railway capacity allocation

Revenue from railway capacity allocation [CZK]	Costs of railway capacity allocation [CZK]	Difference between revenue and costs [%]
109.632.303	**106.780.000**	**2.67**

Source: Authors

4 Summary

Current model for calculation of the price of railway capacity allocation using coefficients K1, K2, and K3 in accordance with Table 8. In 2016, the revenue from railway capacity allocation was 2.67% higher than the total direct costs of railway capacity allocation.

In this paper, the authors defined the requirements on target groups of requests so that they can be attributed to individual products either in the mode of allocating railway capacity within the annual timetable, or in the ad hoc mode. This definition is vital for proper regulation of the demand of railway undertakings for individual products. The aim of the proposed model is for a part of the requests to move from the ad hoc mode to the annual timetable mode.

The work was supported from ERDF/ESF "Project PosiTrans - University of Pardubice cooperation and application sphere in application oriented research of localization, detection and simulation systems for transport processes" (No. CZ.02.1.01/0.0/0.0/17_049/0008394).

References

1. Directive no. 266/1994, Czech Republic, from 14th December 1994, on railways.
2. Derective 2012/34/EU of the European Parliament and of the council of 21 November 2012, establishing a single European railway area.
3. Directive no. 319/2016, Czech Republic, from 6th September 2016, on railways.
4. SŽDC, Network Statement (2015)
5. SŽDC, D1 - Transport and Signal Directive, (2012)
6. Directive. 526/1990, Czech Republic, from 27th November 1990, about prices, actual version 2015
7. Price Bulletin No. 12/2016 of 29th November 2016, Czech Republic
8. Ministry of Finance of Czech Republic, Decree No. 01/2017, of 25 November 2016, listing the regulated price list
9. AŽD Praha, Network Statement, available at: http://www.azd.cz/admin/files/Dokumenty/pdf/Kompletni-zneni-Prohlaseni-o-draze-2017-Cizkovice-Obrnice-Kopidlno-Dolni-Bousov.pdf (2016)
10. J. Gašparík; B. Abramović, M. Halás Promet Traffic & Transportation. 27, 4. 283-290. (2015)
11. Network Statement, Jindřichohradecké místní dráhy, a.s., available at: http://docplayer.cz/1333471-Prohlaseni-o-draze-regionalni.html (2017)
12. Network Statement, Moravskoslezský kraj. available at: http://www.szdc.cz/soubory/prohlaseni-o-draze/2017/01-04-17-prohlaseni-o-draze-2017-1zmena-mosnov.pdf (2017)

* Corresponding author: jaromir.siroky@upce.cz

13. KŽC, Regional Network Statement, available at: http://www.kzc.cz/muzejni-zeleznice-senovka/ prohlaseni-o-draze (2014)
14. Železnice Desná, Network Statement, available at: http://www.sart.cz/wpcontent/uploads/Prohlasení-o-draze-aktualizace-2016.pdf (2015)
15. H. Saláková, *Bachelor thesis: Sledování ekonomických faktorů při přidělování kapacity železniční dopravní cesty*. Univerzity of Pardubice (2015)
16. Decree No. 501/2005, Czech Republic, on the determination of the costs of the operator of the railway connected with the operation and ensuring of operability, modernization and development of the railway infrastructure, (2006)
17. Network Statement. DB NETZE. available at: http://fahrweg.dbnetze.com/fahrwegen/network_access/network_statement/2017.html (2016)
18. Network Statement. PKP PLK. available at: http://en.plk_sa.pl/files/public/user_upload/pdf/Reg_przydzielania_tras/Regulamin_2016_2017/Network_Statement_2016-2017_v11.pdf (2016)
19. Network Statement. ÖBB Infrastruktur AG. available at: http://www.oebb.at/infrastruktur/en/Network_Access//NetworkStatement/2017.jsp (2016)
20. Product Catalogue 2017 Train Path. ÖBB Infrastruktur AG. available at: http://www.oebb.at/infrastruktur//en/Network_Access/Product_ Train_Path.jsp (2016)
21. ŽSR, Conditions for using the railway network for GVD 2016/2017. available at: http://www.zsr.sk//Marketing/PodmPouzivaniaZeleznicnejSiete2017.pdf (2016)
22. J. Gašparík, V. Zitrický. Promet-Traffic and Transport. 27, 4, 283-290. (2015)
23. V. Zitrický, L. Černá, B. Abramović. Procedia Engineering. 192. 994-999. 2017.
24. SŽDC, Annual report 2016, available at: http://www.szdc.cz/soubory/vysledky-hospodareni.pdf (2017)
25. Train Path Pricing System (TPS). DB Netz. [available at: http://fahrweg.dbnetze.com/train_path_pricing_system_2017.pdf (2016)
26. S. Stoilova. International Multidisciplinary Scientific Conferences on Social Sciences and Arts. 627-634. (2016)
27. Gasparik, J.; Majercak, J.; Siroky, J.; Abramovic, B.; Mesko, P.; Nachtigall, P.; Zitricky, V. 2017. Railway Traffic Operation. Žilina: Žilinská univerzita, 292 p.
28. Široký, J.; Cempírek, V.; Gasparik, J. 2014. Transport Technology and Control. Brno: Tribun EU, 238 p.

MATEC Web of Conferences **235**, 00017 (2018) https://doi.org/10.1051/matecconf/201823500017
Horizons of Railway Transport 2018

Capacity range – definition and calculation

Přemysl Šrámek[1], *Jaromír* Široký[*] and *Pavlína* Hlavsová[1]

[1] University of Pardubice, Faculty of Transport Engineering, Department of Transport Technology and Control, Studentska 95, 53210 Pardubice, Czech Republic

Abstract. This article defines a new capacity parameter called capacity range (CR). The definition is explained and then the parameter is calculated for a selected section of Kolín – Choceň (overlapping section of RFC 7 and RFC 9). The calculation is based on timetable 2016. Capacity range, expressed as dependence of the number of trains on average delay increment, is a useful capacity parameter particularly for coherent railway sections with remote traffic control.

1 Introduction

In a doctoral thesis [1], it was found out that it would be appropriate to extend the existing methodologies of calculating capacity parameters (SŽDC D24, UIC 406), particularly by adding parameters allowing for a more network-based calculation of capacity (through several interstation sections). Even though these parameters would be limited by railway network bottlenecks, which they would likely fail to identify, they would still be useful for the evaluation (comparison) of individual operational concepts. At the same time, they would fit the notion that capacity as such actually doesn't exist [2].

With the current implementation of remote traffic control through several interstation sections, it is possible to move trains through bottlenecks more efficiently than before. That's why a new parameter was defined: CR (capacity range).

2 Objectives

The aim of this paper is to define a new capacity parameter called capacity range, including the methodology of calculation. Subsequently, this methodology is applied on the section of Kolín – Choceň.

3 Methodology

To meet the above mentioned objectives, the following methods are to be employed: capacity range calculation, bottleneck identification (operational facility occupancy rate), simulation.

Capacity range calculation – capacity range (hereinafter referred to as CR) definition is shown in Figure 1. CR is defined as an area delimited by 2 lines and 1 curve. Specifically, it is a perpendicular line to the X-axis from the point where N = 1 (N denoting the total number of trains), and the actual X-axis representing the state where ADI = 0. Here, ADI (average delay increment) means average delay increment. The curve then expresses the dependence of ADI on N in a specific section of infrastructure. Mathematically, the relationship is expressed as follows:

$$CR = \left|\int_1^E f(x)\,dx\right| \quad (1)$$

CR is calculated as an integral of a dependence function of ADI on N, delimited by N = 1 on one hand (N can be neither negative nor zero), and a point of balance, where ADI = 0 min/train based on the specified number of trains (based on f(x)), on the other hand. With further increasing the number of trains, ADI becomes positive, leading to reduced stability of the timetable and impossibility to limit or eliminate the stochastically driven delay.

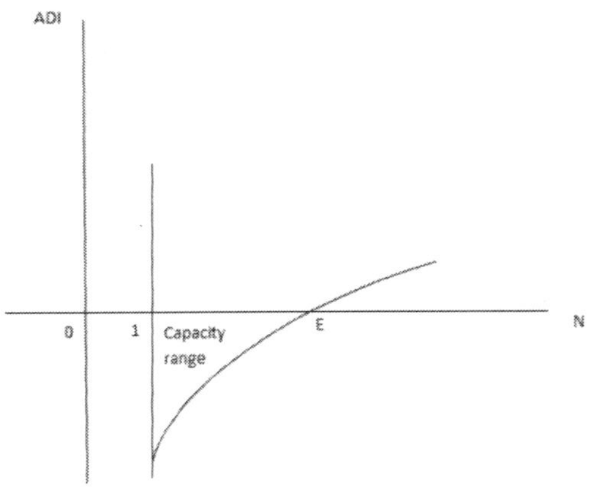

Fig. 1. Capacity range definition.

The average delay increment is calculated according to the following formula:

$$ADI = \frac{z_p exit - z_p entry}{N} \quad (2)$$

[*] Corresponding author: jaromir.siroky@upce.cz

As such, it is a quotient of the difference between the total exit delay and the total entry delay of all trains, and their total number.

Where the railway infrastructure manager is willing to sacrifice timetable stability to a certain extent to increase the number of trains in the specific infrastructure section, enlarged capacity range can be considered as well (CRE – capacity range enlarged).

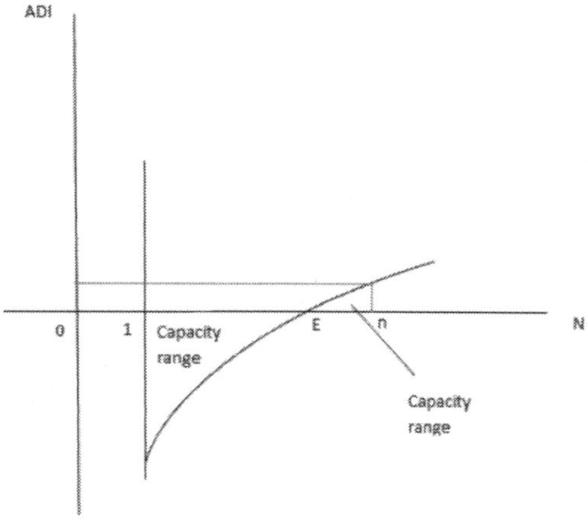

This is shown in Figure 2.
Fig. 2. Enlarged capacity range definition

The capacity range area can optionally be enlarged due to the railway infrastructure manager deciding to operate a higher number of trains at the price of certain instability of the timetable (positive ADI). This actually means increasing quantity at the expense of quality of railway transport. Mathematically, the relationship is expressed as follows:

$$\mathrm{CRE} = \left|\int_1^E f(x)\,dx\right| + \int_E^n f(x)\,dx \quad (3)$$

This gives rise to a new area above the curve (f(x)) delimited by the X-axis and the point of balance, where ADI = 0 min/train, and a set point (either based on the maximum allowable ADI or the required number of trains N).

The greatest challenge in calculating capacity range is the calculation of the dependence function f(x) of ADI on N, this function being different for different railway infrastructure sections. The calculation is described below in the specific application section.

Identifying a bottleneck – since capacity range is mainly useful for evaluating operational concepts, or lines, it cannot primarily be used to identify a railway network bottleneck. Therefore, operational facility occupancy rate is calculated as an additional parameter, according to the Railway Infrastructure Administration (SŽDC) (ČD) regulation No. D24. This is expressed as the ratio of the total time of occupancy of the operational facility with the regular scope of railway transport to the calculation period reduced by the total time of closures and permanent manipulations. A facility is considered sufficiently occupied having a rate of occupancy within a range of S_o = 0.5 to 0.67 (1). Mathematically expressed:

$$S_o = \frac{N \cdot t_{obs}}{T - (\sum t_{výl} + \sum t_{stál})} \quad (4)$$

S_o operational facility occupancy rate [-]
T calculation period [min]
N regular scope of transport [number of trains / calculation period]
t_{obs} technological time of occupancy of a specific operational facility by one train [min/train]
$\sum t_{výl}$ total time of closures of operational facility for repairs and maintenance [min]
$\sum t_{stál}$ total time of occupancy of operational facility due to permanent manipulations [min]

Simulation – as a simulation program, SimuT was chosen mainly due to the simple determination of timetable stability based on the average delay increment calculated as a quotient of the difference between the exit delay and the entry delay, and the number of trains. Entry delay is entered by the simulation program for individual trains based on exponential probability distribution. Furthermore, SimuT allows for the calculation of capacity parameters according to the SŽDC (ČD) regulation No. D24 [3].

4 Application of capacity range on the section of Kolín – Choceň

In practical application, it is necessary to enter the required data concerning the infrastructure and scope of operation into the programme. Capacity range calculation was performed on part of the 010 line, specifically on the section between Kolín (incl.) and Choceň (incl.) with a total number of 10 railway stations, with 9 of them operated remotely from the Central Dispatching Site (Centrální dispečerské pracoviště) in Prague (station Pardubice hl. n. is not yet operated remotely). Using the GVD 2016 analysis, the average daily scope of railway transport was identified as 384 trains, 171 of which were long-distance passenger service trains (Ex, R), 62 of which were regional passenger service trains (Sp, Os), and 151 of which were freight service trains.

The dependence function (f(x)) of ADI on N is determined based on the maximum reliability function created by fitting the curve to individual points. These individual points represent particular ADI values for the given number of trains. The more points are identified, the more reliable will the determination of the function f(x) be. It would thus be optimal to create the individual points by always adding one train at a time. On the other hand, this solution would be very time consuming and would make the calculation too complex. Therefore it is necessary to set individual steps in the simulation programme to specify the number of trains by which the individual points would differ.

So as not to encumber the steps with a different number of trains of different categories, the trains need to be added in "packages". These packages always need to be adapted to the operation on the particular line. As for the section of Kolín – Choceň, where the ratio of trains is 171:62:151 (long-distance passenger service : regional passenger service : freight service), the function f(x) can be determined using 6 points with one step representing the addition of 30 long-distance passenger service trains, 10 regional passenger service trains, and 25 freight service trains. The overall results for the individual steps including bottleneck identification (operational facility occupancy rate) are shown in Table 1. These are outputs from the SimuT simulation program.

Table 1. Setting Word's margins.

Number of trains	ADI (min/train)	Bottleneck	S_o
65 (30+10+25)	-2.86	Pardubice hl. n. – Přelouč, TK 1	0.10
130 (60+20+50)	-2.48	Pardubice hl. n. – Přelouč, TK 1	0.20
195 (90+30+75)	-1.87	Pardubice hl. n. – Přelouč, TK 1	0.28
260 (120+40+100)	-0.99	Pardubice hl. n. – Přelouč, TK 1	0.39
325 (150+50+125)	0.37	Pardubice hl. n. – Přelouč, TK 1	0.48
390 (180+60+150)	2.17	Pardubice hl. n. – Přelouč, TK 1	0.57

As can be seen from Table 1, ADI increases with the individual steps. In each of the steps, the interstation section Pardubice hl. n. – Přelouč (the first track rail thereof) was identified as a bottleneck. This fact reflects the overall consistency of the calculation and shows that the operational facility occupancy rate increases with the individual steps as well.

The dependence function f(x) of ADI on N can also be expressed as follows:

$$ADI = f(N) \tag{5}$$

The results of the individual steps corresponding to the above mentioned equation are shown in Figure 3.

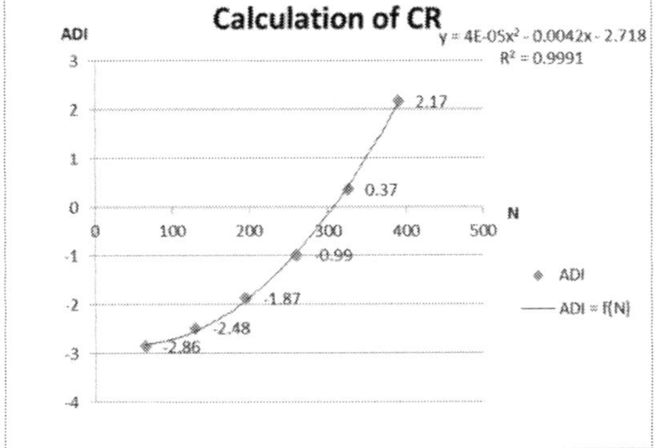

Fig. 2. Dependence of ADI on N – specific CR

Individual points were fitted to a trendline with the highest reliability; the dependence of ADI on N is polynomial, and in this case quadratic. Or:

$$ADI = 0.00004 \cdot N^2 - 0.0042 \cdot N - 2.718 \tag{6}$$

The capacity range formula can also be expressed as follows:

$$CR = \left|\int_1^E f(N)\,dN\right| \tag{7}$$

Point E can be calculated by inserting ADI = 0 min/train into the above mentioned equation, which gives rise to a standard quadratic equation:

$$0 = 0.00004 \cdot N^2 - 0.0042 \cdot N - 2.718 \tag{8}$$

Based on the rules for calculating quadratic equations, the result is as follows:

$$E_1 = 318.4 \tag{9}$$

$$E_2 = -213.4 \tag{10}$$

As it is a total number of trains, the number must be positive and integral. To maintain the timetable stability, E is always rounded down. The resulting point of balance is then E = 318 trains. Given the original ratio 3:1:2.5, one of the possible optimum solutions of transport structure on the given section would be 147 long-distance passenger service trains, 49 regional passenger service trains, and 122 freight service trains. With this transport structure, the timetable is still stable, i.e. able to reduce random delay. Now we can calculate capacity range:

$$CR = \left|\int_1^{318} (0.00004 \cdot N^2 - 0.0042 \cdot N - 2.718)\,dN\right| \tag{11}$$

$$CR = \left|[0.00001333 \cdot N^3 - 0.0021 \cdot N^2 - 2.718 \cdot N]_1^{318}\right| \tag{12}$$

$$CR = 645.2 \tag{13}$$

Capacity range of the section Kolín (incl.) – Choceň (incl.) for further use (comparison of operational concepts, modifications of infrastructure) is 645.2.

5 Application of enlarged capacity range on the section of Kolín – Choceň

As described above, capacity range can be increased at the expense of reduced timetable stability (at the price of increasing ADI). In this article, both cases are described below, i.e. CR for the required number of 450 trains, and CR for an allowable ADI of 5 min/train.

Enlarged capacity range for N = 450 – where as requested by railway undertakings, the railway infrastructure manager comes to a conclusion that it is necessary to operate 450 trains a day, while maintaining the current transport structure (the ratio of 3:1:2.5, i.e. in this case 208 long-distance passenger service trains, 69 regional passenger service trains, and 173 freight service trains) on the existing infrastructure, ADI can be predicted using the identified function as follows:

$$ADI = 0.00004 \cdot N^2 - 0.0042 \cdot N - 2.718 \tag{14}$$

$$ADI = 0.00004 \cdot 450^2 - 0.0042 \cdot 450 - 2.718 \tag{15}$$

$$ADI = 3.492 \text{ min/train} \tag{16}$$

The operation of 450 trains while maintaining the current transport structure would probably lead to an increase of average delay increment by 3.492 min/train. Capacity range will be enlarged using the formula for CRE:

$$\mathrm{CRE} = \left| \int_1^E f(N)\,dN \right| + \int_E^n f(N)\,dN \quad (17)$$

$$\mathrm{CRE} = 645.2 + \left[0.00001333 \cdot N^3 - 0.0021 \cdot N^2 - 2.718 \cdot N\right]_{318}^{450} \quad (18)$$

$$\mathrm{CRE} = 645.2 + 214.6 \quad (19)$$

$$\mathrm{CRE} = 859.8 \quad (20)$$

Capacity range was enlarged by an optional part (214.6) to 859.8.

Enlarged capacity range for an allowable ADI of 5 min/train – the railway infrastructure manager can also accept a timetable with a certain positive average delay increment to be able to offer a higher number of train routes. In this case, the following applies:

$$5 = 0.00004 \cdot N^2 - 0.0042 \cdot N - 2.718 \quad (21)$$

$$0 = 0.00004 \cdot N^2 - 0.0042 \cdot N - 7.718 \quad (22)$$

$$N1 = 494.9 \quad (23)$$

$$N2 = -389.9 \quad (24)$$

As it is a total number of trains, the number must be positive and integral, and always rounded down, as described above. The total number of trains for an allowable ADI of 5 min/train is N = 494 trains. Given the original ratio 3:1:2.5 (maintaining the current transport structure), it is possible to operate 228 long-distance passenger service trains, 76 regional passenger service trains, and 190 freight service trains on this section and with this allowable ADI. Capacity range will be enlarged using the formula for CRE:

$$\mathrm{CRE} = \left| \int_1^E f(N)\,dN \right| + \int_E^n f(N)\,dN \quad (25)$$

$$\mathrm{CRE} = 645.2 + \left[0.00001333 \cdot N^3 - 0.0021 \cdot N^2 - 2.718 \cdot N\right]_{318}^{494} \quad (26)$$

$$\mathrm{CRE} = 645.2 + 400.1 \quad (27)$$

$$\mathrm{CRE} = 1{,}045.3 \quad (28)$$

Capacity range was enlarged by an optional part (400.1) to 1,045.3.

6 Conclusion

With the current implementation of remote traffic control through several interstation sections, it is possible to move trains through bottlenecks more efficiently than before. In this article, capacity range was thus defined analysing the function of average delay increment in a given infrastructure section, allowing for a more network-based evaluation of capacity.

On the other hand, the use of this methodology is problematic mainly in relation to integral calculus, or the need to phase the calculation depending on the local course of the function.

In addition to the possibility of network-based capacity evaluation, another advantage of capacity range is the possibility to predict the delay increment based on a set number of trains or the number of trains according to the maximum allowable delay increment.

The work was supported from ERDF/ESF "Project PosiTrans - University of Pardubice cooperation and application sphere in application oriented research of localization, detection and simulation systems for transport processes" (No. CZ.02.1.01/0.0/0.0/17_049/0008394) and "Student grant competition“ (No. SGS_2018_023).

References

1. P. Sramek, Vliv periodického jízdního řádu na kapacitu trati. Pardubice: doctoral thesis, 2017.
2. International Union of Railways (UIC). UIC CODE 406, 2nd edition. Paris: UIC, 2013. ISBN 978-2-7461-2159-1.
3. P. Sramek, T. Molkova, Horizons of Railway Transport 2016 – proceeding, 207–213. (2016)
4. J. Siroky, et al. *Transport Technology and Control*. Tribun EU. Brno. pp. 238. (2014).
5. J. Gasparik, J. Majercak, J. Siroky, B. Abramović, P. Mesko, P. Nachtigall, V. Zitricky. *Railway Traffic Operation*. 292 p. 2017
6. J. Gasparik, B. Abramovic, V. Zitricky, Technicki Vjesnik – Technical Gazette. **25, 4** 1190-1195, (2018)
7. J. Gasparik, J. Siroky, L. Peceny, M. Halas. Communications: scientific letters of the University of Zilina. **2**. 25-30.(2014)
8. P. Sramek T. Molkova,. *Proceedings of the Third International Conference on Traffic and Transport Engineering (ICTTE)*. 949 – 954. (2016)

Cost Shock in Public Tenders Impacting the Railway Operator

Špetík Ondřej[1*]

[1]Department of Economics, Faculty of Economics and Administration, Masaryk University, Brno, Czech Republic

Abstract. The aim of this paper is to find out whether the cost shocks occur in the passenger rail transport, whether it can cause problems of financing as well as to evaluate the ways which are used by the purchaser of public transport to change the subsidy for carriers in case of cost shock.
Results have shown that costs shocks can negatively impact financing of the subsidised passenger railway transport, so purchasers of the transport services must be extremely cautious in setting the tender documentation. So that purchasers avoid time-inconsistency solution, it is necessary to set the change in the subsidy in the contract. None of the tested clauses used by purchasers is ideal. We can see over-/under-compensation. The clause with 55 % of the consumer price index and 35 % of the change in labour costs in the transport sector reaches better results especially in contracts with vehicles purchased by the carrier.

1 Introduction

Functioning of subsidised passenger railway transport is undergoing changes not only in the Czech Republic but also in some countries of the European Union. Some routes are tendered or are awarded directly to the private transport operator. From 2019, public tenders be will obligatory with some exceptions [14].

The experience has shown that the call for tender brings certain problems. Purchasers of transport services either do not have any experience or have little experience with it. Criteria can be settled in such a way that they will not optimise the results of the tender. Passenger´s preferences or economic possibilities of carriers may not be taken into account. Nevertheless, the criteria must be carefully settled in the tender documentation to reach efficient public spending.

There is only little experience with the call for tender in the railway sector, and most of them ended up unsuccessfully in the Czech Republic (e.g. a tender on the route R 16 Plzeň-Most was rescinded by the Office for the Protection of Competition due to discrimination occurring on the side of some tenderers [20]). There are two successful examples: a tender for three lines in the South Bohemian Region and a tender in the South Moravian Region.

One of the problems which may occur is called "cost shock". It has not been defined in the literature yet. For this paper, we can define it as a significant and unexcepted change of certain costs. Unexcepted means that carriers cannot count with it in competing for tender. The increase of minimum wage for bus drivers in 2017 is a typical cost shock (i.e. real wages increased by 19 %). The increase of minimum wage caused difficulties in some regions (e.g. some bus carriers had to terminate the contract [8]).

From the perspective of legislation, it is possible to resolve this situation in two ways: ex ante approach when the contract can be changed by the inflation or another clause or ex post approach when the contract can be ad hoc changed after the cost shock. Ex post solution may bring problems with time-inconsistency. Coming back to the example of the increase of minimum wage, purchasers of transport services might not address the problem. It also increases insecurity for carriers.

The topic is also important due to its fiscal consequences. Passenger rail transport is annually subsidised by approximately 0.6 billion Euros [17]. Furthermore, public authorities have a contract by due diligence.

The objective of this paper is to find out whether the cost shocks occur in the passenger rail transport, whether it can cause problems of financing as well as to evaluate the ways which are used by the purchaser of public transport in order to change the subsidy for carriers in case of cost shock.

The structure of the paper corresponds to the objective. After the literature review, we will deal with the possible existence of cost shock in the passenger rail transport. Afterwards, we will focus on its influence on financing of the contract, and then we will evaluate the way the purchasers of the transport change the subsidy.

The methodology is to compare changes in real costs with the changes in the subsidy which is presumed in the contract. We have conducted this analysis on the not electrified, regional routes.

The amount of the costs is based on Šíp [18]. Prices of fuel, the consumer price index (CPI), average wages

* Corresponding author: 405154@mail.muni.cz

are based on the statistics of the Czech Statistical Office and dates of other average wages derive from the Average Earnings Information System [7].

2 Literature Review

Time inconsistency was firstly described by Kydland and Prescott [4, 10]. Time-inconsistent politics is such politics whose ex-ante optimum differ from ex-post optimum [14]. An example of time-inconsistent politics is the patent policy. The government in the time t declares the patent´s protection, but in the time $t+s$ does not protect the intellectual property of patents.

In the literature, we can find papers dealing with time inconsistency by the protection of investments. Dastan [3] and Bondarev and Weigt [21] focused on time inconsistency in the energy sector. Nevertheless, the literature has not focused on time inconsistency in public tenders.

The term "cost shock" is a wide term overlapping public transport sector. It is mentioned in relation to the exchange rate risk of multinational corporations. According to these papers, exchange rate movements influence the profit. Several papers deal with cost shock of commodities. They are primarily focused rather than commodity consumers to the mining companies [12].

Regarding the public tenders, literature deals with revenue risks (e.g. Hunold and Wolf [6] or Nigrin [13]). Smith and Wheat (2016) mention cost shocks: they settled them as an important factor influencing the whole railway industry. Uguccioni [19] has focused on costs in the rail freight transport and their structure changes: according to him, wage costs and costs for vehicles have decreased by the time, whereas fuel and the charge of the transport route have increased. He has explicitly mentioned the shock of increased labour costs because of the trade unions pressure, and he has noted that this growth can exceed an increase of work productivity.

Hammoudeh and Li [5] studied the influence of increasing oil prices on return on capital investment in the American transport industry. Nandha and Brooks [11] have investigated the relationship between changes in oil prices and return in the transport industry in 38 countries. They have found out that in developed counties there is a negative impact on return on capital investment in the transport sector. The impact of oil shock can differ from a company to company with regard to the fact whether companies can pass on these costs to their customers [1, 15].

Velde and Beck [22] analysed why carriers in Germany do not sign up to the public tender even though they collect the tender documentation. He has found out that higher risk relates to lower willingness to participate in the tender. Firstly, there is a cost risk and secondly the revenue risk. According to Hunold and Wolf [6], a suitable setting of the tender conditions has influenced not only the sustainability of the project but also the participation in the public tender and consequently its price, too.

3 The Existence of the Cost Shock in the Transport

From the literature review, we can say that the cost shock can influence the profitability of the contract. For our paper, it is necessary to verify the hypothesis of the cost shock existence in the railway passenger sector. Firstly, it is necessary to provide a cost analysis of railway companies. Moreover, the costs can differ from line to line. We have settled the analysis of the line Bratislava-Komárno in Slovakia by Šíp [18] as a measure because of the data availability. This analysis was carried out to award the contract to a private carrier. The analysis calculates with new vehicles on the not-electrified line. We can generalise the results on lines with the similar cost structure.

In the Table no. 1, we can see the costs. The price level is settled in the year 2010, and the net profit is set to 2.8 % in both situations.

Table 1 : Estimated costs on the Bratislava-Komárno line

Item	Thousands EUR	EUR/km	Share	Share without charge of the transport route	Share without charge of the transport route and leasing
Vehicle leasing	3061	2.45	33.2 %	41.3 %	X
Depreciation	172	0.14	2 %	2.4 %	4.2 %
Fuel	1273	1.02	13.8 %	17.2 %	30.3 %
Maintenance of vehicles	907	0.73	9.9 %	12.3 %	21.7 %
Wages	800	0.64	8.7 %	10.8 %	19 %
Social security and health insurance payments	272	0.22	3 %	3.7 %	6.5 %
Other costs and administrative costs	652	0.52	7.1 %	8.8 %	15.4 %
Profit	269	0.21	2.8 %	3.5 %	2.8 %
Total costs (without charge of the transport route)	7405	5.93	80.4 %	100 %	100 %
Estimation of charge of the transport route (2011)	1818	1.45	19.6 %	X	X

Source: Šíp [18]

We will concentrate on the proportion of the costs in the total costs. Some regions (e.g. the South Moravian Region) are going to call for tender of the operation and vehicle separately, we will also count with the contract without vehicles [9]. We will focus on costs which can significantly fluctuate. In our case, they are fuel and labour costs (i.e. the sum of wages and social security and health insurance payments).

Looking at the trend in fuel prices, we can see significant fluctuations in both directions.

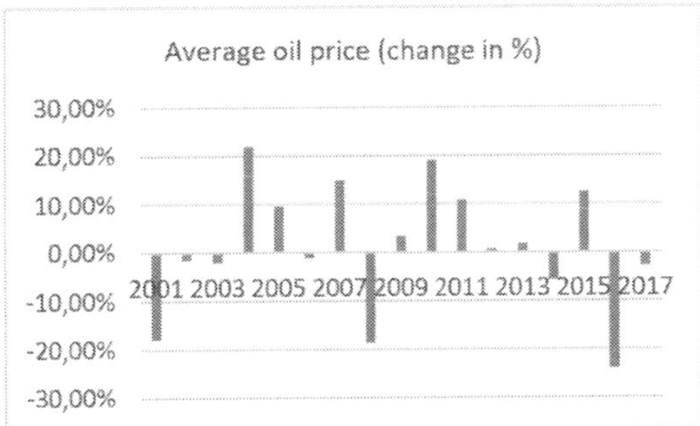

Figure 1: Average oil price
Source: Czech Statistical Office 2018

If we analyse changes in labour costs, we must set up a group of workers whose wage will be examined. The passenger railway transport requires specific employees whose wage development can differ from development in the whole transport industry. Structure of specific employees is indicated in the Table no. 2.

Table 2: Share of the profession on the labour costs

Assistant in the transport	12.16%
Cabin crew	35.81%
Driver	45.95%
Traffic controller	6.08%
Assistant in the transport	12.16%

Source: Šip [18]

We have taken their average wages from the Average Earnings Information System:

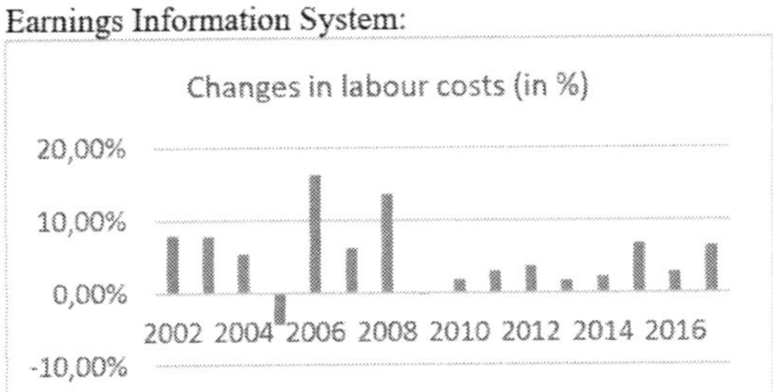

Figure 2: Changes in labour costs
Source: ispv.cz [7]

From the information above, the conclusion is that cost shocks are present. We can see unexpected cost changes especially in the oil prices. In addition, changes in labour cost are significant. Both can negatively influence financing of the subsidised transport services.

4 Influence of Cost Shocks on Financing of the Transport Services

Change of the costs as such would not cause the problem of financing the subsidised transport services. It is also important how the change in the costs will influence the total financing of the services: we are looking for the proportion of the changes in the total costs.

In the next charts, the real change in the subsidy is outlined because of the cost increase. Contracts for the passenger rail transport can be concluded for 15 years; we indicate the real change in the subsidy in two-, three-, five-, ten- and fifteen-year moving averages.

We provide all changes for both situations - when vehicles are purchased by the carrier and by a public authority.

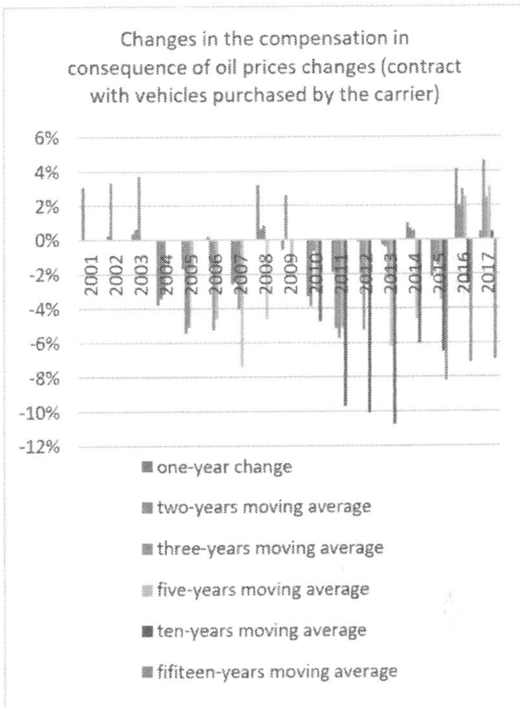

Figure 3: Changes in the compensation in consequence of oil prices changes (contract with vehicles purchased by the carrier) [2, 18]

Figure 4: Changes in the compensation in consequence of oil prices changes (contract with vehicles purchased by the public authority) [2, 18]

In order to analyse the labour costs, we have used costs for workers necessary for transport services.

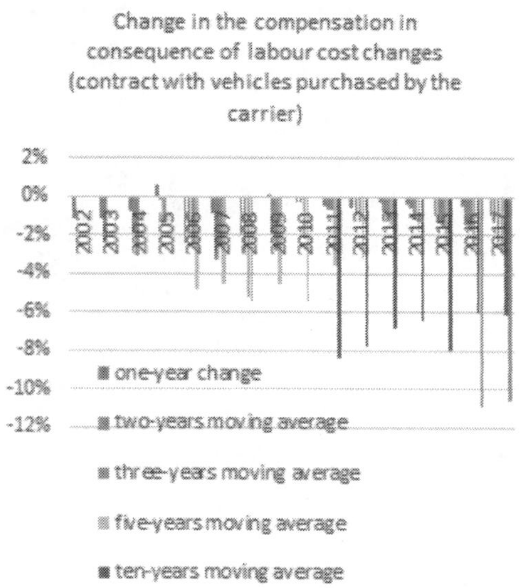

Figure 5 : Change in the compensation in consequence of labour cost changes (contract with vehicles purchased by the carrier) [2, 18]

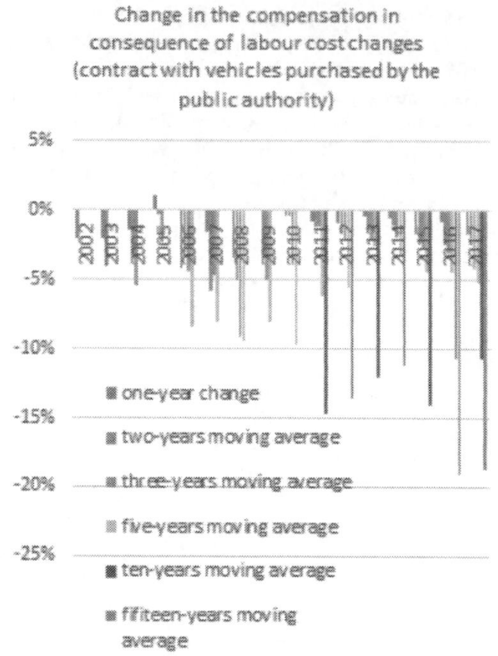

Figure 6: Change in the compensation in consequence of labour cost changes (contract with vehicles purchased by the public authority) [2, 18]

We have summed up oil price changes and labour cost changes for analysing the real impact on the carrier. The consumer price index has changed other variable costs.

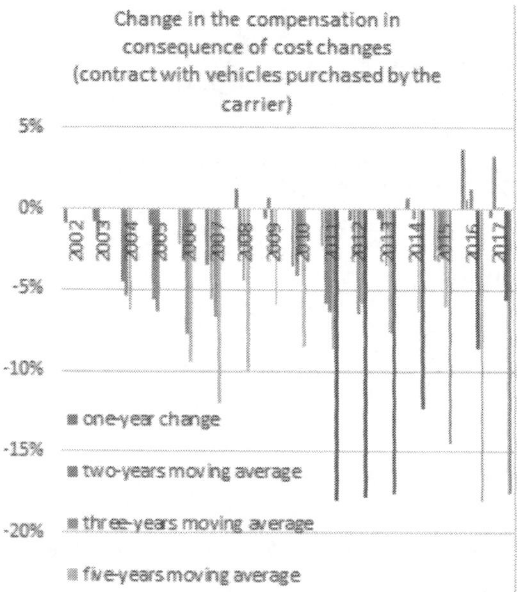

Figure 7: Change in the compensation in consequence of cost changes (contract with vehicles purchased by the carrier) [2, 18]

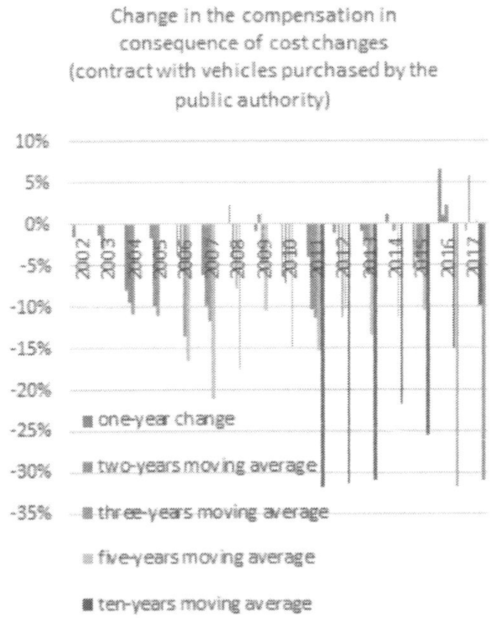

Figure 8 : Change in the compensation in consequence of cost changes (contract with vehicles purchased by the public authority [2, 18]

We can see that changes in costs can have a significant impact on financial sustainability. In case of the contract when the carrier purchases the vehicle, there is lower fluctuation than in case of the contract when the public authority purchases the vehicles. We can see bigger fluctuation in longer period. In case of the contract for ten years when carrier purchases the vehicles, we can see fluctuation from 5.6 % to 18 %. In comparison with that, in case of the contract when the public authority purchases the vehicles, we can see fluctuation from 9.9 % to 31.8 %. Carriers face difficult estimation of the future development. To sum it up, changes in costs may negatively impact financing of public transport.

5 Change in the Subsidy

We will focus on possible ways to change the subsidy and the established case practice. When the contract is directly awarded, there are no limits for changing the contract. It is possible to approve the amendment and raise the subsidy (there are no limits).

When the contract is tendered, the situation is much more complicated. There are two ways [23]: firstly, in the contract can be a clause (it can be linked to the consumer price index – inflation clause – index of wages, etc.). Secondly, the contract can be modified where the value of the modification is below both of the following values: the thresholds set in the Directive 172/2016 and 10 % of the initial contract value. If the value does not fulfil the thresholds, it is necessary to launch a new procurement procedure [23].

In the next chapters, we will find a solution when the contract may not be terminated due to changes in costs.

5.1 The Subsidy Changes Set in the Contract

Inflation clause ranks among the most frequent methods of rising subsidy. Changing subsidy about oil price changes is another way how contracts try to solve cost shocks. We can also find changes in labour costs in the transport sector or changes in wages in the whole economy. In the next table, we can see examples from the bus sector.

Table 3: Change in the subsidy in contracts with bus companies [16]

Contract	Inflation clause	Changes in oil price	Changes in the labour costs in the transport sector	Changes in the average wages
Zdar	66 %	24 %		
BORS	30 %	28 %	27 %	
VYDOS (Jihovýchod)	30 %	28 %	27 %	
VYDOS (Brněnsko)	66 %	24 %		
Chomutovsko		30 %		30 %
Litoměřicko		32 %		32 %
Podbořansko		31 %		31 %

Directly awarded contracts do not usually have any clauses (e. g. contract no. S-OD/2009/05-VI-IDS).

In the next table, we can see the situation in the railway sector.

Table 4: Change in the subsidy in contracts with railway companies [16]

Contract	Inflation clause	Changes in VAT	Changes in oil price	Changes in the labour cost in the transport sector	Transport route
GW Train Regio (R 25)	100 % minus amortisation				The full costs
GW Train Regio Pošumaví	100 %	50 %	Changes in costs for oil minus CPI		The full costs
RegioJet (SK)	100 %				The full

					costs
JMK	55 %			35 %	The full costs
GW Train Regio (R 25)	100 % minus amortisation				The full costs
GW Train Regio Pošumaví	100 %	50 %	Changes in costs for oil minus CPI		The full costs
RegioJet (SK)	100 %				The full costs

The period for which the increase occurs is also important for the financial sustainability of the contact. One year is the most frequent solution (e. g. Contract no. 20016155). The South Moravian Region has chosen two years in the new contract with Czech Railways.

5.2 The Subsidy Rising by Modification of the Contract

Modification of the contract is the second solution. All modifications must fulfil legal requirements. All additional subsidies may not exceed 10 % of the contract, and the amount may not exceed 5.7 mil. CZK [24].

In this case, carriers face the risk of time-inconsistency behaviour of purchasers. If the purchasers target is to minimise the budgetary impact, they do not want to increase the subsidy because of the cost shock. Furthermore, the railway sector is rather specific: it is much more difficult to transfer fix costs to another business than in the bus sector (e.g. it is more difficult to sell a locomotive than a bus).

6 Setting the Subsidy

We will deal with how the purchaser should set the tender documentation so that the contract is financially sustainable and the carriers do not face time inconsistency. We will test the influence of different types of clauses on the financial sustainability of the contract.

We will assume that purchasers can identify the structure of the costs. It is a reasonable assumption because purchasers must estimate the value of the contracts.

Time consistency is such a solution which presumes a change in the subsidy set in the contract. Therefore, we will seek a solution with subsidy changes set in the contract, not in the form of a modification of the contract.

Firstly, we will test the subsidy change according to the consumer price index rise delayed by one year.

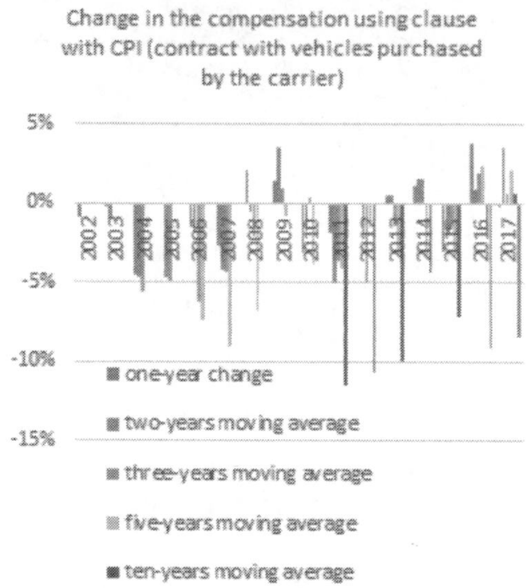

Figure 9: Change in the compensation using clause with CPI (contract with vehicles purchased by the carrier) [2, 18]

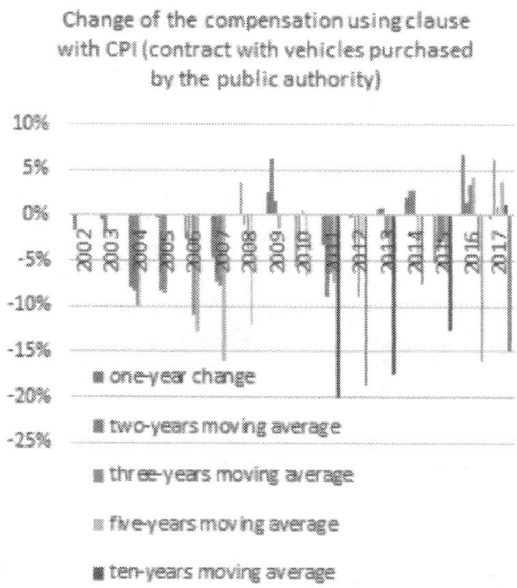

Figure 10: Change of the compensation using clause with CPI (contract with vehicles purchased by the public authority) [2, 18]

Figures provide an overview of different conclusion with utilization of the consumer price index. In case when carriers purchase the vehicles, it is possible to say that the real value of subsidy increases up to 12 %. The case when public authorities purchase the vehicles allows arriving at a different conclusion. The real value of the subsidy is reduced to the extent of up to 20 % in 10 years.

In the next chart, we have included the change in labour costs in the transport sector. The Czech Statistical Office has published this statistic. The chart below includes 55 % of the consumer price index and 35 % of the change in labour costs in the transport sector.

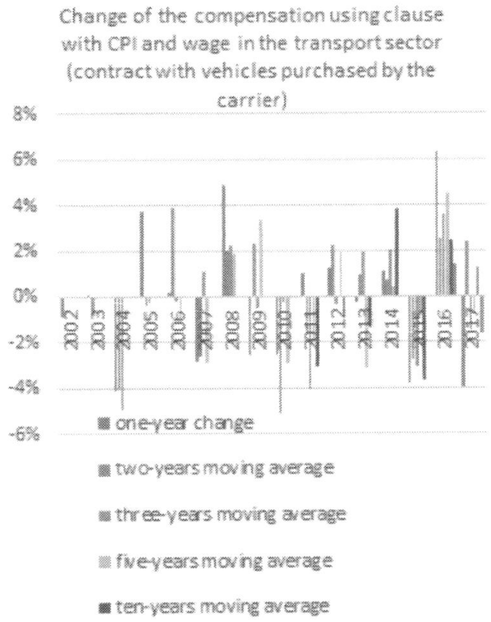

Figure 11: Change of the compensation using clause with CPI and wage in the transport sector (contract with vehicles purchased by the carrier) [2, 18]

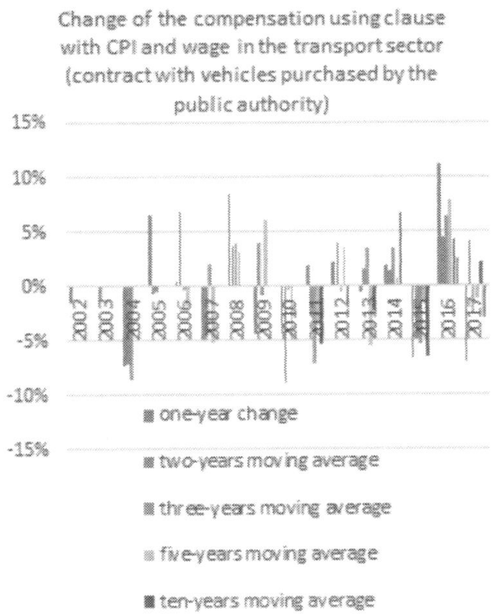

Figure 12: Change of the compensation using clause with CPI and wage in the transport sector (contract with vehicles purchased by the public authority) [2, 18]

When the carriers purchase vehicles, contracts reach better results than with using only consumer price index, the 10-year's increasing of subsidy may be up to 4 %. For contracts when public authorities purchase the vehicles also reach better results. Although, the results are not as good as in the first mentioned contracts, volatility is up to 10 % with one exemption in 2016.

7 Conclusion

Cost shocks can negatively impact financing of the subsidised passenger railway transport. We can observe fluctuations especially in the oil prices in both directions. Labour costs fluctuations are not so noticeable as oil prices, but they can also negatively impact financing. Purchasers of the transport services have to be extremely cautious in setting the tender documentation.

So that purchasers avoid time-inconsistency solution, it is necessary to set the change in the subsidy in the contract. None of the tested clauses used by purchasers is ideal. We can see over-/under-compensation. The inflation clause does not reach good results. In the contract with vehicles purchased by the carrier we can see real decreasing of the subsidy by almost 12 % in ten years. The situation in the contracts with vehicles purchased by the public authority we can see much worse results. The real value of the subsidy has diminished by 20 % in ten years.

The clause with 55 % of the consumer price index and 35 % of the change in labour costs in the transport sector reaches better results especially in contracts with vehicles purchased by the carrier (10-year's increasing of subsidy may be up to 4 %). In the contracts with vehicles purchased by the public authority we can see 10-year's increasing of subsidy up to 6 %.

This article is the output of the project called "New Mobility - High-Speed Transport Systems and Transport-Related Human Behaviour", Reg. No. CCZ.02.1.01/0.0/0.0/16_026/0008430, co-financed by the "Operational Programme Research, Development and Education" and it is output of the project called "Specific Research, no. MUNI/A/1133/2017.

References

1. M. Boyer, M. Martin, D. Filion, *Energy Enonomics,* **29**, 428-453 (2007)
2. ČSÚ [Czech Statistical Office]. *ČR od roku 1989 v číslech* [online]. 2018[cit. 2018-10-01]. Available from: https://www.czso.cz/csu/czso/ceska-republika-od-roku-1989-v-cislech-2017-8jcopi31rm#03
3. S. Dastan, *Renewable and Sustainable Energy Reviews,* **59**, 958-997(2016)
4. DRAZEN, Allan. *Political economy in macroeconomics*. Princeton, (2002).
5. S. Hammoudeh, H. Li, *Journal of Economics and Business*,**57**, 1-21 (2005)
6. M. Hunold, Ch. Wolf, *Competitive procurement design*. ZEW discussion papers (2013)

7. ISPV.CZ. Quarterly Survey of Average Earnings [online]. 2002-2017 [cit. 2018-10-01]. Available from: https://www.ispv.cz/en/ispv-results/archive.aspx
8. Z. Jarolim. E-mail communication 2018.
9. JIHOMORAVSKÝ KRAJ. Veřejná zakázka: Poskytování veřejných služeb v přepravě cestujících v regionální železniční osobní dopravě [online]. 2018 Available from: https://zakazky.krajbezkorupce.cz/contract_display_12923.html
10. F. Kydland, E. Prescott, *Journal of Political Economy,* **85,** 473-491 (1977).
11. N. Mohan, R. Brooks, *Review of Quantitative Finance and Accounting,* **33,** 393-409 (2009)
12. L. Cerna, A. Dolinayova, J. Danis. *Transport means 2016 : proceedings of the 20th international scientific conference.* 414-419 (2016).
13. S. Mohanty, M. Nandha, *Financial Review,* **46,** 465-191, 2011
14. T. Nigrin, Z. Tomeš, D. Seidenglanz, P. Mlsna, J. Dujka, *Železniční reforma v Německu.* (2017).
15. M. Petr, *European Networks Law and Regulation Quarterly,* **4,** 88-97 (2016)
16. S. Raigopal, *The Accounting Review,* **74,** 251-280, 1999.
17. Registry of contracts. Registr smluv [online]. 2018. Available from: https://smlouvy.gov.cz/
18. The Ministry of Transport. *Transport Statistics.* 2016 Available from: https://www.mdcr.cz/Statistiky/Souhrnne-ukazatele/Souhrnne-ukazatele-ekonomiky-dopravniho-sektoru/Dotace-do-pravidelne-verejne-prepravy-osob
19. E. Šíp. *Analýza nákladov prevádzky vlakov verejnej regionálnej koľajovej dopravy na trati Bratislava hlavná stanice – Komárno.* (2010).
20. J. Uguccioni, UGUCCIONI, *International Productivity Monitor* (2016)
21. UOHS. *Decision S0585/2015/VZ-37580/2015/532/KST,* (2016) Available from: http://www.uohs.cz/cs/verejne-zakazky/sbirky-rozhodnuti/detail-13933.html
22. A. Bondarev, H. Weigt, *Energy Policy,* **119,** 196-205, (2018)
23. A. Dolinayova. *Sustainable rail transport: proceedings of railNewcastle talks 2016.* 31-44 (2018).
24. D. Velde, A. Beck, *Research in Transportation Economics,* **29,** 145-151, (2010)
25. Directive 2014/24/EU of 26 February 2014 on public procurement. Available from: https://eur-lex.europa.eu/legal-content/EN/TXT/?uri=celex%3A32014L0024
26. Nařízení vlády č. 172/2016 Sb., o stanovení finančních limitů a částek pro účely zákona o zadávání veřejných zakázek. In: Zákony pro lidi.cz. Available from: https://www.zakonyprolidi.cz/cs/2016-172#f5824225

A FUZZY GROUP DECISION MAKING FOR A RAIL-ROAD TRANSSHIPMENT YARD MICRO LOCATON PROBLEM

Ivan Belošević[1*], *Sanjin* Milinković[2], *Peter* Marton[3], *Slavko* Vesković[4] and *Miloš* Ivić[5]

[1, 2, 4, 5] University of Belgrade, Faculty of Transport Traffic and Engineering, Serbia

[3] University of Žilina, Faculty of Management Science and Informatics, Slovakia

Abstract. Rail-road transhipment yards serve as interchange points between rail and road providing transhipment of loading units. These yards are of great importance for smooth functioning not only of railways but also of the whole intermodal transportation chains. The problem of choosing location of railway facilities, such as yards and depots, has great practical importance and is widely considered in the literature. In this paper, we present a multi criteria decision making (MCMD) model for selecting the most suitable micro location of a rail-road transhipment yard. The developed MCMD model is based on combining a compromise programming technique with the theory of fuzzy numbers. This approach was chosen to solve the considered location problem due to the specific decision-making environment that reflects in the presence of non-commensurable and conflicting criteria. As a case study we analysed several alternatives for locating rail-road transhipment facilities in the area of the rail freight yard in Vršac. The results confirm validity and usefulness of this approach.

1. Introduction

Rail-road combined transport refers to the movement of freight from an origin to a destination relying on railways and trucks. Railways performs long distance hauling supported by trucks that provide first/last mile transport. Complementing these two divergent transport modes, rail-road combined transport enables a sustainable modal share. From the aspect of sustainability, the main contributions of railways refer to high capacity's reserves, reliability and safety, reduction in energy consumption and pollution.
The high increase of container transportation emphasises the role of interchange points. Rail-road transhipment yards serve as interchange points between rail and road where loading units (such as containers, swap bodies or semi-trailers) change transportation mode. Wide range of decision problems arise in the area of managing transhipment yards hierarchically arranged at different planning levels. Decision problems referring micro location, layout design and operational process of a rail-road transhipment yard influence on the local infrastructure and organization of the yard itself. The final results of appropriate decisions on these problems reflect in the reduction of container transhipment costs. Decision problems in the aspect of layout design and operational process have been discussed in papers [1-2].

In contrast to the previous research, this paper considers the location problem of a rail-road transhipment yard. Selecting location is one of the complex problems and requires an extensive multi-criteria evaluation process. The location of transhipment yards has to be considered carefully due to the impacts on different stakeholders involved in the whole transportation chain. These impacts mainly refer to the financial, operational, economic, spatial and environmental aspects among investors, transhipment yard managers, freight and logistic operators, industry and the local community. Due to the such complex circumstances, we propose the multi-criteria decision making (MCDM) approach for solving transhipment yard location problem. We develop MCMD model based on combining a compromise programming technique with the theory of fuzzy numbers. This approach is chosen to solve the considered micro location problem due to the specific decision-making environment that reflects in the presence of non-commensurable and conflicting criteria. The model is based on linguistic variables and extends the evaluation problem to a Group Decision Making (GDM) procedure aggregating individual expert opinions. As a case study we analyse several alternatives for locating rail-road transhipment facilities in the area of the rail freight yard in Vršac. The results confirm validity and usefulness of this model that could serve as a tool for assisting decision makers in selecting the most suitable transhipment yard location.

The paper is structured as follows. Section 2 presents compromise programming approach using VIKOR method. Section 3 determines the proposed framework for selecting location of transhipment yard. The case study is provided in Section 4 and concluding remarks are summarized in Section 5.

* Corresponding author: i.belosevic@sf.bg.ac.rs

2. VIKOR method: a compromise programming tool

The program package VIKOR, based on compromise programming, has been developed by Opricović [3,4] for the multicriteria ranking of alternative solutions. The use of VIKOR method for the evaluation of railway infrastructure alternatives was proposed in papers [5, 6].

The method focuses on the ranking and selection of alternatives in the presence of conflicting criteria. The ideal point is used as the reference point in the area covered by criterial functions. Commonly, there does not exist an alternative that meet all criteria at the same time. The solution closest to the ideal point is proposed for the compromise solution following the Lp metric adopted from compromise programming:

$$L_i^p = \left\{\sum_{j=1}^{n}\left[\frac{w_j(f_j^* - f_{ij})}{(f_j^* - f_j^-)}\right]^p\right\}^{1/p}$$

$$1 \le p \le \infty, i = 1, \dots, m \qquad (1)$$

where m is the number of feasible alternatives $A_i (i = 1, \dots, m)$; f_{ij} is the value of the jth criterion function for alternative A_i; f_j^* and f_j^- are the best and the worst values of the jth criterion function, respectively; and $w_j (j = 1, \dots, n)$ are weighting coefficients (weights).

In the VIKOR method, S_i and R_i stand for boundary ranking measures (see Eq. 2 and Eq. 3) . The measure S_i presents the lower bound when p takes value 1 in Eq. 1 , while the measure R_i presents the upper bound when p takes value ∞.

$$S_i = \sum_{j=1}^{n} w_j(f_j^* - f_{ij})/(f_j^* - f_j^-) \qquad (2)$$

$$R_i = \max_j\left[w_j(f_j^* - f_{ij})/(f_j^* - f_j^-)\right] \qquad (3)$$

The main compromise ranking measure Q_i aggregates S_i and R_i providing a balance between these two divergent ranking measures:

$$Q_i = \frac{v(S_i - S^*)}{(S^- - S^*)} + (1 - v)(R_i - R^*)/(R^- - R^*) \quad (4)$$

where $S^- = \max_i S_i$, $S^* = \min_i S_i$, $R^- = \max_i R_i$, $R^* = \min_i R_i$ and $v \in [0, 1]$. The parameter v is introduced as the weight of the decision-making strategy.

3. The proposed model

The proposed model evaluates alternatives applying VIKOR method extended with fuzzy set theory. The proposed method uses fuzzy sets in describing uncertainties for different criteria involved in the location problem. In this way, Fuzzy VIKOR method finds compromise solution which is closest to the ideal scheme under fuzzy environment. The proposed framework for ranking alternatives and selecting the most suitable micro location of a rail-road transhipment yard is presented in Figure 1.

The method is very simple for practical application and gives an opportunity for decision makers to test the stability of the final ranking results. The applied methodology assumes that experts determine the weights of criteria and evaluate alternatives using linguistic variables. Thereafter, linguistic variables are transformed into the fuzzy numbers to reduce the subjectivity and its impact on the evaluation.

The Fuzzy VIKOR method can be summarized as the following steps:

Step 1: Generate set of alternatives for the reconstruction of the observed railway line A_i $(A_1, A_2, \dots, A_m)$.

Step 2: Create team of experts with K decision makers DM_k $(k = 1, 2, \dots, K)$ and adopt set of criteria C_j $(C_1, C_2, \dots, C_n)$.

Step 3: Create starting decision matrix using linguistic variables to evaluate ratings of criteria and alternatives concerning the adopted criteria. These linguistic variables are pre-defined and expressed in fuzzy numbers in range [0, 1] (Fig. 1).

Step 4: Defuzzificate starting decision matrix.

Step 5: Determine the best value (f_j^*) and the worst value (f_j^-) of all criteria ratings and compute the values S_i and R_i using Eq. (1) and Eq. (2).

Step 6: Compute the values Q_i for alternatives using Eq. (3).

Step 7: Rank the alternatives using three different ranking lists in descend order by the values S_i, R_i and Q_i.

Step 8: Determine the best ranked alternative by the measure Q_i and propose it as a compromise solution if the following two conditions are satisfied [3]:

Acceptable Advantage: $Q(A^I) - Q(A^{II}) \ge DQ, DQ = \min 0.25$

where: A^I and A^{II} are alternatives ranked on the first and second positions in the set of micro location alternatives and DQ is the threshold.

Acceptable Stability: The alternative A^I has to be also the best ranked by QS or QR.

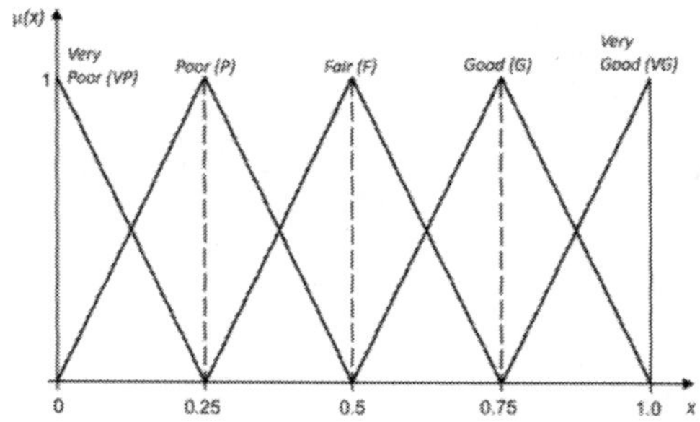

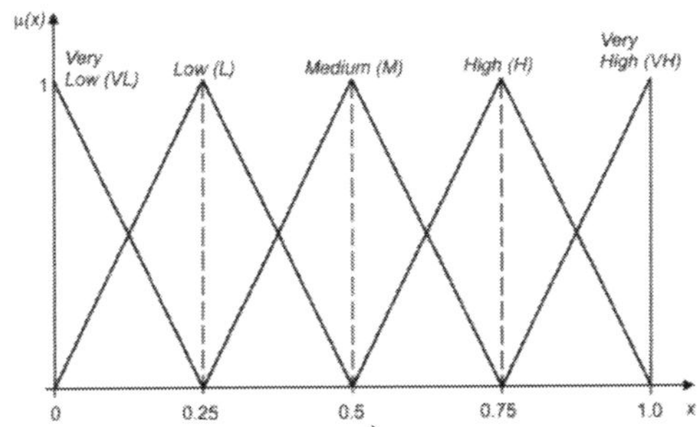

a)

Fig. 1 a) Linguistic variable of criteria weights
b) Linguistic variable of alternative ratings

4. Illustrative example

As a case study we present a several alternatives for locating rail-road transhipment facilities in the area of the rail freight yard in Vršac. The terminal will be located on an important national rail and road routes, and, also, international connection between Corridor IV and Corridor X, situated near river Danube. The intermodal terminal will be located on the outskirts of the city, near to the existing railway station Vršac. The initial framework for connecting intermodal terminal to the public railway network is based on the guidelines to retain existing location of Vršac railway station, as proposed in the Master Plan of Vršac and in the Development Program of Serbian Railways [7].

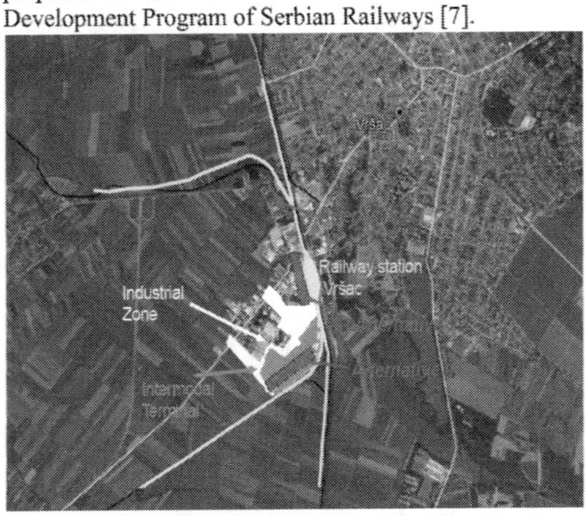

Fig. 2 Intermodal terminal in Vršac with alternatives for road-rail transhipment facilities

Three alternatives were considered as a potential location for road-rail transhipment facilities as presented of Figure 2:

- A_1 – location of rail tracks within intermodal terminal and logistic centre;
- A_2 – building new sidings parallel to existing station tracks in station Vršac;
- A_3 – location of new sidings on the area next to the existing depot.

Team of three experts (DO_1, DO_2 and DO_3) in transport economics, railway infrastructure and operation is formed in the second step. Also, the list of criteria (C_1 – Investment and land acquisition costs, C_2 – Efficiency and quality of rail operations, C_3 - Accessibility and quality of logistic services, C_4 - Availability of road connections, C_5 - Availability of land for further development, C_6 - Environmental restrictions) is adopted to evaluate proposed micro location alternatives.

In the third step, team of experts firstly weights set of criteria using linguistic variables with triangular fuzzy numbers (Fig. 1a). The membership functions of these linguistic variables are presented in Tab. 1 while the assigned weights are presented in Tab. 2.

Table 1 Linguistic scale to evaluate the weight of the criteria

Linguistic scale		Fuzzy number
Very Low	VL	(0,0,0.25)
Low	L	(0,0.25,0.5)
Medium	M	(0.25,0.5,0.75)
High	H	(0.5,0.75,1)
Very High	VH	(0.75,1,1)

Table 2 Linguistic weights of the criteria

Criteria	DM1	DM2	DM3
C1	VH	H	VH
C2	M	VH	H
C3	M	L	H
C4	H	H	M
C5	M	M	H
C6	H	M	L

Secondly, the team of experts evaluates generated alternatives concerning the adopted set of criteria using linguistic variables with triangular fuzzy numbers (Fig. 1b). The membership functions of these linguistic variables are presented in Tab. 3. Tab. 4 presents assigned ratings by decision makers for each alternative concerning criterion. The starting decision matrix is formed combing linguistic variables from Tab. 2 and Tab. 4.

Table 3 Linguistic scale to evaluate the ratings of the criteria

Linguistic scale		Fuzzy number
Very Poor	VP	(0,0,0.25)
Poor	P	(0,0.25,0.5)
Fair	F	(0.25,0.5,0.75)
Good	G	(0.5,0.75,1)
Very Good	VG	(0.75,1,1)

Table 4 Linguistic ratings of the alternatives

	Alternatives	C_1	C_2	C_3	C_4	C_5	C_6
DM_1	A_1	G	F	VG	F	G	VG
	A_2	F	G	G	P	VP	P
	A_3	VG	P	VP	VP	F	F
DM_2	A_1	F	F	G	G	G	VG
	A_2	P	VG	G	F	P	P
	A_3	P	VP	VP	P	F	P
DM_3	A_1	G	F	VG	G	G	G
	A_2	F	G	G	F	P	P
	A_3	VG	P	P	VP	F	F

In the fourth step, the starting decision matrix is defuzzificated. In the fifth and sixed step, the best and worst values of all criteria ratings (f_j^* and f_j^-) are computed. Thereafter, the values S_i, R_i and Q_i are calculated based on equations (2), (3) and (4). After obtaining values for S_i, R_i and Q_i, ranking list is computed for each of these values. Based on these rank lists it is possible to determine a compromise solution. Rank list and values S_i, R_i and Q_i for each alternative are demonstrated in Tab. 5.

Table 5 Rank list of alternatives based on values **S**, **R** and **Q**

	Alternatives		
	A_1	A_2	A_3
S	0.09	0.63	0.8
R	0.09	0.22	0.18
Q	0.0	0.88	0.85

	Alternatives		
	1	**2**	**3**
S	A_1	A_2	A_3
R	A_1	A_3	A_2
Q	A_1	A_2	A_3

Results of multi-criteria evaluation using Fuzzy VIKTOR method and the rank of alternatives show that the alternative A_1 is the compromise solution due to the first position in the rank list and satisfaction of conditions:

Acceptable Advantage:
$Q(A_3) - Q(A_1) = 0.85, DQ > 0{,}25$ *and*
Acceptable Stability: the alternative A_1 *is the first ranked in the lists based on measures S and R.*

In addition, we consider the stability of ranking results considering changes in the values of the parameter *v*. The parameter *v* plays important role in the ranking procedure and reflects a balance between two strategies: a maximum group utility and a minimum of individual regret of the opponent. Figure 3 shows priority ranking comparisons considering different values of the parameter *v* from 0 to 1. It shows that the best ranked alternative is the same in terms of both strategies and satisfies the condition of the acceptable advantage over the entire interval *v*=[0,1] while some changes appear in ranking other alternatives. In terms of a minimum individual regret of the opponent (*v* equal or lower than 0.5), the alternatives are in order A_1, A_3, A_2. When considering a maximum group utility to be more important (*v* equal or higher then 0.6), alternatives A_2 and A_3 switch ranking positions.

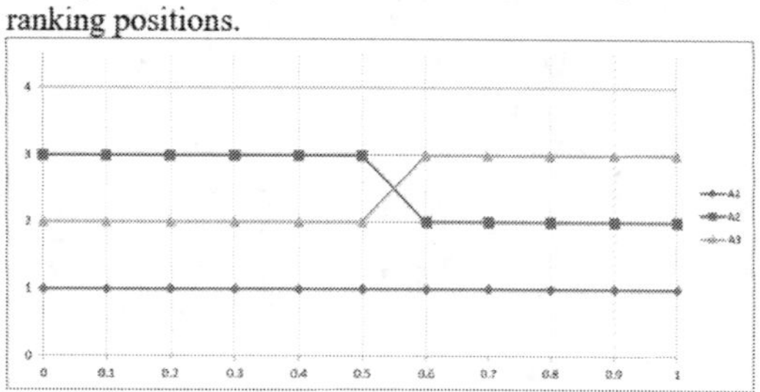

Fig. 3 Sensitivity analysis of the value *v*

5. Conclusion

The proposed MCDM model provides overall and systemic solution to the problem of selecting the most favourable micro location for the rail-road transhipment yard. In this model, proposed location alternatives are compared using different criteria (costs, rail operations, logistic services, road connections, spatial and environmental restrictions). We propose the use of fuzzy VIKOR method to evaluate location alternatives based on linguistic variables used to evaluate weights of criteria and alternatives concerning to the adopted set of criteria. These linguistic values are pre-defined and expressed in fuzzy numbers. The model extends the evaluation problem to a Group Decision Making (GDM) procedure. The final result of the multi-criteria decision-making analysis is the selection of the most suitable micro location. The most suitable location is the best solution from the set of defined alternatives in accordance with the adopted criteria. The result in the presented example shows that this model is practical and represents helpful tool for assisting decision makers in selecting the transhipment yard location that satisfy the often-conflicting criteria.

(C_1 – Investment and land acquisition costs, C_2 – Efficiency and quality of rail operations, C_3 - Accessibility and quality of logistic services, C_4 - Availability of road connections, C_5 - Availability of land for further development, C_6 - Environmental restrictions) is adopted to evaluate proposed micro location alternatives.

This work was supported by the Slovak Research and Development Agency under the contract No. SK-SRB-2016-0024. This work is supported by The Ministry of Education, Science and Technological Development of the Republic of Serbia, within the project No. 36012.

References

1. Belosević, M. Ivić, M. Kosijer, N. Pavlović, S. Aćimović *2nd Logistics International Conference, Belgrade, Serbia*, 62-67. (2015).
2. I. Belosević, M. Ivić, M. Kosijer, N. Pavlović, S. Aćimović. *Hrizons*. 463-469. (2016).
3. S. Opricović, G.H. Tzeng: European Journal of Operational Research, **2(178)**: 514–529, (2007).
4. S. Opricović: Fuzzy Expert Systems with Applications, **10(38)**: 12983-12990, (2011).
5. I. Belosević, M. Kosijer, M. Ivić, N. *European Transport Research Review* 10:43. (2018) https://doi.org/10.1186/s12544-018-0318-4
6. M. Kosijer, M. Ivić, M. Marković, I. Belosević. Građevinar, **3(64)**: 195-205, (2012).
7. J. Masek, M. Kendra, J. Camaj, S. Milinković, S., Vesković, S.,. Transport Problems. **10 (Issue 4)**, 107-119 (2015)

Position of railway passenger transport companies on current liberalized transport market

Zdenka Záhumenská[1,*], *Martin* Vojtek[1] and *Jozef* Gašparík[1]
[1]University of Žilina, Department of Railway Transport, Univerzitná 1, 01026 Žilina, Slovakia

Abstract. Market liberalization is a great contemporary trend in many fields of economy. Liberalized transport market, where the transport demand is meeting with transport offer, is evolving dynamically. Private railway passenger operators want to increase their market share together with national railway passenger transport companies therefore the quality of passenger transportation is getting higher, which positively influences the attractiveness of railway passenger transport. The article is focused on these current trends of liberalization in railway passenger transport market. From operational and economical point of view, there are described some ways, how to make the railway passenger transport system more effective and make the entire transport system more attractive for traveling public.

1 Introduction

Passenger transport is generally considered as an activity, which arises as the consequence of spatial division of places, where people are in exact time and their need to move. Requirements for transport of passengers originate in their need to move, while the passenger transport is dependent on the willingness of travelling. [1]

In passenger transport, there are mostly individual passengers, so it is difficult to determine all transport requirements. Passenger transport is divided into individual and public. Individual passenger transport includes walking, cycling and car transport. Public passenger transport includes railway, road, water, air, city and unconventional transport. From spatial point of view, passenger transport is divided into local, regional and long-haul, which is then divided into interregional transport or international transport. [2]

From economical point of view, passenger transport is classified into tertiary sphere – services. It means that there are not any material production values, but it is reflected in costs. In general, passenger transport has got a great social and political importance. Primary function of the transport system is providing transport for passengers on regional, national-interregional and international level. [3]

2 Railway passenger transport in general

Key element in railway passenger transport is a customer – passenger, who requires the transport from one place to another. A basic precondition for accomplishing the main requirement – transport, is making the complete offer which provides not only transport, but also other associated services. From economical point of view, very important criterion is the price for transport and it mostly depends on economic indicators. In market economy, there are three factors: costs, demand and competition. [3]

Current trend is to optimize connectivity in railway passenger transport by reduction of transfer time – starting at home and finishing in the transport destination. It is important to synchronize arrivals and departures of all connected trains in all points, where passengers can get on, get off and change the vehicle. Minimizing of waiting time will increase quality of passenger transport in general. [4]

Entire quality is defined as an ability to satisfy all requirements of customers. Specific signs for services in transport are insubstantiality, impossibility to store, inseparability, variability, complexity and uniqueness. Level of service quality can be perceived as a disharmony among expectation and perception. Customers – passengers have got different priorities which are connected with quality of service. They usually remember low quality and high quality is a standard for them. The main challenge is to identify the passengers' needs and satisfy them in all cases, because every transport is realized in different conditions. [2]

3 Liberalization process of railway passenger transport market in the EU

A basic step in the liberalisation of the railway passenger transport market has been the separation of railway infrastructure managers from railway transport companies (operators respectively carriers) and consequently providing non-discriminatory access of the railway companies to railway infrastructure in all Member States. The partial liberalisation of the railway passenger transport market in the European Union (EU)

* Corresponding author: zdenka.zahumenska@fpedas.uniza.sk

was already underway in 2010, when all European railway companies gained access to all railway infrastructures in the EU. Opening up the market to new private railway operators means that several transport operators can compete directly for selected lines at the same time. Given that passenger transport is a State instrument of transport, regional and social policy to provide transport services, it is still largely controlled by the State. [5]

The opening up process of domestic railway passenger transport markets according to the fourth railway package has variety of level in Member States. This process is requested to done not later than 2019, while making public tenders for transport service contracts compulsory, in the public interest. The idea is to meet the objectives of the EU White Paper on transport and The Regulation (EC) No 1370/2007 of the European Parliament and of the Council of 23 October 2007 on public passenger transport services by rail and by road and repealing Council Regulations (EEC) No. 1191/69 and 1107/70. [6]

The current EU transport policy is based on the 'White Paper on the Single European Transport Area – Creating a competitive, resource-efficient transport system that includes forty initiatives to create an efficient transport system. The area of the allocation of public services includes an initiative to open up the domestic railway passenger market to competition, including the compulsory award of public service contracts through competitive tendering. The railway transport market has undergone several transformations. Those achieved economic efficiency of rail transport, reducing the financial burden on the state budget, streamlining the management of the railway company and create the conditions for a market orientation. The aim was to increase the flexibility of railway transport to the needs of the customer. The requirements of liberalizing the railway passenger market and harmonizing the conditions for access to this market resulted in the adoption of the so called Fourth Rail Package. [7]

This Regulation lays down the conditions under which competent authorities when imposing or contracting obligations arising from the public service provided by the suppliers pay for costs incurred and/or grant exclusive rights in return for fulfilling the obligations arising from the public service. The Regulation applies to national and international public passenger transport services in rail passenger transport, other passenger rail and road passenger transport. Exceptions are services which are provided mainly for their historical or tourist value. [8-9]

The amended version of the Decree from December 2016 introduced the definition of public service. It is done a public railway passenger transport where they concern transport by rail, with the exception of other track-based modes such as metro or tram. Public service performances are provided by a public or private undertaking for passengers or carriers on the basis of a public service contract concluded between the railway undertaking and the competent public authority or authority with such competence. [8-10]

Contracts are awarded either directly to a selected railway undertaking over which a competent authority control or another possible ways. The competent authority decides itself to provide services in the context of public services in railway passenger transport, and these performances are entrusted to a third party by tender. [11]
The procedure adopted for competitive tendering shall be open to all operators, shall be fair and shall observe the principles of transparency and non-discrimination. Following the submission of tenders and any selection, the procedure may involve negotiations in accordance with these principles in order to determine how best to meet specific or complex requirements. If subcontracting takes place, the operator entrusted with the administration and performance of public passenger transport services in accordance with this Regulation shall be required to perform a major part of the public passenger transport services itself. [12]

A public service contract covering at the same time design, construction and operation of public passenger transport services may allow full subcontracting for the operation of those services. The public service contract shall, in accordance with national and Community law, determine the conditions applicable to subcontracting. Before a public tender is started a decision may be taken by the competent authorities limiting the number of contracts with the same railway undertaking. [8-10]

4 Comparison of railway passenger transport markets

The comparison of the basic requirements railway passenger transport contracts according to the way they are entered is dealt with in Table 1.

Table 1. Comparison of the essential requirements according to their assignment.

Reqiurements		How to enter the contract	
		Direct award	Public tender
Maximum average annual value of performance		to 7.5 mil. €	Not specified
Maximum annual amount of performance		to 500,000 kilometres	Not specified
Maximum time of the contract		10 years	15 years
Opportunity and reason for the extension of the Contract	Depreciation of assets	yes, max. to 50 %	yes, max. to 50 %
	Geographical location	no	yes, max. to 50 %
	Amortization of capital	no	yes, necessary justification
Obligation to publish a pre-announcement in the EU Bulletin for an output range of 50,000 kilometres or more		yes, 1 year before direct awarding	yes, 1 year before starts public tendering

Secondary law of EU affects the allocation of public service performance, in particular in terms of the level of quality of service provided, are the Technical Standard EN 13816 'Transport, Logistics and services, Public passenger transport, Definitions, objectives and measurement of service quality' as well EN 15140 'Public passenger transport, Essential requirements and recommendations for systems to measure service quality'.

Regulation No. 1370/2007 EC governs the general conditions applicable to the Public Service Contract – the procurement, content and reimbursement of services provided under the Treaties. The precise public procurement rules for the award of public service contracts are not specified in the Regulation and remain within the competence of the Member States. [11]

Examples of countries conducting tenders for the allocation of performance in rail passenger transport are shown in Table 2. [15]

Table 2. European countries, that realized public tenders for allocation of performance in railway passenger transport.

Country	Public Tender	Direct contracting	Administration monopoly
Belgium		x	
Bulgaria		x	
Czechia		x	
Denmark	x	x	
Germany	x	x	
Estonia		x	
Ireland		x	
Greece		x	
Spain		x	x
France		x	x
Italy	x	x	
Latvia		x	
Lithuania		x	
Luxembourg		x	
Hungary		x	
Netherlands	x		x
Austria		x	
Poland	x	x	x
Portugal	x	x	
Romania		x	
Slovenia		x	
Slovakia		x	
Finland		x	x
Sweden	x		
Great Britain	x		

Outsourcing tenders in the Member States may be at varying levels of state administration as well as can vary the conditions of operation of public tenders. Requirements for safety and interoperability must be common to all Member States with regard to the common EU legislation and technical specifications for interoperability (TSI).

5 Economic aspects of railway passenger transport

Railway passenger transport has to be evaluated from economical point of view. Basic evaluation method is operating costs calculation. Costs are financial representation of company sources consumption for realizing services per time. Internal costs of the transport company arise from operation of trains on railways. Thanks to calculation, the exact amount of these costs is known. In railway passenger transport, the calculation unit is the service – transporting of passengers. It can be defined by quantity (number of trains, vehicles), time (staff working time, time of traveling) or other way (passenger-kilometers, train-kilometers). [16]

In general, there are these costs: vehicle costs (price for vehicle, repairs and maintenance, insurance, operational cleaning), railway infrastructure access, staff costs (wages of vehicle-drivers and stewards), traction energy consumption and other indirect costs (management, marketing, travel ticket selling system, information system etc.). Sum of all costs, which are converted to one typified train on the route, is the base for making the tariff charges. [14]

Railway vehicle costs are calculated this way:

$$r^{RV}_{trkm} = \frac{D_Y + \Sigma RM_Y + OC_Y + INS_Y}{\text{Ø annual vehicle kilometrage}} \quad (1)$$

where: r^{RV}_{trkm} – railway vehicle costs rate for train-kilometre [€/trkm],
D_Y – depreciation of vehicle per year [€],
ΣRM_Y – entire costs for repairs and maintenance of vehicle per year [€],
OC_Y – entire costs for operational cleaning of vehicle per year [€],
INS_Y – entire costs for vehicle insurance per year [€],
Ø annual vehicle kilometrage – average kilometrage of railway vehicle per year [km].

$$C_{RV} = \Sigma trkm \, . r^{RV}_{trkm} \, . NRV_{tr} \quad (2)$$

where: C_{RV} – entire railway vehicle costs per route [€],
Σtrkm – sum of train-kilometres per route,
r^{RV}_{trkm} – railway vehicle costs rate for train-kilometre [€/trkm],
NRV_{tr} – number of railway vehicles in the train on the route [vehicles].

Staff costs are calculated this way [14]:

$$r^{S}_{emph} = \frac{\text{price for working} + \text{equipment}}{\Sigma \text{ work time}} \quad (3)$$

where: r^{S}_{emph} – staff costs rate for employee-hour [€/emph],
price for working – all month company's costs for the employee [€],
equipment – month costs for equipment of employee [€],
Σ work time – entire month work time of employee [hours].

$$C_S = t_r \cdot CR_S \cdot r^S_{emph} \quad (4)$$

where: C_S – staff costs per route [€],
t_r – train ride time [hours],
CR_S – conversion ratio: train ride time →employee-hour,
r^S_{emph} – staff costs rate for employee-hour [€/emph].

Traction energy consumption costs are calculated this way [14]:

$$C_{TEC} = \frac{\Sigma gtkm \cdot mc_{TE} \cdot r_{TE}}{1000} \quad (5)$$

where: C_{TEC} – entire traction energy consumption costs per route [€],
Σgtkm – gross-tons-kilometres per route,
mc_{TE} – measurable consumption of traction energy per thousand gross-tons-kilometres,
r_{TE} – traction energy rate [€].

From operating costs calculation, tariff rates can be appointed. The tariff reflects valuable relations among the operator and passengers. These rates have to include internal goals of the operator (increasing profit, decreasing costs, market share etc.), social sphere (quality and offer of public transport, reducing regional gaps etc.) and environmental aspects. Current transport demand and complementary transport offer are also important part of setting tariff rates.

Application of economic aspects of operating passenger trains from transport company point of view is described in case study of fictional Express train from Bratislava to Bardejov in Slovakia. Firstly, according to previous formulas, there is a summary of operating costs calculation shown in table 3.

Table 3. Operating costs per one fictional train.

COSTS (€)		
vehicle	1051.76	26.61%
infrastructure	649.18	16.42%
vehicle-driver	190.88	4.83%
stewards	184.38	4.66%
traction energy	888.74	22.48%
indirect	988.31	25.00%
sum	**3953.24**	100.00%
converted to one seat	12.35	
converted to one placekm	0.023265	

Comparison of costs and revenues are shown in table 4. When the result > 0 → operation of passenger trains is economically effective. On the other hand, when the result ≤ 0 → operation of passenger trains is not economically effective therefore the transport company would find some way how to reduce costs or increase revenues. Some government subsidies are also one of possible ways how to make operation of passenger trains economically effective.

Table 4. Economic effectiveness of fictional train on the route.

Route: BRATISLAVA - BARDEJOV		
Sum of costs per one typified train	3953.24	€ without VAT
Costs converted to one seat	12.35	€ without VAT
Sum of revenues per one typified train	5210.00	€ with VAT
minus VAT	4168.00	€ without VAT
Result (revenues minus costs)	**214.76**	**€**

The case study indicates that it is economically effective to operate trains on this route according to proposed conditions, which are not specified above. It shows the way how to consider operation of trains on new routes. Transport companies must consider enormous number of factors and conditions, which are necessary for objective and accurate assessment of providing railway passenger transport services.

6 Conclusion

The process of railway passenger market liberalization influences contemporary progression in the whole passenger transport system. There are several ways how to evaluate this process, because it has got different consequences to traveling public and transport companies, which also vary in each country according to its legislation and economic conditions. Market share growth of private transport companies is a basic result of liberalization and it is proportionally connected with improvement of transport qualitative criteria and increasing attractiveness of railway passenger transport. Private transport companies are focused on costs optimization and making higher profit therefore railway passenger transport market is significantly affected by economical aspects. These aspects are mainly internal costs, transport demand and level of competitiveness. In central Europe, there are some private railway passenger transport companies with different strategies. The most famous are WEST Bahn (Austria), Regio Jet (Czech Republic, Slovakia, Austria), LEO Express (Czech Republic, Slovakia), ARRIVA vlaky (Czech Republic, Slovakia). These companies provide also bus services, which is partially complement to their trains.

The paper is supported by the VEGA Agency under Project 1/0095/16, 'Assessment of the quality of connections on the transport network as a tool to enhance the competitiveness of public passenger transport system'

References

1. Z. Záhumenská, V. Gáborová, J. Gašparík, Perner´s contacts,**12, 1.** 174-185 (2017)
2. J. Gašparík, O. Stopka, L. Pečený, Our sea, **62.3** SI. 114-118 (2015)
3. L. Černá, A. Dolinayová, J. Daniš, *Transport Means,* 414-419 (2016)
4. J. Gašparík, V. Ľupták, P. Meško, *ICTTE*, 388-392 (2016)
5. L. Pečený, J. Gašparík, V. Gáborová, *ICTTE*, 462-466 (2016)
6. J. Gašparík, P. Meško, Z. Záhumenská, Periodica Polytechnica Transportation Engineering **Onlinefirst**. 6 (2018)
7. R. Kampf, O. Stopka, I. Kubasakova, V. *Zitricky. World Multidisciplinary Civil Engineering-Architecture-Urban Planning Symposium.* (2016) DOI: 10.1016/j.proeng.2016.08.623.
8. A. Dolinayová, M. Ľoch, J. Čamaj, *Transportation research procedia*, 916-925 (2016)
9. REGULATION (EC) No 1370/2007 OF THE EUROPEAN PARLIAMENT AND OF THE COUNCIL No. 1370/2007 of 23 October 2007 on public passenger transport services by rail and by road and repealing Council Regulations (EEC) Nos 1191/69 and 1107/70, 13 (2007)
10. P. Blaho, P. Sulko, L. Cerna. LOGI proceeding, MATEC, **134.** (2017
11. M. Poliak. Ekonomicky casopis. **61, 2**. 206-220. (2013)
12. REGULATION (EU) 2016/2338 OF THE EUROPEAN PARLIAMENT AND OF THE COUNCIL of 14 December 2016 amending Regulation (EC) No 1370/2007 concerning the opening of the market for domestic passenger transport services by rail, (2016)
13. COMMISSION STAFF WORKING DOCUMENT Accompanying the White Paper - Roadmap to a Single European Transport Area – Towards a competitive and resource efficient transport system, 32 (2011)
14. J. Čamaj, J. Mašek, M, Kendra, *Transport Means*, 147-151 (2015)
15. L. Černá, V. Zitrický, J. Ponický, Mathematical methods in economics, 126-131 (2016)
16. M. Kvizda, Rewiew of Economick Perspectives. **16. 4.** 389-394. (2016)
17. A. Dolinayová, E. Nedeliaková, *Controlling in railway transport*, 148 (2015)
18.
19. A. Torok. Procedia Engineering **187**, 722-725, (2017), https://doi.org/10.1016/j.proeng.2017.04.446
20. L. Černá, V. Zitrický, J. Ponický.. *Mathematical methods in economics = MME*. 126-131.(2016).

Proposal international transport chain during the transport of tires

Lenka Černá[1*], *Vladislav* Zitrický[1], *Peter Blaho*[2] and *Peter* Šulko[3]

[1] Department of Railway Transport, Faculty of Operation and Economics of Transport and Communications, University of Žilina, Žilina 010 26, Slovak Republic
[2] Železnice Slovenskej republiky, 1. Mája 34, 010 01 Žilina, Slovak Republic
[3] Železnice Slovenskej republiky, Klemensova 8, 813 61 Bratislava, Slovak Republic

Abstract. In view of a continuously growing freight turnover, the road transport has been increasingly developing not only in the Slovak Republic, but also in the European Union, as a whole. The European Union, by applying legal instruments, seeks to change the situation in favour of safer, more efficient and environmentally friendly modes of transport, such as railways and inland water transport. The railway scientists and experts from the Slovak Republic compare the transport and tariff conditions for the carriage of automobile tires by railway and road transport on the route Puchov - Hannover in a real environment. Based on the comparative analysis, they assess the time and economic indicators of complexity and efficiency in the transportation of a particular type of cargo on a particular route.

1 Introduction

The international transport of goods and the transportation contracts under which they are carried out are the most commonly used type of contract, in addition to the international agreements for the purchase and sale, and create a kind of bridge between the different transport organisations. International freight transport can be considered secondary neither from the legal nor from the commercial points of view.

One of the most important criteria is the cost of transportation, which can significantly affect the final price of goods on the market and, therefore, the consumer choice. In addition, this the criterion that must be taken into account by business partners.

The transportation of goods and cargoes in international traffic depends on a number of important criteria in the field of transport security (including legal, economic, and contractual and transport security). [1,2] All types of processing and handling can be carried out by a specific freight forwarding company, whose task is to provide the most suitable vehicles and to ensure the integrated transport Services. This is due to a high degree of inaccessibility of some railway infrastructure objects and their depreciation. In the current economic conditions within the EU, it is very difficult to forecast the development of freight railway transport in Slovakia. Due to certain instability in the Steel and automotive industries, significant decline in the productivity of railway transport may be observed as well. On the contrary, given the growing trend of globalization, there will most likely be further positive developments in intermodal transport. Apparently, intermodal transport is an important lever for the development of railways. To ensure further sustainable development, it is essential to create all necessary conditions, including a reliable railway infrastructure, flexible Services in accordance with customer requirements and an optimized terminal network. [3]

Road transport provides such Services, because of which "door-to-door" transportation is ensured almost anywhere in the world. Road transport is flexible, with minimal administrative delays associated with loading, and the driver throughout the transportation controls the cargo. Fast delivery of goods for short distances has no competition. Another advantage of road transport is the low cost of the Services it provides, which is beneficial even for carrying lightweight and special cargoes, accompanied by less damage during the transportation. [4]For longer transport distances, the delivery time is limited to the AETR (European Agreement on the Work of Crews in International Road Transport).

2 Analysis of freight transport development in the EU

The main transport problems in the European Union include the unbalanced development of various modes of transport, congestions of road infrastructure, as well as of the airspace, and the negative impact of transport on the environment. [5, 6]

The efforts undertaken to limit the efficiency of road transport and its reorientation onto the railway transport have been developed by the EU in a number of regulations and national documents. First, the White Paper and the Lisbon Strategy are meant. [7]

* Corresponding author: lenka.cerna@fpedas.uniza.sk

It should be noted that not only in the Slovak Republic, but also in other EU countries, road freight transport has the largest representation. This is due to the fact that road transport is cheaper for the client, more attractive, more adaptable to the customers' specific requirements, and, because of the high competition, carriers are ready to significantly reduce the cost of transportation (provide discounts), which ultimately becomes unrealistically realistic. One of the most important criteria is the speed of delivery. From the point of view of the share of road freight transport in the EU the first place is occupied by Germany (36% of the total volume of goods transported), then comes France (22%), and the UK (18%) in third place (according to the Statistical Office of the Slovak Republic). [8]

By railways, the speed of transportation is higher, especially when it goes to the block trains, and the cost of carriage is higher, too, since it includes other costs associated with the transportation itself, including the costs for weighing, handling, sealing or loading and unloading. For this reason, railway transport for the customer may be less attractive. Basing on the analysis of railway transport in the European Union, it can be concluded that the largest share of freight railway traffic is again in Germany (37%), second in Austria (20%), and third in the UK (11%) (According to the data of the Statistical Office of the Slovak Republic).

3 The initial parameters for the transportation of tires at the railway section Puchov - Hannover

On the Puchov-Hannover railway route, automobile tires with the dimensions of 385/55R 22.5 for trucks are delivered. The reason why the loading of a batch of tires is performed at the Puchov station is that a railway siding to the main network connects the manufacturing enterprise producing the tires. [9]

Railways in a special transport tires. Transport swap body belonging to the Deutsche Bahn AG (the so-called Container or transport unit) - "DB-Megabox", which was specially designed for the transportation of automobile parts. These swap bodies are also used in road transport using a special combination with a hydraulic lifter.

In railway traffic, wagons of "Sgkkms" type are used for the carriage of "DB-Megabox" transport units. Since "DB-Megabox" containers are special swap bodies, they are classified as special cargo consignments when transported by railway. On the other hand, trailers are used for the carriage by road. Railways deliver the cargo through the border crossing points of Lúky pod Makytou - Horní-Lideč, Děčín - Bad-Schandau, and by road - through the border crossing points of Drietoma and Ústi nad Labem. [10]

Basic parameters for railway transportation:

- Type of transportation: transportation under special conditions.
- Goods type: tires 385/55R 22.5.
- Number of tires in one Container (swap body): 120 pcs.
- Transport unit (TE): "DB-Megabox" swap bodies;
- Wagon type: "Sgkkms".
- Number of transport units: Two swap bodies on one flat wagon;
- Auxiliary transport equipment: Two fixing rods.
- Cargo weight: 9600 kg.
- Loading station: Puchov.
- Unloading station: Hannover Linden.
- Border Crossing points: Slovakia - Lúky pod Makytou, Czech Republic - Horní Lideč, Děčín, Germany - Bad Schandau.
- Distance: 891 km.
- Delivery time: 4 days.

3.1. Railway delivery route plan

Figure 1 shows the route plan of goods railway delivery on the railways of Slovakia, Czech Republic and Germany.

Characteristics of DB-Megabox transport unit (TU)

The "DB-Megabox" swap bodies are used for the transportation of tires on the Puchov-Hannover railway route, referred to as transport units (TU). The special transport unit of "DB-Mega- box" type was developed jointly by DB Cargo, DB Technical Centrum Minden and Continental AG. They represent a combination of large containers with open top. They are all-metal boxes on four legs, which are suitable for the transportation by using special road vehicles equipped with a lifting platform, as well as for the transportation on railway flat wagons. The main characteristics are shown in Table 1.

Fig. 1. Railway delivery route plan

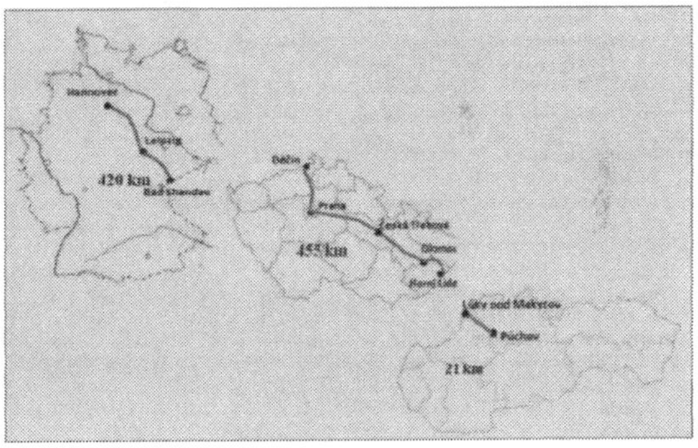

Table 1. Characteristics "DB-Megabox "transport.

External dimensions (m)	7,82 x 2,60 x 3,29
Internal dimensions (m)	7,67 x 2,52 x 3,03
Loading capacity (m^3)	58,5
Maximum load (kg)	11255

Container tare weight (kg)	4745
Total weight (kg)	16000

These open-top containers - swap bodies are transported by railway mainly on special railway fiat wagons, since their width exceeds the maximum permitted by Regulation 349/2009 Coll. in relation to the maximum dimensions and mass of vehicles. Figure 2 shows a "DB-Megabox" transport unit and its loading onto the "Sgkkms" fiat wagon by using a special loader.

In railway transport these open-top transport units do not correspond to the maximum permissible overall dimensions in height, therefore specially designed and modified "Sgkkms" fiat wagons are used for their transportation by rail. [11] The "Sgkkms" fiat wagon used to provide the carriage of tires in "DB-Megabox" container type swap bodies is a modified fiat wagon with the loading surface level lowered 33 cm below the standard (845 mm above the rail) used in the "classical" freight wagons, which corresponds to the permitted dimensions for the transport of "DB-Megabox" TUs. The basic data of the fiat wagon is shown in Table 2. Two containers can be transported on one fiat wagon at the same time. For loading and unloading the Container onto/from the fiat wagon a special loading and unloading device with clamp-like cramps (the so-called ‚spreadrom') is used.

Table 2. Characteristics of "Sgklans " flat wagon

Wagon length with buffers (mm)	19130
Length of the loading surface (mm)	15890
Floor height above the rail (mm)	854
Maximum load weight (kg)	46000
Maximum permissible speed (km/h)	120

Two "DB-Megabox" TUs can be loaded onto one "Sgkkms" fiat wagon. Figure 3 shows special wagons with lowered floor along with the loaded "DB-Megabox" transport unit. [12]

3.2. Loading of tires for railway transportation

The cargo inside the open-top container-type swap body should be evenly distributed, fixed sufficiently to avoid the displacement, sheltered and protected from damage by the appropriate material. Before being loaded, the "DB-Mega- box" transport units are cleaned and checked. [13] The consignor with the use of his own loading equipment provides the loading. It takes an average of 90 minutes to load one Container (distribution of tires by type, preparation of the ramp, loading of the cargo itself). Tires through the open top are lowered onto special tube-type rods (Figure 4). 8 truck tires with the dimensions of982 x 212 mm and with the weight of 40 kg of one tire are placed on one rod. Tires are located next to each other and, thus, are fixed from over- timing. Two fixing rods are fitted on a TU's body to prevent the load from overtiming when the front door is opened. After loading and securing the cargo, the top is closed, locked and sealed with three seals. Then the data on the seals are entered into the wagon sheet, which is attached to the consignment load.

Fig. 2. Transport unit (TU) "DB-Megabox" and its loading onto the fiat wagon "Sgkkms" by means of a special loader

Fig. 3. A flat wagon of "Sgkkms" type loaded with two "DB-Megabox" container-type transport units

The railway carrier provides the delivery of goods to the industrial siding tracks. Unloading is provided by the consigner, which in this time is covered by a special procedure. In accordance with the UIC Loading Regulations, a cargo is considered to be a special cargo when its external dimensions, weight or nature cause special difficulties when it is processed by railway equipment and when transported by railway rolling stock of one of the participating railways and, therefore, can only be accepted for the transportation under special technical or operating conditions. Specific goods can be accepted for the transportation only under special conditions, which must be agreed in advance between the carrier and the railway undertaking (UIC Loading Regulations). [14]

3.3. Technology of cargo handling on railway siding tracks (loading tires in a transport unit)

Comparing the carriage conditions of tires by railway and road transport, the technology of their unloading and processing can be also analysed, i.e. the process of handing over a 'relay baton' from the road carrier to the railway carrier at the place where the tires were loaded into the railway containers before they were sent from the siding tracks. [15]

The analysis shows that 20 standard operations are required for loading and carriage by rail, and only seven - for road transport.

3.4. Analysis of tariff conditions for the carriage of tires by rail on the route Puchov - Hannover

In the context of studying of the cost of railway

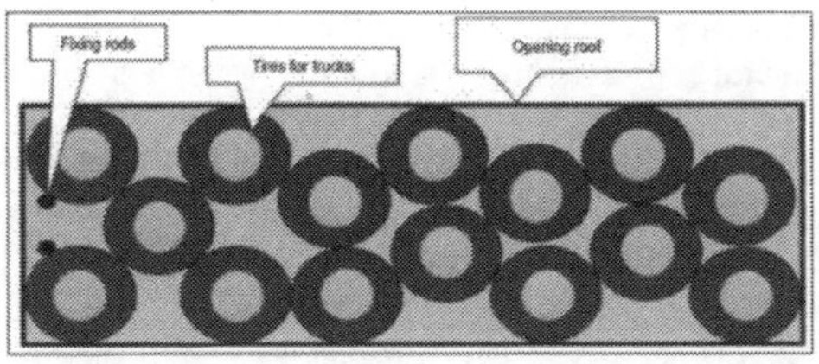

transport, ten carriers were considered who provide transport Services in Slovakia and provide the delivery by rail. The information provided by three of them is given in Table 3. The tariffs for carriage of goods in the Puchov-Hannover traffic are presented in two versions: when transporting the same number of tires in one large Container and in two swap bodies. [16, 17]

Fig. 4. Stowing of tires in a Container

Table 3. Examples of tariff conditions for the carriage of tires by railway on the route Puchov — Hannover

Information, received from selected railway carriers		
	Price for a carrier's wagon in euros	Price for a consignor's wagon in euros
Carrier 1	2138	1932
Carrier 2	2300	2070
Carrier 3	3902	3175,30

3.5. Basic parameters for transportation by road

Transportation of tires by road differs from that by rail, mainly by the choice of the vehicle used to carry out this transportation. While in railway transport container-type swap bodies are used as transport units, in road freight traffic a special semitrailer with reinforced sideboards is used for the carriage of tires. This is because "DB-Megabox" transport units do not meet the maximum width requirements for road transport. [18]

- Type of transport: international freight trans-port.
- Goods: tires 385/55 R 22,5.
- Number of tires in one TU: 240 pes.
- Transport unit: Fliegel semitrailer with rein- forced sideboards.
- Axillary equipment: slings.
- Freight weight: 9600 kg.
- Loading station: Puchov.
- Unloading station: Hannover Linden.
- Border Crossing points: Slovakia - Drietoma, Czech Republic - Ústi nad Labem.
- Distance: 880 km.
- Time of delivery: 2 days.

3.5.1. Characteristics of transportation by road

Figure 5 shows the route of cargo delivery by road passing through Slovakia, Czech Republic and Germany, together with the marked border Crossing points. Border crossings through which the road transport route passes are: in the territory of Slovakia — Drietoma, and in the Czech Republic - Ústi nad Labem. [10]

Fig. 5. Road delivery route plan

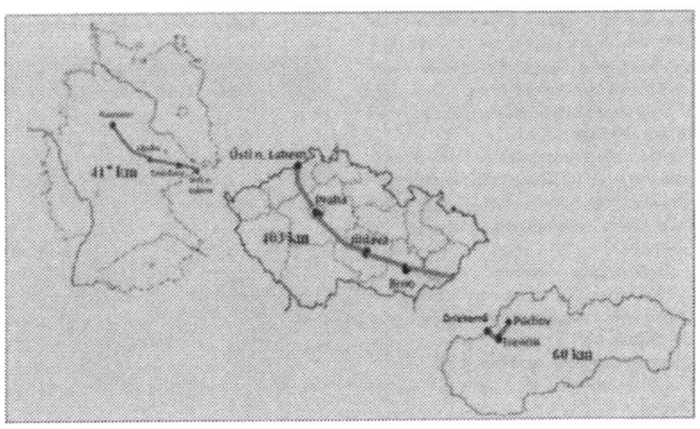

3.5.2. Characteristics of an automobile semi-trailer

When transporting tires by road, a special semi-trailer was chosen for the transportation of tires with reinforced side boards, as shown in Figure 6.

Fig. 6. Semitrailer used f or road transport

In the time of road transport, the carrier submits for transportation a vehicle authorized for transportation. The driver shall take part in the loading and unloading process of the vehicle, but does not performs either loading or unloading himself. The driver must check the external condition of the cargo, its quantity and packaging, as well as identify the damages, if there are any, to indicate them in the consignment note, and notify the consignor in time of improper loading and securing the cargo. The semitrailer used for the carriage of tires shall be equipped with horizontal crossbars on both sides of the semitrailer with a maximum gap of 35 cm between them, which must be fixed transversally by at least nine lines. Figure 7 shows a semi-trailer with loaded tires fixed with slings. [7, 9]

Figure 7 shows the loaded tires in the "Fliegl"semi-trailer, loaded manually. In the semitrailer the same number of tires of the same type as in the two "DB-Megabox" swap bodies is placed.

Technology of transportation of goods by road (loading of tires into a semi-trailer)

Transport technologies for railway and road freight transport are very different. On the railways, the technology of transport processing is much more complicated than in road transport. Trains run in accordance with the exact traffic timetable. [10]

The so-called, hierarchyl of traffic exists on the railways. In most cases, priority is given to passenger trains, and then preference is given to regular (route or block) freight trains, and, in the end, trains with irregular traffic follow. In road transport, the carrier is guided by the terms of transportation and is limited only by the ban on traffic during public holidays and on weekends or by the provisions of the European agreement concerning the work of crews of vehicles engaged in international road transport. [3,5,9]

Fig. 7. Example of stowed tires and their fastening with slings

4 Analysis of tariff conditions for the transportation of tires by road on the route Puchov – Hannover

The cost of transportation of goods by road from Puchov to Hannover was analysed using the Internet resources of individual road carriers. The data of five road carriers were analysed. Table 4 shows the information of three out of five carriers.

Carriers provided information about the cost of transportation per 1 km. Travel costs for the driver amounted to 56.11 euros for the mentioned countries, whereas travel expenses amount to 219.63 euros. Of these values, the total cost of transportation was calculated, including a calculated profit margin of 10%. Thus, the final cost of transportation ranges from 1077.72 euros to 1222.92 euros for each load carried.

Table 4. Tariff conditions for the transportation of tires by road

Information provided by selected automobile carriers		
	Price per 1 km in euro/km	Cost of transportation in euros
Carrier 1	0.80	1077.72
Carrier 2	0.87	1145.45
Carrier 3	0.95	1222.92

Evaluation of selected conditions for the carriage of tires by railways and by road transport

Table 5. Compares tariffs for road and railway freight traffic on the mentioned route.

Comparison of the cost of transportation of tires by railways and by road transport		
	Cost of transportation by rail	Cost of transportation by road
Carrier 1	1932	1077.72
Carrier 2	2070	1145.45
Carrier 3	3175.30	1222.92

This table shows that the cost of freight transportation by road is about half of the cost of the transportation by railway. Therefore, for a Client, automobile transportation is more economically sound than transportation by railway.

The cost of transportation by road on the route Puchov - Hannover is 1077.72 euros and 1222.92 euros for a load shipment. In all cases, carriers calculated the cost of international transportation performed within the period of one day. Swap semi-trailers and wagons belong to the customer. The cost of railway freight transportation on the same route amounts from 1,932 euros to 3,175.30 euros for one fiat wagon sent, i.e. two removable Container bodies belonging to the customer. If the fiat wagon or transport unit do not belong to the consignor, the cost of transportation would be 2138 - 3902 euros for a load shipment. In most cases, carriers could provide tariffs for the transportation within a period of 3 days. The indicated cost of delivery is calculated based on the tariff. If it were a normal delivery, the delivery cost could be reduced by 10-15%.

The conditions for the transportation by railway and by road are very different. Table 6 presents a brief comparison of transport and tariff conditions. [2,11]

The responsibility of the carrier for the transportation by road is 8.33 XDR/kg gross weight of a damaged or lost consignment, and in railway transport - 17 XDR/kg, which demonstrates the carrier's greater responsibility for the damage or loss of goods in the transportation by rail. In time of delay, the carrier is liable: when transporting by road - up to the price of transportation, when carrying by rail - up to four times of the cost of transportation. The delivery time in both cases is agreed in the contract between the consignor and the carrier. If the delivery time is not agreed by the contract, the time of delivery by rail shall not exceed the specified time of delivery and handling of the goods. In road transportation, the delivery time cannot be more than the actual time of transportation taking into account the circumstances and time necessary for loading and preparation of the vehicle.

The tariff/price conditions for railway transportation are deter- mined in different ways. In this particular time, the method for determining the price was chosen on the basis of the "Client Agreement", because the fare for transportation on this route does not include the cost for the transportation of an intermodal transport unit (swap body). Selected railway and road carriers were chosen. In road transport, the cost of transportation is calculated using individual calculations of carriers. On the chosen route and with the selected transport criteria, the most economical is road transport.

Fig. 8. Comparison of transport and tariff conditions

TRANSPORT CONDITIONS		
	Railway transport	**Road transport**
Responsibility in accordance with	COTIF Convention, JPP CIM	CMR Convention
Responsibility for providing data	Consignor	Consignor
Liability of carrier in case of loss or damage	17 XDR/kg	8,33 XDR/kg
Liability of carrier in case delay	4 times of the cost of transportation	Up to the cost of transportation
Number of copies of consignment notes	5	4
Responsibility for measuring the weight of load	Consignor	Consignor
Departure control	Carrier	Carrier
Estimated delivery time	According to the contract but not longer than the time required for handling and delivering	According to the contract but not longer than the actual time of delivery
Claims	In accordance with the contract	Within 7 days after the goods have been received – for an obvious damage, 7 days after the moment a hidden damage has been detected
Statute of limitations	1 year	1 year, intentional damage – 3 years
TARIFF CONDITIONS		
Method of pricing	Client agreement	Individual calculation
Unit of calculation	Price / 1 wagon	Price / 1 km

3 Conclusion

The conditions for the transportation in international and domestic freight traffic are analysed taking into account the time and cost- effectiveness. For the customer ordering the transportation, the economic component consists of the total cost spent for transportation. The time of delivery is the time required for transportation.

The reasons why the road freight transport in the European Union has the largest share in the transport market is not only the price, but also the delivery time. When comparing transportation by road and rail, the delivery time for most customers is determinative. In this time, road freight transportation was twice as fast as compared to the railways. The duration of transportation is determined taking into account the time of delivery and handling of the goods, which are only indicative. If it were certified regular transports, then the transportation time would be 24 hours. Comparing the tariff and time conditions of these transportations, it can be mentioned that railway transportation in this time is not competitive with respect to the road transport.

The purpose of this article was to compare the transportation and tariff conditions in railway and road transport, taking into account the selected criteria on the route Puchov – Hannover, basing on the analysis of the relevant provisions of the Convention on the Contract for the International Carriage of Goods - CMR and the Uniform Rules of COTIF Convention, regulating the conditions for the transportation of goods by rail and by road. The results of the analysis can be used in practice to select a more advantageous option. Both railway and automobile carriers provided data on delivery tariffs and conditions.

This research was financially supported by the project VEGA No. 1/0019/17 Evaluation of regional rail transport in the context of the region's economic potential focusing on the efficient use of public resources and social costs of transport.

References

1. A. Dolinayová E. Nedeliaková. *Proceedings of the 16th International Symposium Eurnex – Žel*, 101 – 108, (2008)
2. G. Fedorko, V. Molnár, M. Kopas,, Tem journal - technology education management informatics, **7, 3** (2008) DOI: 10.18421/TEM73-02
3. V. Ľupták, J. Gašparík, M. Chovancová, MATEC Web of Conferences.**134**, 00033, (2017)
4. A. Dolinayová, J. Čamaj, J. Daniš, J. *Globalization and its socio-economic consequences, 16th International scientific conference proceedings*, 383-390 (2016)
5. M. Zoldy, A. Torok. Period. Polytech. Transp. Eng., **43**, **4**, 172-176 (2015),
6. L. Černá, V. Klapita, J. Mašek, J. Technická základňa intermodálnej prepravy. 1st edition, (2015)
7. J. Ližbetin, O. Stopka, F. Nemec. *Proceedings of the 20th international scientific conference Transport Means 2016*. 151 – 155. (2016)
8. M. Poliak. Transport and communications, **1** (190- 201. (2010).
9. J. Královenský, J. Gnap, M. Poliak, V. Konečný. Ekonomika cestnej a mestskej dopravy. **1**, University of Žilina, ISBN 978-80-8070-831-3 (2008)
10. Transport development strategy for 2020 [on-line], (cited 12.02.2017)
11. G. Fedorko, V. Molnár, J. Strohmandl et al. *Transport Means 2015 – Proceedings of the International Conference*. 466-+ (2015)
12. T. Skrúcaný, J. Ponický, M. Kendra, J. Gnap *Proceedings of the 3rd International Conference on Traffic and Transport Engineering ICTTE* 2016, 744-749. (2016).
13. M. Chovancová, V. Klapita. Open Engineering. **7. 1**, 50-54 (2017) DOI: 10.1515/eng-2017-0009
14. J. Majerčák, M. Plíhal, P. Majerčák. Modelovanie prepravy vybranej komodity v medzinárodnej preprave tovaru, 1st edition. (2016)
15. B. Buková, E. Brumerčíková, P. Kondek. Open Engineering. **7**, **1**. 14-18. (2017) DOI: 10.1515/eng-2017-0003
16. Siroky, J. LOGI proceeding, MATEC, **134** (2017)
17. M. Poliak. EKONOMICKY CASOPIS. 61, 2. 206-220. (2013)
18. L. Bartuska, O. Stopka, M. Chovancova, J. Lizbetin. *Transport Means - Proceedings of the 20th International Conference*. 107-111 (2016)
19. A. Plakhin, R. Kampf, E. Ogorodnikova, A. Kokovikhin. MATEC Web of Conferences. **134**, 00047 (2017) DOI: 10.1051/matecconf/201713400047

Cancellation of delayed trains: passengers' and capacity points of view

Josef Buliček [1,*]

[1] University of Pardubice, Faculty of Transport Engineering, Department of Transport Technology and Control, Studentská 95, 53210 Pardubice, Czech Republic

Abstract. The paper is focused on interface between passenger transport demand after utilization of individual trains influenced by current operational situation (delays, overtaking of trains etc.) and on aspects of capacity utilization of railway line. The research is inspired by an idea (based on UIC 406) to operate rush lines on higher level of capacity utilization due to possibility to cancel delayed trains and make the time schedule more stable in this way. Simulation model of distribution of passengers between trains according to current operational situation is modified and applied for this assessment. The main result is that the effectiveness of this measure (especially in passengers' point of view) is depended not only on high capacity utilization of railway line, but also on time schedule and a complex of other operational conditions.

Introduction

This paper refers about the results of preliminary research works realized in the field of railway infrastructure capacity within the project PosiTrans solved at the University of Pardubice.
There is a tendency to shorten headway times between trains on railway especially in suburban areas. Effort to create railway more attractive for passengers can be a reason. The aim is to create transport system with similar features like in urban public transport. Short headway times belong to these features. Capacity utilization of railway infrastructure is increased due to this tendency.
The impact of possible delay can be a question in the case of shortened headway times. Possible delay can cause redistribution of passengers between individual trains in this case, so that the demand after delayed train can be seriously reduced. This leads to an idea to cancel such delayed trains [1]. Train cancellation can be helpful for operational stability and can be used as a tool allowing designing of time schedules with higher capacity utilization (occupancy rate) than in the case when redistribution of passengers' demand and related train cancellation are not possible. This can be description of an invited effect. The interface between passengers' point of view and capacity is researched as a scope of this paper. For some cases. This possibility is also considered by and based on the UIC 406 Codex – Capacity [1] for estimation of capacity of railway transport infrastructure.
Simulation model designed for exploitation of train passenger capacity, taking into account potential delays, is applied. This model was developed by the Department of Transport Technology and Control at the Faculty of Transport Engineering of the University of Pardubice. [2]
Simulation in field of railway capacity estimation is also international accepted approach, e.g. [3]. Consideration of robustness within planning of railway operation is accented by [4] and [5]. Interconnection of quality, capacity, dispatching and performance is creating a scope of the research presented in [6]. As follows from this brief analysis, this theme is topical on international scientific level at present time.
Potential delay of individual trains can lead to a change of demand with passengers moving to another train. Problem is that this effect can be occurred with no regard if delayed train is operated or cancelled. In some cases, passenger capacity might be insufficient, while in other cases it can be unexploited. This effect is the focus of the model as well.
In terms of the approach to transport demand, the model can be classified as a microscopic simulation model. As for the approach to traffic, macroscopic modelling is usually sufficient. Overall, the model can thus be classified as mesoscopic.
Simulation models are generally descriptive. It follows that naturally, the quality of outputs depends on the quality of inputs, specific set of realized replications and the scope of simulation experiments (replications) carried out. [7]
It is also to be noted that the aim of this paper is to present the situation on a methodological and general level. That's why the individual inputs and outputs of the model are characterized by their type and description of their processing, rather than by specific values. The specific values need to be established on a case-by-case basis, as there can be differences in practice. These differences can exist on all levels – in the way the

* Corresponding author: josef.bulicek@upce.cz

passengers accumulate, in their decision making regarding the train they would use, and in the railway transport organisation itself (e.g. different priorities in waiting for delayed connecting trains).
The aim of this paper is to identify in which cases is possibly relevant to cancel the trains and which effects influencing this. This will be done next to brief model presentation and although the fact that model results have particular character as it was mentioned above.

1 Problem definition

As indicated above, the model focuses on the changes of use of train passenger capacity considering a potential delay and mutual "interaction" of different trains within a common section or in a part of railway network serving the same transport relation.

In case of a delay, the passenger generally:

a) waits for the train he/she intended to use (even though he could use another (replacement) train),
b) uses a delayed train he didn't intend to use, using which he/she expects to reach the final destination (or interchange station) earlier, as he/she came to the station earlier and caught the delayed train unexpectedly,
c) uses a replacement train where the (delayed) train he/she intended to use hasn't arrived until the departure of the replacement train, using which he/she expects to reach the final destination earlier.

On the other hand, the reaction of individual passengers to a train delay is often ambiguous. The reasons why different passengers wouldn't react in the same way can be the following:

a) the delay length, number and structure of the respective trains,
b) possibilities to reach the final station (including the available information on trains),
c) ticket validity (e.g. different rates for different trains, inclusion of the train in the integrated transport system, obligatory seat reservation, etc.),
d) wheelchair accessibility of the train,
e) other factors – e.g. the possibilities of transport of baggage, bicycles, prams, or direct connection requirements.

These factors often don't apply absolutely and it is more or less possible that in the end, the passenger decides differently than expected. For instance the passenger can give up a reserved seat and use a replacement train. In other cases, he/she can reserve a seat for a train with obligatory seat reservation. This means that the model cannot provide a clear solution to this, but it needs to consider the aspect of "willingness" to these decisions, e.g. to consider their probability.

The purpose of the proposed model is to monitor passengers' demand for a given train collection, including the changes to this demand caused by the effect of potential delays, so that these changes can be predicted and considered in establishing the passenger capacity (number and type of cars, or units) for individual trains.

To meet the above mentioned purposes, it is necessary to create a microscopic model to be able to distinguish between different batches of passengers. This approach is also advantageous if trains with minimum use are included. The principle is basically the same as for the discrete model of transport demand, as presented in [8].

On the other hand, even entering such a large scope of input data can be problematic. For the above mentioned reasons, it is necessary to create support processes to potentially replace this data entering and to function based on the assumed general characteristics. This ultimately means that even the basic input – establishing the transport demand – will be subject to modelling, including the deliberate simplification and generalization. For instance, accumulation of passengers is modelled in uniform manner in model application presented in chapter 8 of this paper.

The structure of the model is modular, meaning that the processes (approaches) involved both in input data entering, and evaluation, can be complemented, combined and used simultaneously or alternatively depending on the given situation. Consideration can be equipped by other methods for capacity estimation, mentioned e.g. in [9].

Specific functions and procedures were added and applied into the model also within the research of train cancellation. Number of trains with decreased transport demand can be evaluated in relation to general indicators reflecting transport situation like average delay, ratio of delayed trains, categories of passengers' demand change etc. It is invited that this will be more related to structure of delayed trains as well. For that reason, possibilities to define structure of delayed trains are followed in the model. Simulation is equipped by an exhaustive-search algorithm evaluating all possible values of delay by selected set of trains as well for this purpose.

Passengers' selection of train to use is improved by an algorithm comparing candidate trains not only by departure times in station of boarding, but also by arrival times to the station of alighting. The necessity of this is caused by the fact that model is designed for comparison of trains using different railway lines between the origin and destination as well. The model is able to work in partial network point of view as well. On the other hand, model does not reflect any condition related to vehicle and staff circulations. Possible delay of each train is generated independently according to defined conditions (valid for this discrete random variables).

The model was created in Microsoft Excel using tables of inputs and outputs. The main part of the model is based on set of macros created in Visual Basic for Applications (VBA). The length of the programme code is more than 2,000 lines.

2 Problem definition

The first input to be entered is the data on the line or operational section (part of the network). The calculations consider each transport direction (direction of travel) separately. This data is entered using the names of railway stations or stops to be included in the future solution. At the same time, it is possible to reduce the number of stops by replacing a homogeneous section, in which all the stations are of similar type, with one stop combining all the stops within the section. This results in a simplification which needs to be approached carefully. It is recommended to enter selected stations or stops representing train journeys and transport demand beyond the area in question. This allows for examining the total number of passengers.

Another key input is the information on the individual trains in question, the stations and stops they serve and their timetable (arrivals/departures to/from these stations). Technically, this involves indicating the train (entering the number or name, or their combination) and the respective arrivals and departures.

The third input is the data on passenger capacity of the individual train units. For simplicity, the same capacity is considered within the entire section. Naturally, it is possible to enter different passenger capacities for different trains in accordance with the (planned) unit structure.

3 Railway traffic simulation

As mentioned above, the simulation of railway traffic (specific train journeys) is approached macroscopically. The aim is to establish a collection of specific values of arrivals and departures of the respective trains to/from the individual stations.

As for stochastic modelling of these train journeys, the same approach is used as in model [10]. Delays (or early arrivals) are approached as a discrete random variable with a precision of 1 minute. The (continuous) variable is discretized to facilitate the distribution of probability of delay using the specific values of delay probability. From a substantive point of view, this distinguishing level is acceptable. Setting parameters of continuous distribution of probability of a random variable would often involve variations and would entail a more complex approach. For a discrete variable, the delay statistics is sufficient with a relatively easy statistical evaluation (establishing the probability).

Furthermore, it is possible to create a set of conditions representing the correlations between arrivals and departures and modelling e.g. the waiting for delayed trains or basic technical-technological relations (e.g. the need to wait for an oncoming train on a single-track line).

On the other hand, if there is a more detailed microsimulation model of railway traffic already available, it is possible to draw on it and adopt its results. In this case, it is necessary to take over the collection of modelled ("real") values of arrivals and departures of individual trains to/from the stations. Only the part of the model presented herein will then be used which deals with the allocation of passengers to individual trains.

4 Railway traffic simulation

When monitoring the use of trains by passengers, it is also necessary to model the way they accumulate at stations in regards to the possibility to use an earlier train, or a train leaving earlier in case of a delay.

Moreover, as already mentioned, not all passengers have the same possibilities or willingness to use them. The train leaving earlier in the case of a delay doesn't need to be a suitable alternative for all passengers originally waiting for the delayed train.

For these reasons and to ensure the most faithful modelling of the above mentioned factors, a microscopic approach is taken. Transport demand is classified in the model according to the intended (demanded) train and the demanded transport relation.

On the other hand, specific transport demand, intensities of passenger transport flows and way of accumulation are applied for theoretical evaluation of train cancellation. It is inspired by real situation. For instance, uniform way of passengers' accumulation confirmed for short headway times by rapid transport surveys made by authors' department [7]; weaker transport flows originating in smaller stations than in rush hubs etc. On the other hand, all of these features were adapted to purposes of theoretical research and they are not related to any real train service.

The transport relations are classified according to the stations and stops served (by the originally demanded train). For simplicity, where the passenger intends to change trains, the interchange station is considered as a final one.

For modelling the passenger accumulation, a similar approach is taken as in the case of delay: time discretization with a precision of one minute. It is assumed that passengers arrive deterministically in accordance with the planned timetable/departure, regardless of a potential delay. It is possible to enter the manner of accumulation, i.e. the time period, in which the passengers accumulate, and the nature of this accumulation (e.g. continuous, batch).

The final train choice of passengers is modelled using four key conditions:

a. the "replacement" train must serve the final (interchange) station or stop,
b. "replacement" train is selected between all trains departing in defined time period (7 min period is applied) according to the first arrival (e.g. passengers can wait on a faster train departing later, but arriving earlier),
c. for transfer between individual trains, a probability matrix is created for the choice of this train as a replacement train (depending on the intended and replacement trains), passenger flow can be divided between individual trains according to defined probabilities,
d. additional conditions need to be met, e.g. that it is expected for the replacement train to reach the final (interchange) station earlier than the originally intended train (taking into account its delay).

Condition (a) is simplifying to a certain extent; for the future, it will be necessary to explore the possibilities of including such alternatives, where the passenger changes the final station or makes an indirect journey, to make the model even more accurate. The change to final station can for instance apply to railway centres, where different stations have similar connections to urban public transport, or to the city infrastructure.

As for (c), probabilistic assessment is performed due to the fact that not all passengers are equally willing to use a replacement train. Moreover, the probability is generally not the same for all pairs of trains. As mentioned above, the reason can be the different rates, obligatory seat reservation, etc. In some cases, this option can even be impossible (e.g. in case of trains solely composed of sleeping cars and couchette coaches). The collection of waiting passengers is then distributed between the intended and replacement trains based on this probability.

The (d) conditions can then be added depending on the specific situation in question.

5 Configuration of model for research of train cancellation

Model railway network is consisted of 8 stations and 8 line segments interconnecting them. The configuration of this network is in the Fig. 1. Considered transport direction is from east to west (in direction from A to C).

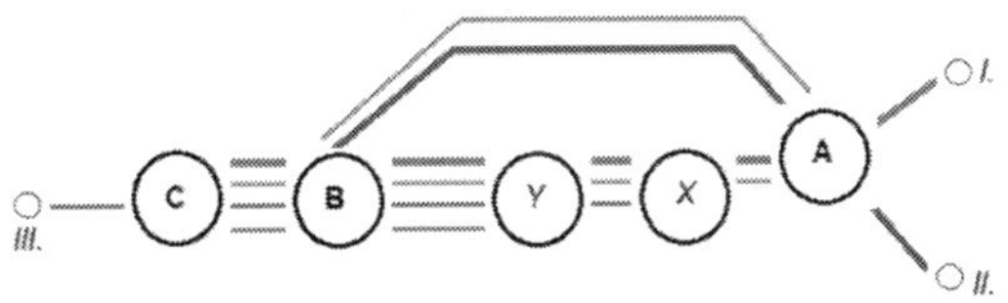

Fig. 1. Model railway network.

The core of transport relations is organized between the stations A, B and C. Stations X and Y are intermediate stations on the railway line A–B(–C). Stations marked as I., II. and III. are stations representing direct leading of trains in outlying areas of primary modelled network between A and C.

Station A is representing a regional hub station. Station B is a station where 2 railway lines are connected. This station is located e.g. on the border of city area and it can be destination for high number of passengers as well. Station C is representing the main destination (e.g. the main station in a city).

Red line I. – A – B – C is a line of long-distance fast trains (with no stop at X and Y intermediate stations). This line is operated with 60 min line headway time.

Blue line II. – A – B – C is also a line of long-distance fast trains, but using different railway line between A and B. Headway time on this line is 120 min.

Other lines are regional (suburban). Lines A–C (green), X–III. (violet) and Y–C (grey) are operated every 30 min. These lines are creating a collective operation between Y and C with collective headway of 10 min between subsequent regional trains.

Regional line A–B (brown) is leaded by alternative railway line. The line is operated with variable headway time of 30 or 60 min. Travel times to regional centre (from station A to B or C) on this alternative line are longer in comparison with lines using the main line (over X and Y).

The model is consisted of 42 trains representing morning and forenoon operation. The length of modelled time period is 6 hours.

6 Parts of network

Each part of the railway network in model represents different aspects of railway public passenger transport.

Station A is representing selection between different kinds of trains (long-distance fast or regional trains) as well as substitution effects of alternative railway line for lines going over X and Y. Network approach is applied.

Stations X and Y represent situation of selection between regional trains of similar type using the same railway line.

Railway line segment Y–C is a segment with rush operation of regional lines comparable with urban public transport (like tram transport) in short headway of 10 min.

Segment to the station III. represents a limitation for passengers travelling behind the station C. For instance, passengers from Y travelling to III. can't simply change the train in the case of delay and to use synergic effects of this operation. It is modelled as a specific step with the model algorithm.

7 Principles of evaluation of redistribution passengers' demand

he situation in the field of redistribution of passengers is evaluated by demand of passengers after utilization of each train in given specific conditions (in competition of other trains possibly going to the same destination). The base for comparison is created by passengers' demand after using of the train in the case of operation with according to time schedule (in time).

Following categories are defined for easier possibility to evaluate changes in demand in comparison with operation according to time schedule (in time):

- Increase – passenger number in specific replication is higher.
- Stagnation – the same number like in the case of operation according to time schedule.
- Small decrease – demand in specific replication is decreased by less than 30%.
- Medium decrease – demand in specific replication is decreased by 30 – 50%.
- Important decrease – demand in specific replication is decreased by more than 50%.

8 Discussion of results

Simulation was realized according 3 different scenarios for possibility to reach more general results. On the other hand, the results of simulation always have particular character in mathematical point of view as it was stated above.

8.1. Scenario 1 – Small and frequently occurred operational delays

This scenario is inspired by a common (normal) operation. Each train has individually simulated delay as integer value from 0 to 9 minutes. These delays are simulated according to probability distribution displayed by the Fig. 2.

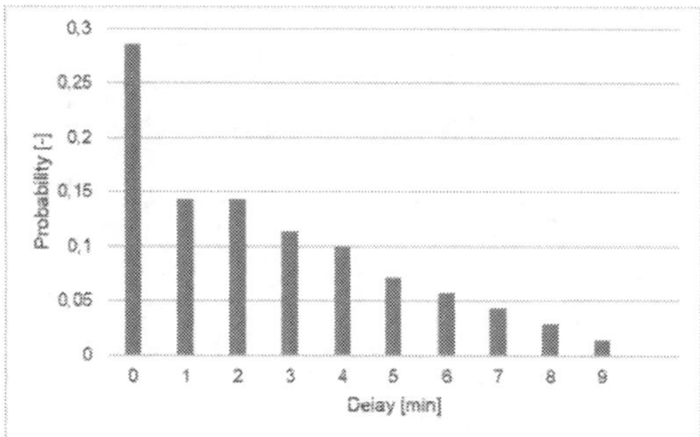

Fig. 2. Scenario 1 – discrete distribution of probability of delays.

Average percentage of delayed trains is 71.02%. Minimum number of delayed trains is 20, maximum is 38 from all 42 simulated trains. Average delay (of all trains) is 2.50 min and average delay of delayed train is 3.52 min. These values are related to 500 simulated replications.

Average results characterizing redistribution of passengers for 500 simulated replications are summarized in the Tab. 1.

Table 1. Scenario 1 – Redistribution of passengers.

All the trains departing from	Increase of demand	Stagnation of demand	Decrease of demand		
			Small	Medium	Important
Station A	38.65%	27.48%	25.88%	1.02%	6.97%
Station X	31.96%	36.73%	14.47%	8.67%	8.17%
Station Y	39.13%	21.23%	21.54%	9.96%	8.14%

There is visible in the Tab.1 that there are some cases of trains, when demand after they can be decreased about more than 50%. These trains can be candidate trains for cancellation. Problem following from more detailed analysis of simulation results, is that the situation is quite different in individual replications. All types of demand change are recorded by the most of trains. Only one train (fast long-distance train from A using longer alternative line) is typically placed in the category of important decrease of demand. On the other hand, this train can't be cancelled due to long-distance character of train.

Train cancellation can be strictly related to operative control of transport. It can't be prepared in a "systematic" way – e.g. with possibility to place such notice about this into time schedules, to reflect it by optimization of vehicle circulations etc. With respect to these facts it can be presupposed that train cancellation can be effective by less than 7% of trains (if all other technological conditions will be fulfilled as well). This number of trains can be lower in practice due to other operational conditions (e.g. circulation of vehicles, necessity to ensure possible interchanges etc.). Mentioned value is estimation of ratio of trains able to be considered in this point of view only.

8.2. Scenario 2 – Higher and less frequently occurred delays

The second simulated scenario is focused on the situation of higher, but less frequent delays. Distribution of probability is in Fig. 3. Delays of 0, 5, 10, 15 and 20 minutes are generated only.

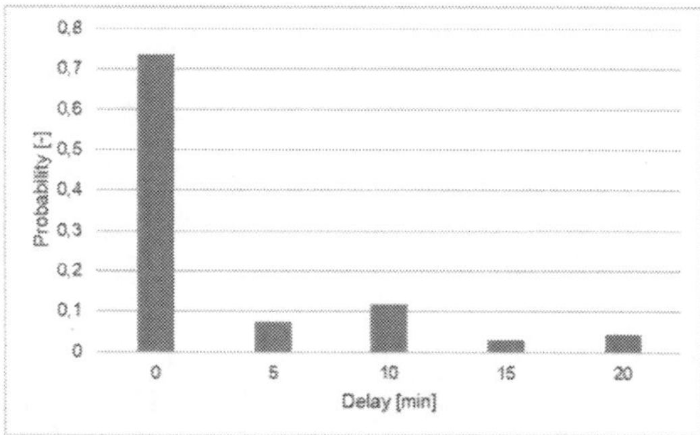

Fig. 3. Scenario 2 – discrete distribution of probability of delays.

Average percentage of delayed trains is 25.85%. Minimum number of delayed trains is 3, maximum is 21 from all 42 simulated trains. Average delay (of all trains) is 2.80 min and average delay of delayed train is 10.83 min. These values are related to 500 simulated replications.

The main difference between scenarios 1 and 2 is illustrated by above mentioned values. Result of evaluation of redistribution of passengers mentioned in the Tab. 2 are corresponding.

Table 2. Scenario 2 – Redistribution of passengers

All the trains departing from	Increase of demand	Stagnation of demand	Decrease of demand		
			Small	Medium	Important
Station A	32.00%	47.60%	6.99%	2.28%	11.13%
Station X	14.86%	70.94%	1.33%	1.15%	11.71%
Station Y	20.16%	58.97%	1.80%	3.24%	15.82%

Dominant result of this scenario 2 is that there is stagnation of passengers' demand by most of trains. It is related to the fact that the most of trains was operated in time. Increased length of delay caused increased ratio of trains in category with important decrease of demand.
Only one train (regional train from A to B using longer alternative line out of X and Y) is typically placed in this category. All other trains are located in categories of increase or stagnation. It means that it is not possible to select a set of trains candidate for possible cancellation. Decrease of demand can be occurred randomly. It must take a part of operational control of railway transport as well.

8.3. Scenario 3 – Relation to value of delay

The last scenario is quite different. It is dedicated to research of relation between the value of delay and passengers' demand after train.

The most of trains is operated according to time schedule (in time) in this scenario. The trains of regional line A–X–Y–B–C are influenced by delay only. Impact of this delay on passengers' demand after one selected train of this line is evaluated. Exhaustive search algorithm is applied. All integer values of delay from interval <1, 60> minutes are applied, simulated and evaluated. The results are in the Fig. 4.

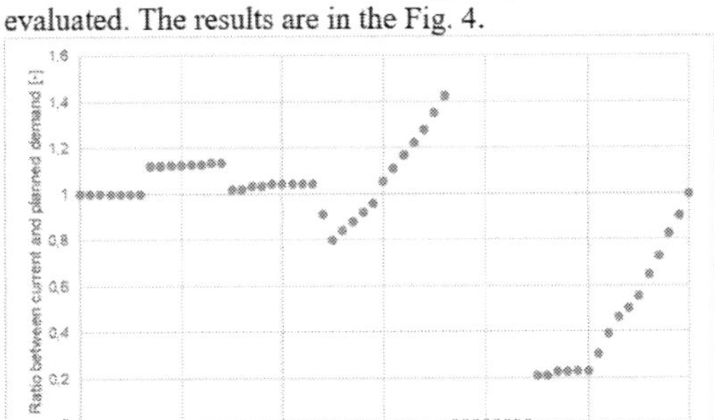

Fig. 4. Scenario 3 – relation between value of delay and ratio of current and planned demand

Planned demand represents demand considered for a situation when considered regional train from A to C is operated according to time schedule (in time). Current demand represents demand identified by simulation model for given (considered) value of delay.

It can be seen in the Fig. 4 that this relation is influenced by departures (from station A) of other trains operated on the same relation. It is possible to be seen that delay can have positive as well as negative impact on demand. As it follows from Fig. 4 it is relative effective to operate this train when the delay is lower than 37 min or higher than 50 min (if the planned occupation is used by 50% or more). When the train will be delayed about 37 – 49 min, train cancellation can be considered as the possibility how to save the capacity of railway infrastructure (line). This decision is strictly related to specific (given) operating conditions (time schedule), Fig. 4 will be different for each specific situation.

Conclusion

It is declared by the research results that application of simulation models in the field of interface between railway infrastructure capacity and transport demand after individual trains is possible.

Train cancellation is not a simply tool. Its application and effectiveness is related to complex of operational conditions and time schedule, not only to capacity occupation rate of give railway line. A lot of questions based on results mentioned in this paper can take a part of future research.

The work was supported from ERDF/ESF "Project PosiTrans - University of Pardubice cooperation and application sphere in application oriented research of localization, detection and simulation systems for transport processes" (No. CZ.02.1.01/0.0/0.0/17_049/0008394).

References

1. Codex UIC 406 – Capacity. 1st ed. Paris: UIC International Union of Railways, 2004
2. J. BULÍČEK, *Nová železniční technika: nové železniční trendy*, **25**, **2** 28-31 (2017)
3. L.W. JENSEN L.W. et al., Transportation Research Part C **74**, 126-149, (2017) http://dx.doi.org/10.1016/j.trc.2016.10.013
4. R.M. LUSBY et al., European Journal of Operational Research (2017), http://dx.doi.org/10.1016/j.ejor.2017.07.044
5. J.W. ANDERSON et al., Journal of Rail Transport Planning & Management **3** 95-110, (2013),. http://dx.doi.org/10.1016/j.jrtpm.2013.12.002
6. J. LIANG, et al. Journal of Rail Transport Planning & Management (2017), http://dx.doi.org/10.1016/j.jrtpm.2017.08.002
7. J. BULÍČEK, V. MOJŽÍŠ, et al. *Modelování technologických procesů v dopravě.* Pardubice: Univerzita Pardubice, , 223 pp. (2011)
8. A. ČERNÁ, – J. ČERNÝ, *Manažerské rozhodování o dopravních systémech.*, 230 pp. (2014)

9. J. GAŠPARÍK, J. KOLÁŘ, *Železniční doprava – technologie, řízení, grafikony a dalších 100 zajímavostí.*, 432 (2017)
10. J. BULÍČEK, Vědecko-technický sborník ČD No. **40** (2015). [online], https://vts.cd.cz/documents/51448/51619/4015.pdf/

Draft of strategy of the European rail freight corridor Amber

Juraj Kanis[1], *Anna* Dolinayova[2*] and *Matej* Kmetik[3]

[1]VVUŽ Žilina, Section of Expert Activities, 010 02 Žilina, Slovakia
[2]University of Žilina, Faculty of Operation and Economics of Transport and Communications, 010 26 Žilina, Slovakia
3University of Pardubice, Faculty of Transport Engineering, 532 10 Pardubice, Czech Republic

Abstract. An increase in rail freight transport performances is affected by the quality of transport services, which is influenced to the great extent by the transport infrastructure and infrastructure manager services. An increase in the quality of services provided by railway infrastructure has also been achieved through the establishment of the European rail freight corridors. At the same time, the trends in routing of goods flows and demand for transport services have contributed to the establishment of the new European rail freight corridor "Amber". For this reason, an implementation strategy of the Amber corridor is proposed using the SWOT method and Balanced Score Card method. The draft of strategy is based on the findings contained in the study of transport and market environment of the Amber corridor.

1 Introduction

The current economic development in EU countries has an impact on continuous increase in demand for transport services. The continuous increase in demand for transport services results from a higher consumption of EU population and a higher production manufacturing enterprises. The demand is directly influencing also by the need to transport, the final products and the intermediate products from East Europe and Asia [1]. Several European companies cooperate with the companies in Asia and their income, level of innovations and social benefits depend on their cooperation. This demand then creates an offer that results in a market for transport services. There are many offers from several modes of transport in this market, where each mode of transport has its advantages and disadvantages for the transport process, the society and environment [2].

Rail freight is currently considered to be the most environmentally friendly mode of transport of goods, with an important role in the freight transport market. It contributes to the development of human society and combines economic and social progress while respecting the development [3]. Due to exogenous (e.g. entry of competition in road and air transport, technological innovations oriented to other modes of transport, change in transport requirements) and endogenous (e.g. inefficiency, overemployment, low level of innovations and modernization, technological lag) factors, rail freight lost the competitiveness in the transport services market resulting in decrease in the transport performances of rail sector [4]. At the same time a shift of transport performances to other more environmentally demanding modes of transport has occurred. This shift leads to a higher production of the negative external costs of transport and need for higher state subsidies to the transport from public funds. This unfavorable condition has to be addressed by individual states and EU.

EU, to promote the competitiveness of rail freight transport, in particular in the field of infrastructure quality, safety, time and administrative effectiveness, international cooperation, has established the European Rail Freight Corridors [5]. The establishment of the European Rail Freight Corridors should bring, in particular, better, more complete, more reliable and less expensive services to railway undertakings. Such services of the single European railway infrastructure consequently contribute to the better services of the railway undertakings providing freight services [6]. Increased commercial activity, reliable, fast and safe and cost competitive service lead to a shift of transport performances from more environmentally demanding modes of transport to rail freight transport. In addition to its environmental advantage, rail freight transport can provide more reliable, safer, less expensive and faster transport service in case of harmonizing the transport and technological processes in comparison with other modes of transport [7]. The shift of transport performances to rail leads to overall decrease in social costs (infrastructure owner costs, carrier costs and negative external costs of transport) generated by transport. Increasing requirements on quality and availability of rail freight service led to intention to establish the new European rail freight corridor Amber. The corridor establishment brings the connection between Adriatic seaports in the Republic of Slovenia and inland ports on the Danube and terminals in Hungary and the Slovak Republic and Poland, but it brings also the perspective of railway transport development with Serbia and the improvement of the railway transport in Europe – Asia direction. The Amber RFC implementation plan requires a draft of strategy.

* Corresponding author: anna.dolinayova@fpedas.uniza.sk

For this reason, we decided to propose a strategy for Amber RFC, in this paper. The proposed strategy is developed based on the technical and technological statistical data provided by the affected infrastructure managers using the synthetic SWOT analysis method and the Balanced scorecard method.

2 Aspects of establishing the Amber RFC

The main objectives of establishing the rail freight corridors, defined by the European Commission (hereinafter referred to as EC) are:
• Strengthening competitiveness of rail freight transport compared with other modes of transport,
• Effective modal split with an emphasis on environmentally friendly rail freight transport,
• Coordination of investment in more qualitative railway infrastructure with possibility of financial support from EU funds,
• Harmonization and synergy between national rail systems,
• Strengthening cooperation in allocation of railway infrastructure capacity to international freight trains between single infrastructure managers,
• Conformity with existing objectives of other specific freight corridors [8].

The establishment of the Amber RFC is to lead to the fulfilment of the partial objectives that can be summarized in the following points:
• General growth of transit rail freight performances,
• General growth of international rail freight performances,
• General growth of intermodal transport performances.
• Improve the interconnection of the main intermodal transport terminals,
• Facilitate the interconnection between the Adriatic Sea Port in the Republic of Slovenia and the inland ports on the Danube in Hungary and the Slovak Republic,
• Promote the railway transport development with Serbia,
• Improve, potentially, the railway transport across EU Eastern border and on the land bridge between Europe and Asia,
• Better services of infrastructure managers provided to railway undertakings,
• Better services provided by railway undertakings to customers,
• Shift of transport performances from environmentally demanding modes to rail freight – change in modal split in favour of rail freight,
• Increase in reliability and decrease in transport time,
• Decrease in railway undertaking costs,
• Growth of socio-economic efficiency of the railway system [9].

In addition to the partial objectives mentioned above, the establishment of the Amber RFC also brings particular benefits to railway undertakings and terminals:
• Participation in the process of creating a corridor offer,
• Overview concerning the capacity included in the corridor, including the capacity provided with priority,
• Better services in terms of transit time, regularity, reliability and information,
• Strengthening customer approach,
• Information on investment projects in railway infrastructure between railway administrations,
• Information about operating restriction,
• Harmonization of infrastructure technical transport parameters,
• Harmonization of track possessions between individual railway infrastructure managers,
• Better possibility to improve the infrastructure included in the corridor,
• Support for remove of technical and technological barriers,
• Chance to strengthen priority rules in operative traffic control for freight trains carrying out transport performances on the corridor,
• Possibility to express the opinion of carriers on the quality of Amber corridor services.

The defined objectives and benefits of the Amber RFC establishment are, in particular, to increase the competitiveness of rail freight services compared with other modes of freight transport. The expected objectives and benefits are to contribute to better, more reliable and more available rail freight services and harmonization technological and operational conditions. The fulfilment of Amber corridor´s objectives requires the cooperation of all stakeholders: state, government, ministries concerned, infrastructure managers, carriers, intermodal operators and external suppliers of the railway sector.

3 Characteristics of Amber RFC

The Amber rail freight corridor No 11 is being established based on Commission Implementing Decision (EU) no. 2017/177 of 31 January 2017, that was issued of the basis of "Letter of intent" as request of 4 Ministries competent for Rail Transport of Hungary, Republic of Poland, Slovak Republic and Republic of Slovenia. The establishment of Amber rail freight corridor is on the compliance with Article 5 of Regulation (EU) No 913/2010 of the European Parliament and of the Council of 22 September 2010 concerning a European rail network for competitive freight. This Regulation lays down rules for the establishment and organization of international rail freight corridors with a view to the development of a European rail network for competitive freight [8], [9].

RFC Amber routing: Koper–Ljubljana/Zalaszentivan –Sopron/Csorna/(Hungarian/Serbian border) – Kelebia – Budapest– Komárom– Leopoldov/Rajka –Bratislava–Žilina–Katowice/Kraków–Warszawa/Łuków – Terespol – (Polish/Belorusian border) [9].

Member states: Republic of Poland, Slovak Republic, Hungary, Republic of Slovenia.

Length of railway lines: approximately 3 400 km.

Seaports: 1

Inland ports: 3

Intermodal transport terminals: 25

Deadline for making Amber RFC operational: by 30.01.2019

Seat of Corridor One Stop Shop (C-OSS): Poland
Equations should be centred and should be numbered with the number on the right-hand side.

Fig. 1. Graphical representation of Amber RFC routing

4 Draft of the Amber corridor strategy

We suggested the Amber corridor strategy using two methods: SWOT analysis and Balanced Scorecard method. Draft takes into account comments and suggestions Infrastructure managers of MAV, GYSEV, VPE, SŽ-I and PKP.

4.1 SWOT analysis

Method of SWOT analysis consists in identifying the internal environment of the studied subject using its strengths and weakness and in identifying the impact of external environment using opportunities and threats. Based on recognized results a review of internal and external environment analysis will be obtained, while the most appropriate strategy for the studied subject will be made up based on given scores. Elaboration SWOT analysis is conditioned by completion of collection and subsequent evaluation of all available data collected. Then, the created basis of SWOT analysis is qualitatively and quantitatively assessed by independent experts and stakeholders, in this case by individual members of Amber RFC. Without assessment of several experts and stakeholders [10], SWOT analysis has only subjective character of its maker and it is inconsistent for the adoption of strategic direction and decision-making.

Strengths and weakness:

• Importance. Importance shows how important or a weakness is for organization as some strengths (weakness) might be more important than others might. A number from 0,01 (not important) to 0,99 (very important). The sum of all weights should equal 1,0.

• Rating. A score from 1 to 6 is given to each factor to indicate whether it is a major (6) or a minor (1) strength for the organization. The same rating should be assigned to the weaknesses where -1 would mean a minor weakness and -6 major weakness.

• Score. Score is result of importance multiplied by rating. It allows prioritizing the strengths and weakness. You should rely on your most important strengths and try to convert or defend your weakest parts of the organization.

Opportunities and threats:

• Importance. It shows to what extent the external factor might influence the business. Again, the numbers from 0,01 (no impact to 0,99 (very high impact) should be assigned to each item. The sum of all weights should equal 1,0.

• Probability. Probability of occurrence is showing how likely the opportunity or threat will have any impact on business. It should be rated from 1 (low probability) to 6 (high probability). For threats -1 (low probability) to -6 (high probability).

• Score. Importance multiplied by probability will give a score by which you´ll be able to prioritize opportunities and threats. Pay attention to the factors having the highest score and ignore the factors that will not likely affect your business.

The following four tables give strength, weaknesses, opportunities and threats of internal and external environment of Amber RFC [12, 13].

Table 1. Strengths of Amber RFC

S (Strengths)	Impor-tance	Ra-ting	Score
Interconnection of railway infrastructure within the countries included in Amber RFC	0,07	5	0,35
Railway system reliability	0,08	5	0,41
Available information on technical specification of corridor railway lines	0,04	5	0,18
Access to the important seaport Koper in the Republic of Slovenia	0,10	5	0,51
Thanks to the corridor strategic location and routing, good connection with other RFC corridors is guaranteed	0,08	5	0,41
Existing cooperation between individual infrastructure managers within Amber RFC countries	0,08	5	0,40
Railway infrastructure safety	0,10	6	0,54
Good technical conditions of railway infrastructure	0,08	5	0,41
Available free capacity	0,07	5	0,39
Connection by rail with countries outside the EU through BY/PL (Brest/Terespol) railway border crossing	0,10	6	0,60
Flexibility of railway infrastructure (e.g. suitable alternative routes)	0,05	6	0,28
Schengen area	0,03	6	0,21
Procurement of railway infrastructure capacity from one place C-OSS	0,05	4	0,19
Connection of railway transport with terminals within Amber RFC	0,06	5	0,31
TOTAL	**1**	-	**5,19**

In tables, there are assigned importance to each indicator and scores achieved. These two figures are then multiplied, while their product determines the final evaluation of indicator. [14] The data presented in the tables are the resulting average values obtained from the infrastructure managers affected by the Amber RFC, the TMS elaborator and the academic environment.

Table 2. Weaknesses of Amber RFC

W (Weaknesses)	Importance	Rating	Score
Insufficient implementation of TEN-T infrastructure minimum standards	0,09	-4	-0,38
Enforcement of various interests of infrastructure managers of member states	0,12	-3	-0,34
Traffic restrictions related to possession causing temporary capacity constraint	0,17	-5	-0,78
Reducing the quality of rail freight services provided within Amber RFC	0,14	-3	-0,42
Poor technical condition in some sections of railway lines	0,15	-5	-0,69
Bottlenecks of capacity utilization	0,10	-5	-0,44
Insufficient technical parameters of railway infrastructure – requirements for modernization	0,11	-5	-0,57
Long waiting times at border crossings	0,13	-4	-0,50
TOTAL	**1**	-	**-4,11**

Table 3. Opportunities of Amber RFC

O (Opportunities)	Importance	Rating	Score
Trend of using more environmentally friendly mode of transport (opportunity for rail transport)	0,08	4	0,35
Complete modernization of railway lines which limit the increase of line capacity	0,12	4	0,51
Investment of railway undertakings in sidings and siding operation	0,08	4	0,34
Increase in costs of road goods transport, e.g. toll charges	0,10	5	0,47
Increase in impact of transport policy of individual countries in favour of rail	0,10	5	0,47
Favourable economic growth of countries included in Amber RFC resulting in increase of import / export	0,12	5	0,56
Improving mutual cooperation between RFC corridors	0,06	5	0,30
Potential for corridor extension to the north of the Republic of Poland towards seaports	0,08	4	0,32
Connection of major economic active regions within the Amber RFC	0,09	4	0,38
Investment and modernization (e.g. construction of new line, double-tracking, station upgrade-signalling equipment, etc.)	0,08	3	0,23
Connection between inland ports on the Danube in Hungary and Slovakia	0,05	4	0,21
Connection with the lines in the Czech Republic	0,03	5	0,17
TOTAL	**1**	-	**4,32**

Table 4. Threats of Amber RFC

T (Threats)	Importance	Rating	Score
Building logistic centres without connection to railway infrastructure	0,06	-3	-0,17
Lack of qualified personnel in operation	0,08	-4	-0,37
Insufficient coordination in infrastructure development work	0,09	-4	-0,37
Reducing transport volumes of international freight trains	0,10	-4	-0,34
Tendency of transport policy of individual countries to rail transport disadvantage	0,06	-3	-0,16
Unfavourable economic development within Amber RFC countries	0,07	-3	-0,21
Reducing investment subsidies for rail transport	0,07	-4	-0,30
Reducing non-investment subsidies for rail transport	0,06	-3	-0,19
Higher transport time compared to road goods transport	0,10	-5	-0,44
Lower flexibility compared to road goods transport	0,10	-5	-0,46
Insufficient coverage of railway corridor routes to cover customer needs	0,11	-5	-0,57
Stagnation (unsolved problems) in the field of maintenance and modernization	0,10	-2	-0,25
TOTAL	**1**	-	**-3,82**

The quantitative scores were assigned to strengths, weaknesses, opportunities and threats (risks) in SWOT analysis for the Amber RFC. Quantified assessment of internal and external environment analysis needs to be put in comparison of vectors from which we find a particular position, which represents model strategy for the Amber RFC.

Using quantified evaluation of internal and external environment, it was found by comparison of vectors: Offensive strategy, as model strategy for the Amber RFC. Graphical representation of matrix of model strategies with initial strategy for the Amber RFC is shown in Figure 2.

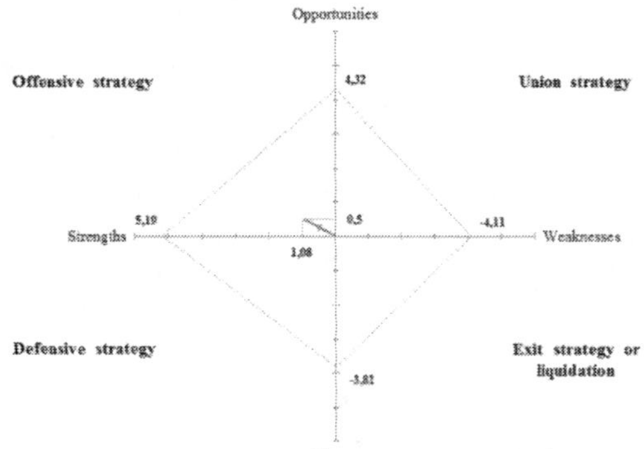

Fig..2. Matrix of model strategies for the Amber RFC

Offensive strategy is considered the most attractive strategic alternative. It can be used by an entity whose position is ideal with the predominant strengths over the weaknesses. Such an entity is able to use its strengths to realize the opportunities offered by the external environment. [15] However, an entity must monitor its

weakness and avoid defined risks. Based on the resultant strategy, it is necessary to take the following measures for the Amber RFC:
• Increase the reliability of rail system services,
• Developing the high-quality and available services of C-OSS,
• Developing the cooperation with other RFC corridors,
• Support for intermodal transport services,
• Reducing the railway infrastructure charges for local service trains– positive impact on last mile,
• In operative transport management, to proceed to prioritize international freight trains,
• Cooperation with seaport and inland port Małaszewicz,
• Close cooperation between infrastructure managers,
• Coordination of investment projects in railway infrastructure within Amber RFC lines,
• Increased awareness of the corridor, its services and perspectives,
• Exchange of information concerning operation, control and possessions,
• Measures to reduce the technological times of operations for transport of goods from/to countries outside the EU,
• Providing the best resource, e.g. human, IT,
• Investments in interoperability,
• Exclusive or dominant access to the most capable suppliers of MB Amber RFC.

These measures result from the strategy and its characteristics. However, the Amber corridor bodies cannot affect all measures. Therefore, it is necessary that the subjects that can influence the individual measures proceed to their implementation, thereby helping to meet the expected goals and benefits of the Amber corridor establishment.

4.2 Balanced Scorecard Method

In order to fulfil the basic objectives of the Amber RFC, it is necessary to set out the strategic steps for their fulfilment. One of the appropriate methods for creating strategic processes is the Balanced Score Card - BSC. BSC is complex strategic method that looks at the subject under consideration through four perspectives and their mutual relationships. I tis a financial, customer, process, learning and growth perspective. BSC is based on the vision and strategy of the object under consideration and on that basis for each perspective the mission and strategic objectives, to which certain metrics and their target values are assigned, will be determined. All perspectives are logically connected and linked and this method, therefore, provides a complex view of the object under consideration and its performance [16].

Amber RFC main vision are:
• Growth of rail freight transport performance,
• Fulfilling the EU transport objectives and reducing the negative external costs of transport,
• Strengthening rail freight position within the individual member states of the Amber RFC,
• Expand cooperation with rail carriers as well as between IM,
• Strengthening and developing the cooperation between RFC corridors,
• Maintaining and developing the rail freight services,
• Developing the services concerning free capacity allocation,
• Fulfilling the basic objectives of the liberalization of rail freight services market [12].

Amber RFC mission consists particularly in:
• Providing and improving the rail freight services,
• Creating a positive perception of rail freight transport and the Amber RFC,
• Development and modernization of railway infrastructure,
• Participation in transport policy development within the individual countries of the Amber RFC as well as at the EU level,
• Promoting the development of rail freight transport as an environmentally friendly and perspective mode of transport compared to road transport,
• Decreasing the transport performances of more environmentally demanding modes of transport,
• Available non-discriminatory access to railway infrastructure and its capacity,
• Effective transport of goods from/to EU, from/to countries outside the EU,
• Reducing public spending,
• High satisfaction of all customers of the Amber RFC [16].

Figure 3 shows the BSC strategic map for Amber corridor. The strategic map is based on the vision and mission of the Amber RFC and its four perspectives.

5. Conclusion

The paper deals with the draft of the Amber corridor strategy. The draft of strategy itself was created using the data and results included in the study of transport and market environment of the Amber corridor elaborated by the Research and Development Institute of Railways Žilina. The strategy is to serve for the implementation plan, support of management and routing of the corridor so that it provides high-quality and available services. The high-quality services should lead to an efficient modal split in the freight transport and so to promote sustainable development.

The paper was supported by the project EUREKA E!11214 EPLOS – "European portal of logistics services" at Faculty of Operations and Economics of Transport and Communication, University of Žilina, Slovakia.

EUREKA Σ! innovation across borders

MINISTERSTVO ŠKOLSTVA, VEDY, VÝSKUMU A ŠPORTU SLOVENSKEJ REPUBLIKY

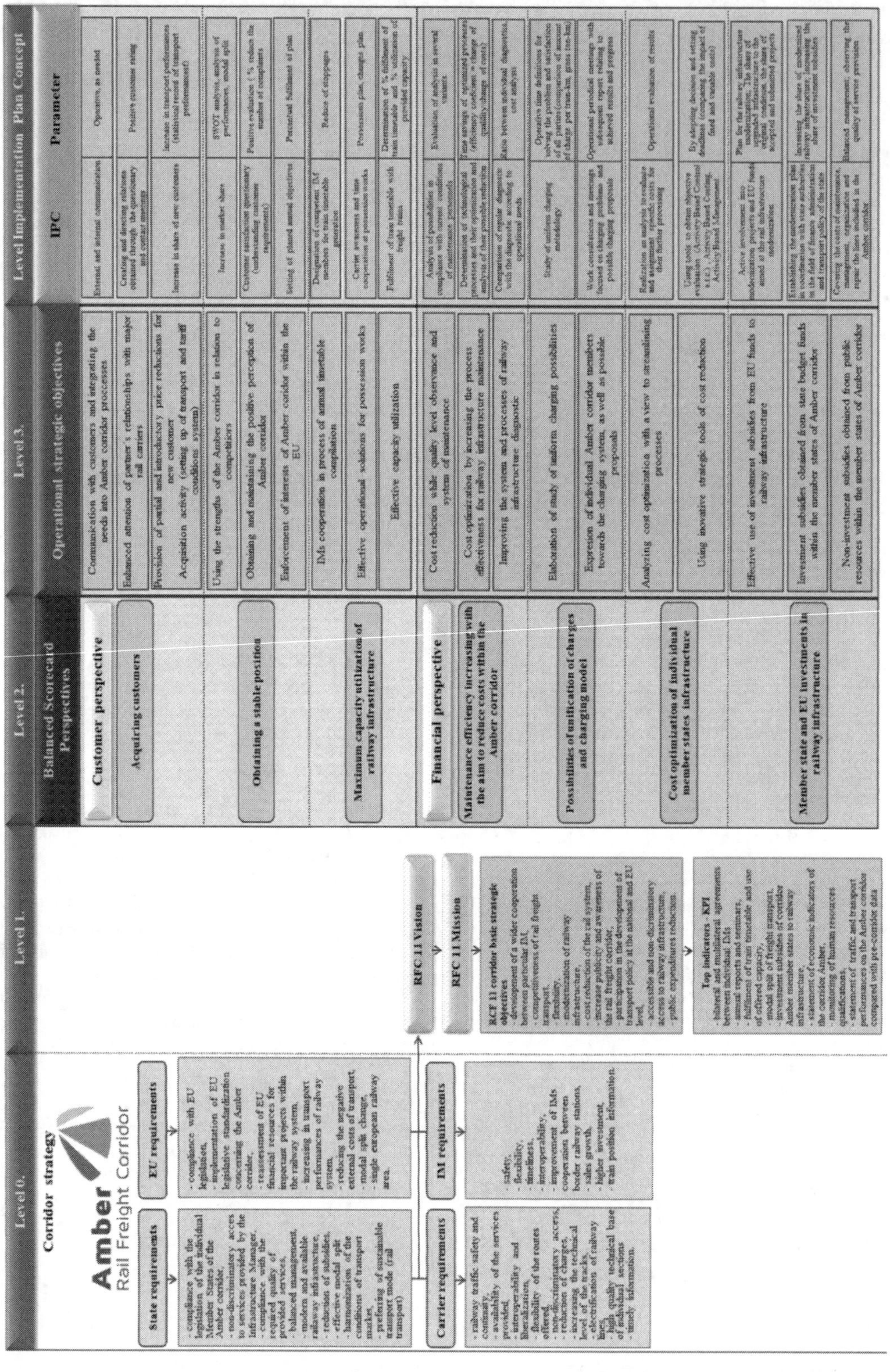

Fig. 3. Map Balanced Score Card of Amber RFC

* Corresponding author: anna.dolinayova@fpedas.uniza.sk

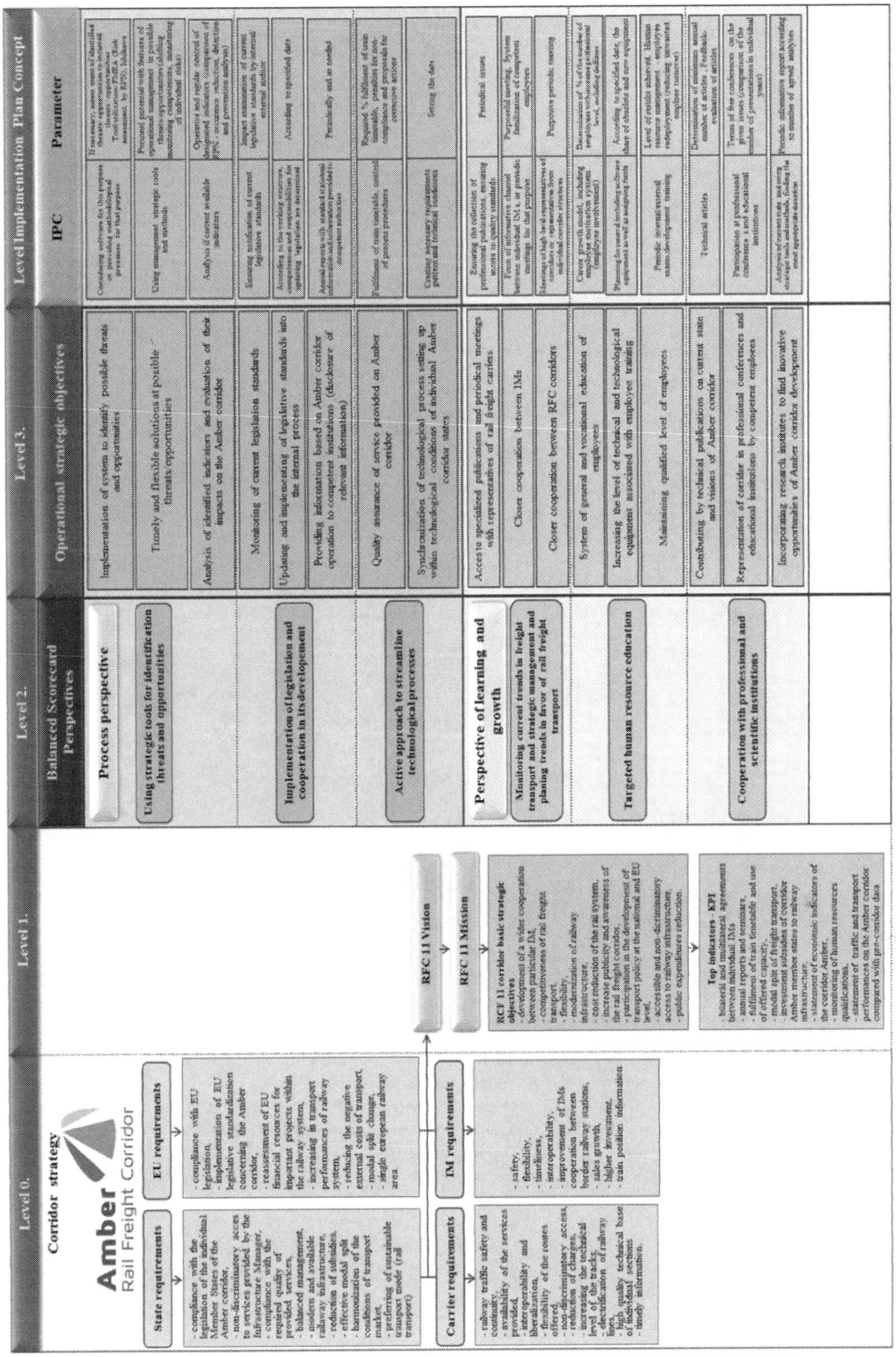

Fig. 3. Map Balanced Score Card of Amber RFC – Continue

References

1. P. Drozdiel, B. Bukova, E. Brumercikova, Transport Problems **10**, 4 (2015)
2. T. Skrucany, M. Kendra, B. Sarkan, J. Gnap, Communications in Computer and Information Science, **531** (2015)
3. P. Mesko, J. Gasparik, V. Gaborova, Proceedings of the 20th International Scientific Conference Transport Means (2016)
4. T. Zunder, DMZ Islam, M. Marinov, Transport problems, **5**, 3 (2010)
5. M. Poliak. EKONOMICKY CASOPIS. 61, 2. 206-220. (2013)
6. V. Zitrický, L. Černá, B. Abramovič, Procedia Engineering, **192** (2017)
7. O. Stopka, R. Kampf. Transport, 33, 280-290. (2018) DOI: 10.3846/16484142.2016.1174882.
8. B. Abramovic, E. Nedeliakova, M. Panak, D. Sipus. Proceedings of International Scientific Conference Business Logistics in Modern Management (2017)
9. E. Nedeliakova, M. Panak, J. Ponický, R. Soušek. AER – Advances in Engineering Research. **62**, 195-198. (2016)
10. R. Kampf, L. Lizbetinova, K. Tislerova, Open Engineering, **7**, 1 (2017)
11. Regulation (EU) No 913/2010 of the European Parliament and of the Council of 22 September 2010 concerning a European rail network for competitive freight. Available at: https://eur-lex.europa.eu/LexUriServ/LexUriServ.do?uri=OJ:L:2010:276:0022:0032:EN:PDF
12. M. Kvizda, Politická ekonomie. **53**. 5. 617-633. (2005)
13. Commission Implementing Decision (EU) 2017/177 of 31 January 2017 on the compliance with Article 5 of Regulation (EU) No 913/2010 of the European Parliament and of the Council of the joint proposal to establish the ‘Amber’ rail freight corridor. Available at: https://eur-lex.europa.eu/legal-content/EN/TXT/?uri= CELEX%3A32017D0177
14. A. Galierikova, J. Sosedova, NASE MORE. **65**, **3**. 8 – 11. 2018
15. A. Torok. Procedia Engineering **187**, 722-725, (2017), https://doi.org/10.1016/j.proeng.2017.04.446
16. Š. Zapletalová. *Business Environment*. [In Czech: Podnikatelské prostředí], (Slezská univerzita, 2016)
17. The study of TMS Corridor Amber, VVUŽ Žilina. Internal material (2018)

* Corresponding author: anna.dolinayova@fpedas.uniza.sk

Current Problems in Extraordinariness Situations with the Impact on the Management of Rail Transport Under the Conditions of Slovak Railways

Juraj Čamaj[1*], *Ján* Ponický[2], *Lumír* Pečený[1], *Pavol* Meško[1] and *Karel* Zeman[3]

[1]Department of Railway Transport, Faculty of Operation and Economics of Transport and Communications, Univerzitná 8215/1, 010 26 Žilina, Slovakia
[2]Železnice Slovenskej republiky, The Regional Directorate Žilina, The Railway Station Dolný Hričov, Slovakia
[3] Institute of Technology and Business in České Budějovice, Faculty of Technology, Department of Informatics and Natural Sciences, Okružní 517/10, 370 01 České Budějovice. Czech republic

Abstract. Rail transport is a dynamic and progressive process. It is important that the requirements of individual carriers be meet as efficiently as possible in accordance with the current train schedule. In the case of extraordinariness situations (delay, reconstructions activities, different routing of a train, etc.), which affect the fulfilment of train schedule, are applying so called "operational traffic management according to the applicable internal regulations". In the event of the occurrence of these extraordinary situations, it is necessary to compile a written injunction in order to inform the train drivers about these extraordinary circumstances. The introduction of an electronic injunction system causes the opposite effect in selected situations, such as rapid, accurate, and effective delivering this paper injunction. The article focuses on the current problem of introducing a new information system for the compilation and creation of written injunctions in relation to selected types of injunctions and situations in the railway transport.

1 The Structure of Operational Management in Railway Transport

Operational transport management on the network of Slovak railways (ŽSR) network is divide into several levels. Within each level of operational management, there is always one responsible manager of ŽSR – leading changes (figure 1), which ensures in its assigned area the operational management of railway transport. [1]

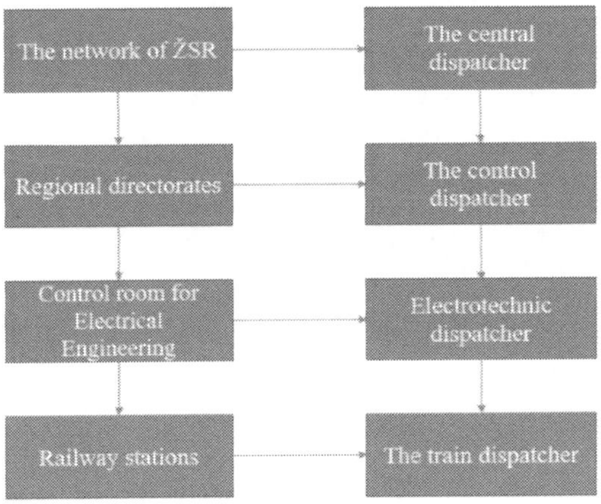

Fig. 1. Head of change in individual circuits.

The individual employees of the dispatching apparatus are all subordinate employees of the ŽSR and the transport dispatchers who participate in the strategic and operational management of the railway transport. [2] In the context of operational traffic management, j. in the case of extraordinary situations (delay, reconstructions activities, different routing of a train, etc.), which have an impact on the fulfilment of the train schedules, proceeds in accordance with regulation SR 1003 – Operational Management of Transport at Slovak railways.[3]

1.1. The managers for operation management of the General Directorate level

The Central dispatcher is the head of change on the entire network of Slovak railways. The central task of the central dispatcher is to manage and coordinate all traffic management activities on the Slovak Railways network in a comprehensive manner [4].

The central dispatcher cooperates with the main dispatcher and the individual dispatchers at the level of the Regional Directorates.

He coordinates activities with foreign railway administrations, carriers, state administration bodies and the control dispatchers of the Regional Directorates (fig. 2). [5]

* Corresponding author: juraj.camaj@fpedas.uniza.sk

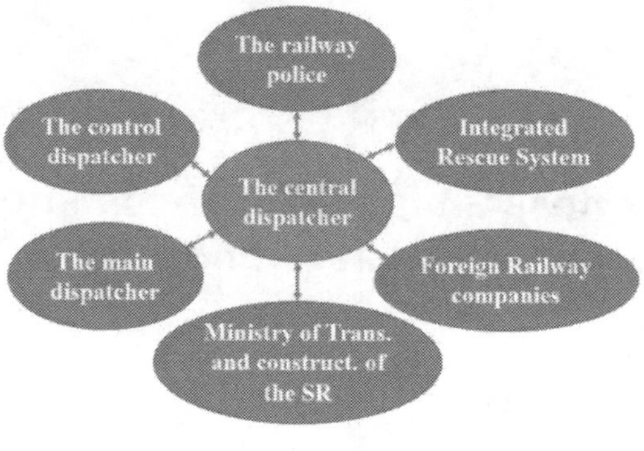

Fig. 2. Links of the central dispatcher to other entities.

The main dispatcher coordinates, in cooperation with the central dispatcher, all activities on the Slovak Railways network (fig. 3). [5]

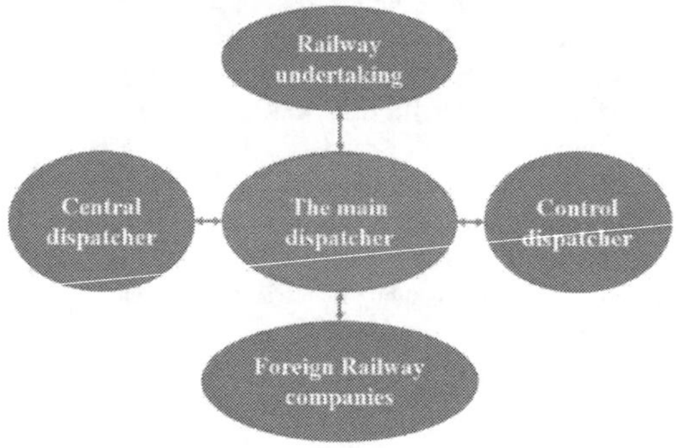

Fig. 3. Links of the main dispatcher to other entities.

1.2. The Managers for Operation Management on the level of the Regional Directorates

The Head of Change in the Regional Directorates (Trnava, Žilina, Zvolen and Košice) is the Control dispatcher.[4]

Its main task is coordinating to operationally and comprehensively traffic in its assigned area, in conjunction with other railway administrations and other Regional Directorates. He is responsible for the work of the operating dispatchers in his specified areas, for the optimal solution of the traffic situation (fig. 4). [5]

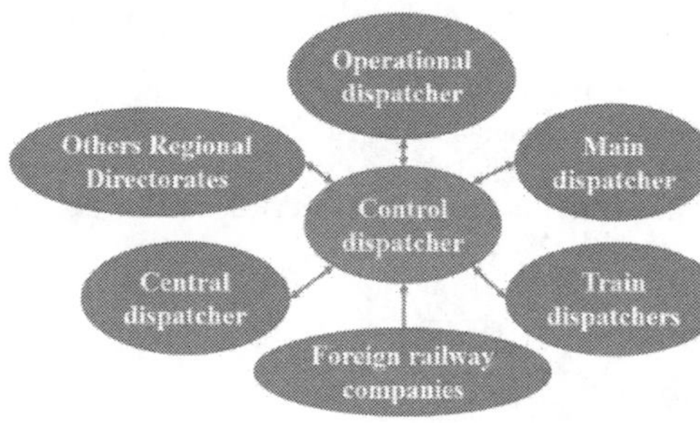

Fig. 4. Links of the control dispatcher to other entities.

The operational dispatcher manages, organizes, coordinates train traffic in its specified areas, and proposes operational measures to address the extraordinariness situations that impact of the train schedule (fig. 5). [5]

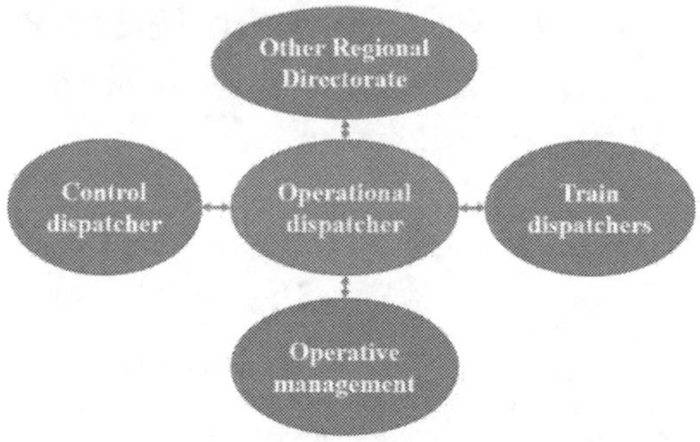

Fig. 5. Links of the operational dispatcher to other entities.

1.3. The Operating Trains Managers on the level of the Regional Directorate

The role of operations staffs, i.e. train dispatcher, operating supervisor, station dispatcher, etc. is to implement a shifts plan in the Slovak railways and the allocated disposition sections (fig 6). [5]

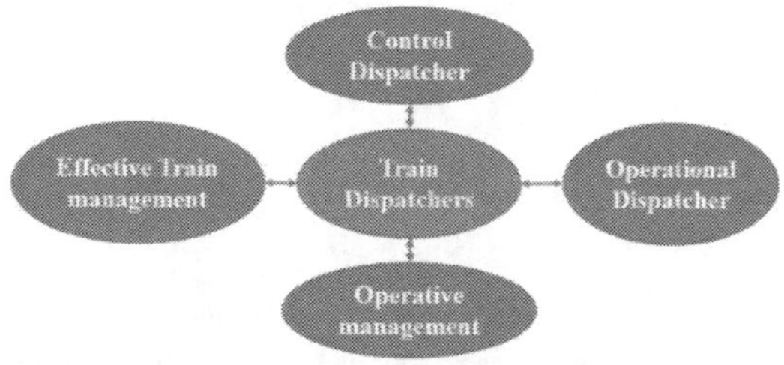

Fig. 6. Links of the train dispatcher to other entities.

2 Extraordinariness situations in field of the railway transport and the written injunctions

For all the extraordinariness situations related to train directions, the train driver (and the other train drivers on the active locomotives) must notified by written injunctions. [2]

If the train dispatcher does not receive a report that the train driver was notify by written injunction or himself did not give a written injunction (in case of extraordinariness situations), he must not allow him to go further. [6]

Each written injunction must include the stamp of the station and the signature of the employee who issued the injunction and who takes the injunction. The train driver (or train drivers) who takes the injunction is obliged to read it before acknowledging the receipt (with signature). [7] The written injunction must be clear, concise and legible.

In the case where the issuing of injunctions is using a computerized technique, the name of the station may be use instead of a stamp, which was place in the same place as a stamp imprint [8]. The font size of the injunction executed on the computing technique must be at least 3 mm.

The train driver must be notifying with the written injunction at the starting or the select station for delivery this injunction or at the last station in front of the place for which the injunction is valid. In the case of regular stopping trains, this is the last station where the train stops regularly.[9]

2.1. The train driving in the event of a breakdown the interlocking system of level crossing

The crossing is a level crossing between the road and the railway track. Under Slovak Railways, we divide level crossings to:

- Secured - equipped with crossover interlocking systems,
- Unsecured - without crossover interlocking systems

The next part of paper will be processed measures while riding trains in the event of failure of the level crossing interlocking systems. It means in the conditions while the train not ride through the level crossing with the prescribed manner warn road users that the rail vehicle approaching the crossing or there is another failure.

For monitor the activity and control of the light-signalling device, a control box is located in the station or track points with indicate the individual states of the level crossings. [10]

In the case of an indication of the fault condition of the level crossing, the train dispatcher is obliged to take the appropriate measures to ensure the train's driving around the fault line - to inform the train driver with the "Injunction for cautious driving - Part A". [3-4] In this section, is for train driver provide all details of the location of the fault level intersection as well as all measures for safe passage of the train through this level crossing. The written Injunction for cautious driving - Part A is delivered to the train driver at the last station where the train dispatcher is located before the level crossing and where the train stops for the last time. In the injunction, the train dispatcher is obliged to fill in the kilometre position of the level crossing to three decimal places (e.g. 174,052 km) and the names of the stations, or stops etc., between or in which places the level crossing is located. [4]

3 The electronic application for the written injunctions in Slovak railways

The application of written injunctions is one of the Modules of the Automated Workstation of the Train Dispatcher. It used to compile the injunctions for the notification of trains and to send them electronically to other workplaces that have the authority to issue the written injunctions. [11] The injunction generated by this application can displayed for a specific track, for a particular train or for unknown trains. It also allows you to create an injunction for handling between stations (HBS). [3]

All necessary data on stations, specified tracks and train timetables are summarizing in the application's root data. The actual application uses the operator to input the data about the trains to notified, i.e., specifies the speed limitation of railway tracks, reconstructions activities, and other extraordinary situations. This information is generally referring to as an instruction. The written injunctions for train notification are generating based on the instructions given (fig. 7). [12]

Creation of injunctions a based on the facts that the operator gradually enters the instructions in the application as they are informing about them in time. At the time when the injunction is needed, the operator enters a command to create one of the types of injunctions (The General injunction, The Injunction for cautious driving – Part A, [4] [13] The Injunction for cautious driving – Part B, The Injunction for handling between stations). The application then selects appropriate instructions and assembles them to one or more injunctions. [9-10]

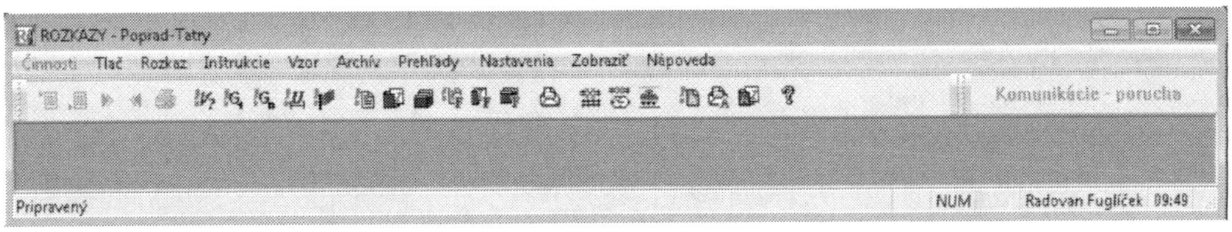

Fig. 7. The main window of Application for the written injunctions.

3.1. The links between templates, instructions and injunctions in the Application

Instruction refers to the part of the injunction that applies to a particular section, multiple sections, or the entire route. In order to work with the instructions, you need to create and see a list of instructions. [4]

To create an instruction, you need to fill in all required fields. In total, these fields are 19 (Fig. 8). [12] [13] Their completion is conditional on the data stored in the application's root data. Each of the fields is bound to others, and some of them cannot omitted.

The creation of individual instructions is very complex. From the point of view of complex reconstructions of railways, instruction creation and forwarding between stations are acceptable. Likewise, the subsequent creation of written injunctions.

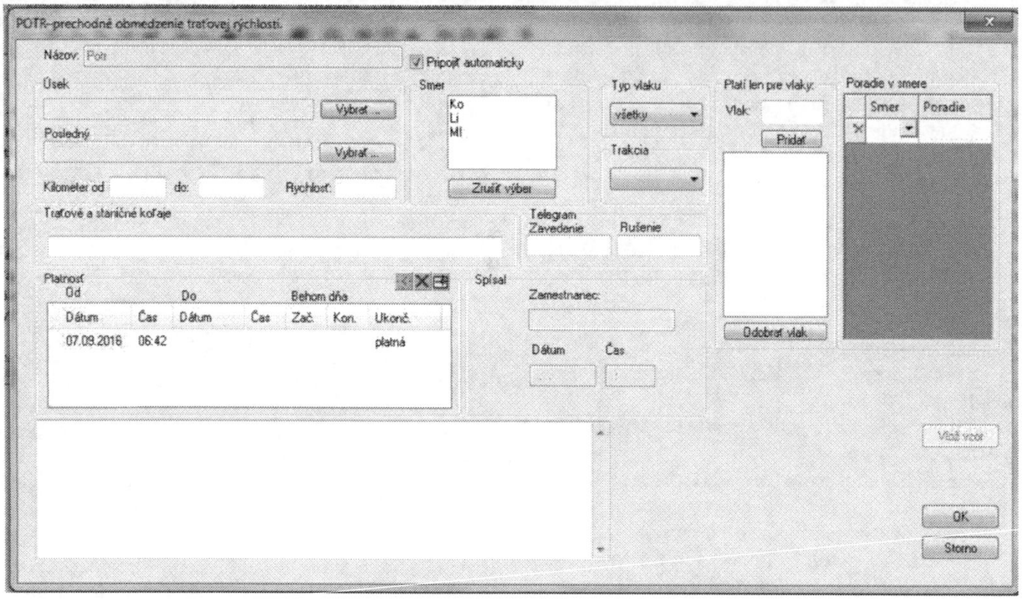

Fig. 8. The main window for creating the instructions in application.

4 Current problems with the application of injunctions and their possible solution

In cases where the train dispatchers deal with individual operating failures (level crossings, signal lights, etc.), the compilation process of writing orders from the instructions is very stressful and lengthy. The train dispatcher must inform all competent departments (operating dispatcher, etc.) from his position, manage the other train traffic, and create instructions for a new written injunction. All these tasks carried out in time stress, which caused precisely with a view to minimizing the delays of the affected trains. [14]

However, the users are also available with template's he previously created in the application. [4, 14] The compilation process of the written injunctions is most visible in the case of the Injunction for cautious driving - Part A and Part B very long and stressful.

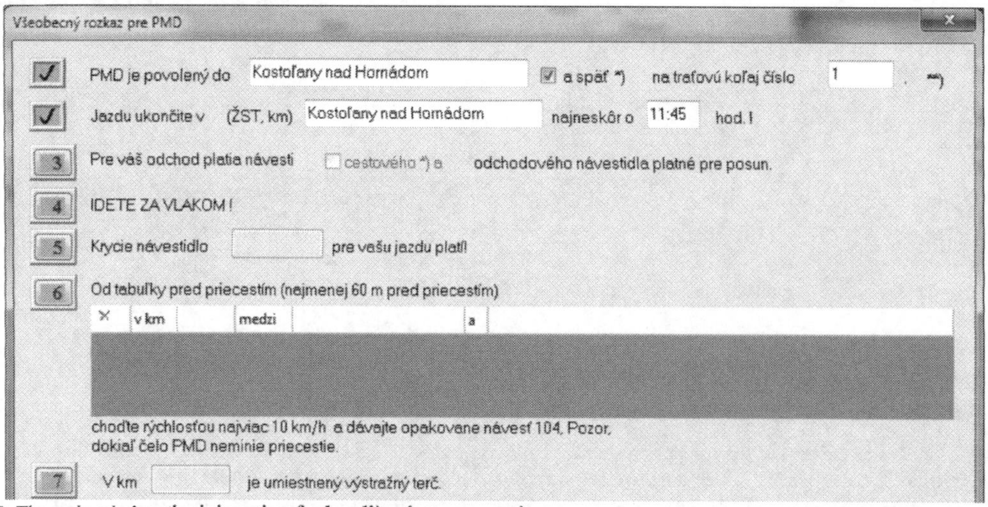

Fig. 9. The main window the injunction for handling between stations

The solution for the Injunction for cautious driving could be another type of the injunction - The injunction for handling between stations (Figure 9). [12] In this case, the operator completes with double-click to mouse the particulars applicable to the handling between stations.

The solution is the possibility that the operator should, in the case of the Injunction for cautious driving, the possibility of the template of this injunction. [7]

In the above-mentioned template, a simple double click would complement the specific kilometres positions of the level crossings. Based on the inserting the kilometres positions of the level crossing, shall be inserted to the injunction obligatory information automatically (from the root data of the application). [2] The compilation process of the Injunction for cautious driving would be very simple, user-friendly and minimize the stress factor too in its compilation technique. [14]

5. Conclusion

In the case of extraordinariness situations (delay, reconstructions activities, different routing of a train, etc.), which affect the fulfilment of train schedule, are applying so called "operational traffic management according to the applicable internal regulations". In the event of the occurrence of these extraordinary situations, it is necessary to compile a written injunction in order to inform the train drivers about these extraordinary circumstances. The compilation process of the written injunction is very complex and also very stressful and lengthy. The compilation process of the written injunctions is most visible in the case of the Injunction for cautious driving - Part A and Part B.

The solution for this Injunction for cautious driving could be another type of the injunction - The injunction for handling between stations. With all knowledge's from other written injunctions, the compilation process of the Injunction for cautious driving would be very simple, user-friendly and minimize the stress factor too in its compilation technique.

This research was financially supported by the project VEGA No. 1/0019/17 Evaluation of regional rail transport in the context of the region's economic potential focusing on the efficient use of public resources and social costs of transport.

References

1. R. Kampf, S. Lorincova, M. Hitka, O. Stopka. Sustainability, 9. (2017) DOI: 10.3390/su9091561.
2. E. Nedeliaková, J. Sekulová, I. Nedeliak. *Transport Means – Proceedings of the International Conference*. 236-239. (2015)
3. ŽSR *Regulation Z1 – Railway Transport Policy*. 523 pages. (2011).
4. O. Stopka, R. Kampf. Transport, 33, 280-290. (2018) DOI: 10.3846/16484142.2016.1174882.
5. ŽSR *Regulation SR 1003 – Operational Traffic Management at the ŽSR*. 27 pp. (2011)
6. B. Abramović, E. Nedeliaková, M. Panák, D. Šipuš. *Proceedings of 7th International Scientific Conference on Business Logistics in Modern Management*. 15-28. (2017)
7. J. Gašparík; B. Abramović, M. Halás Promet Traffic & Transportation. **27, 4**. 283-290. (2015)
8. G. Fedorko, V. Molnár, J. Strohmandl et al. *Transport Means 2015 – Proceedings of the International Conference*. 466-+ (2015)
9. L. Černá, A. Dolinayová, J. Daniš. *Proceedings of the 20th International Scientific Conference Transport Means 2016*. 414-419. (2016).
10. V. Ľupták, J. Gašparík, M. Chovancová. MATEC Web of Conferences. **134**. Article Number: 00033. (2017).
11. A. Dolinayová, J. Kanis, M. Ľoch, Procedia Engineering, **134**. 187-195. (2016).
12. B. Abramović, J. Gašparík, Jozef, V. Zitrický. Tehnički Vjesnik. **25, 4**. 1190-1195. (2018)
13. V. Zitrický, L. Černá, B.Abramović. Procedia Engineering. **192**. 994-999. (2017)
14. ŽSR *User manuals – Applications of Electronic Injunctions*. Version 3.95. 52. (2017)

* Corresponding author: juraj.camaj@fpedas.uniza.sk

Practical application of city logistics in the city of Trenčín

Milan Dedík[1,*], *Tomáš* Čechovič[1] and *Jozef* Gašparík[1]

[1] Department of Railway Transport, Faculty of Operational and Economics of Transport and Communications, University of Žilina, 01026 Žilina, Slovakia

Abstract. At present, mobility is an important part of everyday activities in cities. The increasing number of passenger and freight vehicles on the widening transport infrastructure raises several problems. The range of use of public passenger transport does satisfy enough the citizens needs and its use is constantly decreasing, especially in the new EU countries. One of the reasons is also the growing sale and production of new passenger cars in the EU, resulting in congestion in the road infrastructure of cities. These issues are dealt with the modern City Logistics concept.The paper is aimed to the basic principles of city logistics, the reasons for its development and, in particular, the possibilities of practical use. An example is the city of Trenčín in Slovakia, where a regional rail passenger transport is proposed for city traffic service including Park and Ride system.

Introduction

High traffic intensity level on the roads currently saturates a great demands on individual transportation. The congestion level is also reflected in the accident rate, which is still very high despite the various measures to reduce them compared to other European countries. Many traffic accidents in cities, as well as high traffic intensity, inadequate capacity of urban communications, result in the emergence of crisis situations. One possible tool for preventing crisis situations in cities is the concept ofCity logistics. This part of logistics is very important, because the movement, transport and handling of material are influential part of the logistical chain.[1]

1 City Logistics

The definition of City Logistics follows from the logistics definition: "Logistics is an interdisciplinary science that deals with the coordination, alignment, interconnection and optimization of the materials, information and fiannceflow in terms of customer satisfaction with optimal spending. "City logistics is the process of optimizing logistics and transport activities, involving various companies with the support of advanced information systems in the city with regard to the transport environment and its impact on the emergence of congestion, safety and energy savings. Thereforeone of the tasks of city logistics is to provide comprehensive transport serviceability.[2]

There are cities in Slovakia that are beginning to have problems with the traffic situation, although we can not fully compare the problems of our cities with the problems of major European and world cities. Cities try to handle these situations with different measures and drive as much traffic out of town as possible. Of course, most of these measures, especially construction, have a particular impact on transit traffic, when it is possible to route this kind of out-of-town traffic through the construction of by-pass roads. However, the problem remains with the organization of source and destination traffic that is heading directly to the city. It is precisely the solution to these problems in the cities that gave rise to the emergence of city logistics.[3]

City logistics is part of logistics that deals with issues related to the movement of goods flows and means of transport within the city. Thereforethe primary task is the organization of passenger and freight transport in cities and the organization of goods flows and flows of consignments entering or leaving the city. The main reason for this new direction in logistics was the increase in the number of vehicles in cities.

In the understanding of city logistics, related only to the freight sector, three stakeholders - the carrier, the self-government and the private sector - are involved. It is the triangle of city logistics. If we extend the scope of city logistics to the area of passenger transport, we also need to increase the number of participants who will be involved in the development of solutions. These are, in particular, passenger transport undertakings, police forces and, last but not least, citizens.[4]

The basic principles of city logistics have been elaborated in a number of expert and scientific papers and theses. A comprehensive approach gives Cisarova [5]:"Given that city logistics is perceived in a number of countries to be essentially negative and not conceptually addressed, it is appropriate to define the preparation and phases that should be preceded by the introduction of a quality global concept. First of all, it is necessary to

* Corresponding author: tomas.cechovic@fpedas.uniza.sk

resolve the issue of transit traffic. The city logistics itself solves or should solve only the problems arising from traffic-logistic service, not heavy freight transport. However, there are cities that have major problems with transit, but they are smaller towns with less than 50,000 inhabitants located on busy 1st and 2nd class roads. Transit traffic as well as transit traffic can be guided by traffic restrictions".

This issue is also addressed to a large number of transport experts and scientists. For example Julian Allen, Christian Ambrosini, Michale Browne, Daniele Patier, Jean-Louis Routhier and Alan Woodburn [6] propose a wide review of survey methods that have been carried in Europe, based on the conclusions of the BESTUFS II project. From survey work carried out with experts in 11 European countries, a comparison of urban freight transport data collection efforts is made to better understand what currently takes place and to identify examples of good practice. Authors observe that the extent of urban freight data collection varies significantly between the European countries, as existing urban freight data comes mainly from the disaggregation of national survey results. Finally, authors identify a set of gaps in data collection, as well as the need for greater standardization in data collection methods and in analysis and reporting of this data.

Jesus Gonzalez-Feliu, Routhier, J. L et al [6] propose a statistical-based modeling approach to propose a data estimation tool that can be transposed to different cities, avoiding the needs of making very costly surveys. The proposed framework needs standard inputs able to be obtained by public authorities and/or private stakeholders to make a diagnosis of urban logistics in current situations. The joint process of collecting data and modeling is described, and the different modules of the framework are presented. The authors conclude by presenting the main applications and further developments.

However, at present, it is important to address, in particular, the solution of city logistics in passenger transport. Therefore, it is necessary to be able to correctly define the reasons for the development ofthe city logistics and especially the opportunities of its use in practice.

1.1 Reasons for creating the city logistics

There are a number of reasons why city logistics is becoming more and more important. These include[7]:

- Environmental: the negative impacts of transport on its surroundings are generally known (noise, exhales, soil cover, etc.). By proposing effective solutions to urban and freight transport in cities, these negative impacts can be reduced;
- Transport: the road network in the cities has been diminished during its construction to a lower capacity. Today, however, the traffic flow is far higher than previously thought. Therefore, measures in city logistics can at least partially solve this situation;
- Living standards in the city: the city is a place for its inhabitants to provide not only their living needs but also their social needs. Improving the living standards of city residents can greatly help to improve the traffic situation, mitigate negative environmental impacts, or reduce traffic in the chosen area or time.

1.2 Opportunities for solving urban problems in practice

Solving problems related to increasing the number of vehicles in cities can be solved in a number of ways, but finding an optimal solution is not easy. City Logistics offers a lot of suitable logistics concepts for passenger and freight transport optimization, which usually consist of one or more combinations of the following city logistics approaches[8]:

- Vehicle entry restrictions or permissions; depending on the type of vehicle (size, weight, amount of produced emissions), vehicle entry is permitted or prohibited in a particular area. Mostly, the weight limit applies to the total weight and width of the vehicle or to certain emission limits.
- Paid communications and toll-road fee system; in this case, to introduce tolls for selected types of vehicles that are interested in entering the marked part of the city that is subject to the city toll system [9]. This solution can reduce the number of unacceptable vehicles on toll roadson the one hand, and it can be beneficial for the city budgeton the other hand,
- Night supply; when the supply is done at night, the vehicles are avoided by the daily traffic peaks while not feeding them. This results in benefits such as shorter driving times, the possibility of using larger vehicles to replace several small ones, reducing emissions and fuel consumption, fewer roads,
- Distribution of goods by other means of transport. By introducing this measure, a modal shift of the goods to ecologically acceptable modes of transport should be achieved, and consequently the less number of journeys,
- Making urban transport more attractive; only the high-quality, fast-moving urban public transport at regular intervals should motivate citizens to prefer public passenger transport to individual motoring.
- The use of the urban rail transport kindsas a main means of transport; in large cities with built-up tramways, overground and underground railways (metro), this mode of transport should form a transport system and should provide passengers an attractive concept of transport.
- Utilization of the rail network (or private sidings) for city traffic service. Several railway lines in many cities pass relatively close to the center and there is a low number of trains. There are also unused private sidings. [9] In this case, there is a possibility of the new trains intoduction as public transport, whereby these trains will use the existing regional or main lines orprivate sidings.

- Support for the walking and cycling by building of the paths and cycle tracks.

2 Practical use in the city of Trenčín

The city of Trencin lies in the western part of Slovakia in the Trenčianska highlands, which extends along the valley of the river Váh. I tis divided into four major city districts – the middle, south, north and west, which are further decomposed into smaller units. According to the population survey in January 2018, Trenčín had a total of 54,916 inhabitants. It is possible to rank it on the 8th place among 140 Slovak towns.

There is a problem with an excessive increase individual car transportation in the city of Trenčín and thus also congestion. Instead of creating alternatives for the city to create a parking space on the fringe of the city before entering its center, the large number of cars park directly in its narrow center and core, resulting in an increase in the number of cars in the city center. Although the city charged almost all the parking places in the center and close the city center, the problem has not been solved yet.If there is no option to park the motor vehicle in the parking garages on the periphery, the drivers will have to park their vehicle in the city center.[10]

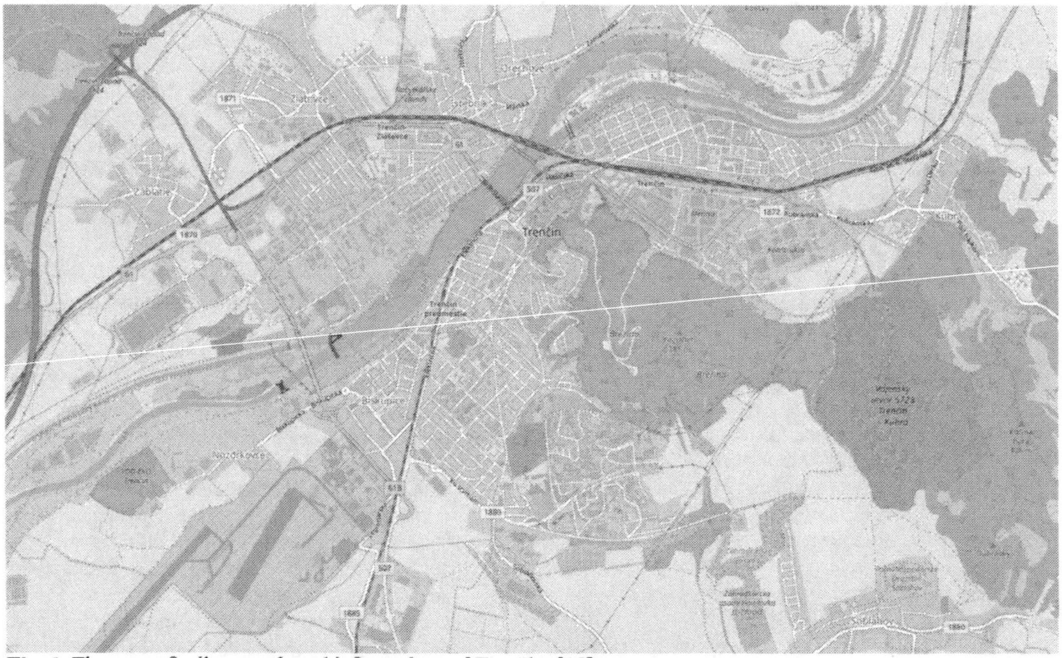

Fig. 1. The map of railway and road infrastruktur of Trenčín: [15]

2.1 The proposal of solution by using rail transport

In accordance to the city logictic concept is proposed to extensive usage of rail transport. A suitable way to solve the present situation is to use the existing railway lines or private sidings passing through the city of Trenčín and ensure their operation at dense intervals, thus creating the "Urban railway transport". This should create an alternative way to go through the city by a private car in order to connect the city center with its other parts more efficiently.

Start of the route is proposed on the rail leading to the area of military siding on the estate housing „Pred Poľom“ and continues through the railway station Trenčín, further down the railway line no.143 through the stop „Trenčín –predmestie“ to„Trenčianska Turná“ as the final stop where the train units will change direction. The private siding belongs to the Slovak Army, there is an average of 1 train per year, and its use is very problematic but possible.

The proposed route should cooperate together with the proposed parking and applying appropriate measures, such as the cost of using these car parks, which should include the ticket to proposed train. In addition to the priority objective, an alternative to public transport buses will also be created along the entire route. Thus, not only the city's inhabitants, but also the entrants can go from one side of Trenčín to another or to the city center and back more faster.

The calculation of the interval run from the new „Pred Poľom“ stop to „Trenčianska Turná“ station at the intended average speed of 35 km per hour and the total distance 6,720 meters, which is considered with 30 seconds additional time for the motor starting run and stopping, and 30 seconds for the dwell timein 10 of 12 stops during the whole ride, the result of running timeis 20.5 minutes. It means the train units should ensure the entire territory traffic servicetheoretically over 20 minutes. Also they would do three backruns during one hour if they did not give priority to any other passengersor freight train.[9]

There are proposed total 12 stops in the given order, with 3 already existing:

Pred poľom(A) – starting or the final stop, which could be built at the estate housing, on the private siding leading to a military area. It should ensure transport services to this city part and a small part of Kubra.

Merina (B) – a stop just before the railway station Trenčín. Its name is derived from the nearby Merina building, but it can ensure transport service in the Sihoť city part. It has an important strategic location because of the big parking place.

Railway station (C) – one of the already existing stops for this route, its strategic location would make it possible to ensure transport service in the most of Sihoť I. part and part of Milan Rastislav Štefánik Park to supermarkets Billa and Lidl as well as the available MERINA area.

Hasičská (D) and Zátoka pokoja (Bay of Peace) (E) – these two stops are the closest and the most accessible to the historical center of Trencin. The entire center and the beginning of Kolar's quarter are available from these stops. There is a subway crossingat the Hasičská streetthat would serve the passengers for easy access to the stop. Similarly, it is also at the stop of the Bay of Peace, where there is already an existing level crossing.

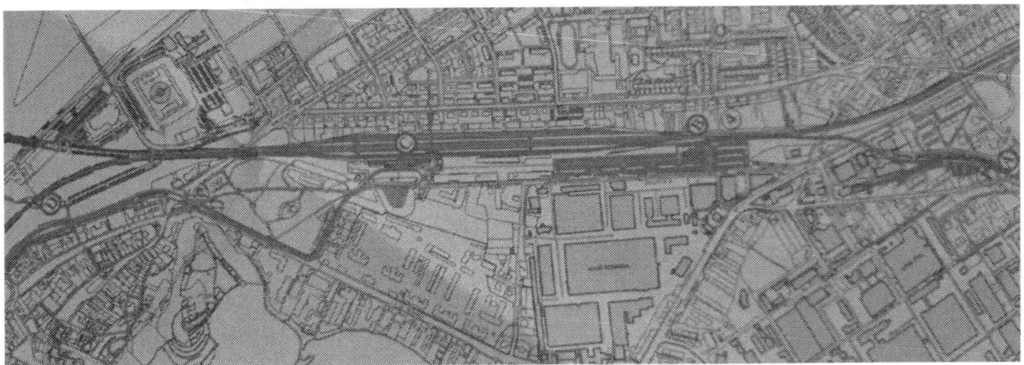

Fig. 2. Stops A – E and car park next to B stop, Source: [9]

Predmestie (Suburb) (F) – one of the existing stops currently used by passenger trains near the clinic and the District Court. There is a subway crossing at the stop. In addition there is proposed to build an overtaking station. There will be ensured safe and crossing of the both trains.

Hospital (G)–the location of this stop is near to the hospital and the city part „Noviny", where is the subway crossing.

Noviny (H) and Kaufland (I) – these proposed stops sholud ensure transport service od the area from the hospital to Kaufland, as well as residential areas in the Legionary Street and Biskupice district.

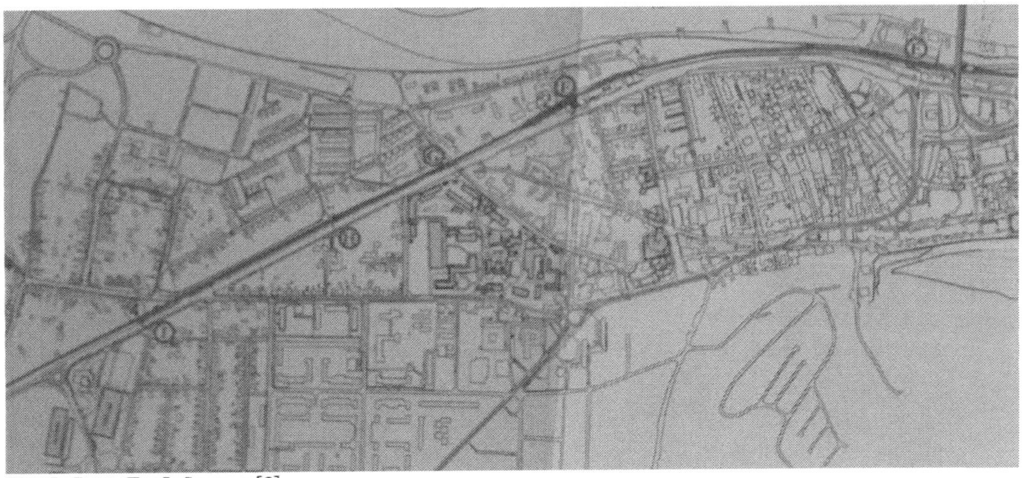

Fig. 3. Stops E – I, Source: [9]

Diaľničný privádzač (Highway feed) (J) – there is no residential area or large residential buildings. Therefore, it is primarily intended for the arrival and exit of passengers who will come from motorway and will parked their vehicles in a parking place and consequently go the city center. There must be build an upperbridge over the route no. 507 because of safe and easy passenger access to the stop.

Tesco (K) – the stop is close to the Tesco and Laugaricio shopping centers.

Trenčianska Turná (I) – the final stop of this planned route in the figure depicted by the letter "L", The train units will change their travel direction at this railway station. It will also ensure transport services of the entire Trenčianska Turná and to transport passengers to and from the city center, who have parked their vehicles at the proposed parking places next to the railway station Trenčianska Turná.[9]

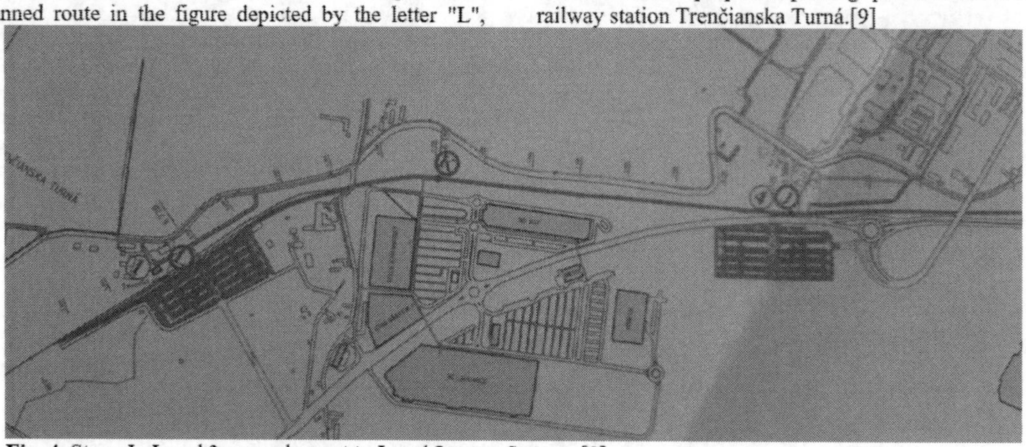

Fig. 4. Stops J –L and 2 car parks next to L and J stops, Source: [9]

2.2 Usage of the Park and Ride transport system

One of the most widespread problem-solving systems in the city is Park and Ride (P & R), "park and go by public transport". The P & R system is a typical application of intelligent transport systems. It includes parking places and other facilities that enable the interconnection of individual and public transport and combine their advantages. The basic task is to enable an advantageous transfer to urban public transport, thus reducing the number of passengers to the city center by individual transport.

On the proposed transport route is important to build three catch car parkings in front of the center of Trenčín. The first car parking (A) would serve for cars coming from Banovce nad Bebravou. The parking is designed in the area between the railway and the existing local road, located at the railway station at the beginning of Trenčianska Turná. The second car parking (B) should be built at a downhill run from the D1 highway, which should primarily serve to shut-down cars from the D1 motorway and to offer them the possibility of parking in the contact with rail transport.

Another proposed car parking (C) should be next to the Trenčín railway station, in the unused warehouse building area. This strategic location of the car parking should enableto be parked the vehicles coming from Trenčianska Tepláand to offer the possibility of parking in the contact with rail transport too. [11, 12]

2.3 The proposal of the possible timetable

When timetable construction,passenger trains for transport services and thus serving as a train service for public transport, it is very important to offer as many connections as possible in dense time intervals. This philosophy is also respected in the construction of passenger trains for transport services in the city of Trenčín. However, in order to maintain the line and station intervals on the line section after the overtaking stationconstruction at the stop, it is possible to ensure a 22-minute full-day interval of the proposed passenger

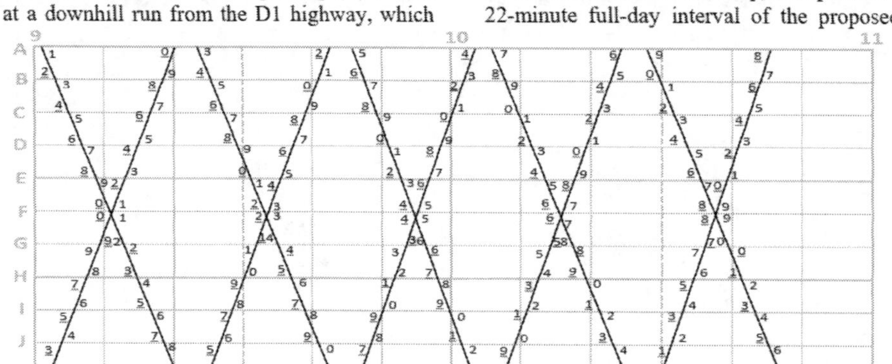

Fig. 5. A preview of two hours period train traffic diagram, Source: [10]

trains. Crossing of the passenger trains is ensured at the overtaking stationTrenčín – Predmestieand at Trenčianska Turná station. Running times on the route are recalculated for the 813/913class of the train unit. A preview of a possible train traffic diagramfor two hours periodis shown on the Figure 4.[13, 14]

2.4 Benefits and risks of the proposal

In the final assessment of the proposal, it was found that the implementation of the proposal should result in a significant reduction in individual car traffic in the city, thereby reducing congestion. Also the cancellation of some car parks in the city center should be replaced by greenery, which would reduce the number of vehicles in the city, reduce greenhouse gas emissions, and thus improve environmental impacts. After building a overtaking station, train units should drive for a minute accurately and on time. It is a very important criterion in terms of time aspect of choosing the right means of transport to the city center.

The only possible risk and collision point within the entire planned section is the level crossing of the line leading to the private siding object, to the „Pred Poľom"stop and Gen. M. R. Štefánik street. It's the only place to meet a passenger car with the railway vehicle. A solution for this risk would be to ensure this collision point by light signaling and barriers. The next collision point is also possible to consider the overtaking station where passenger and freight trains will be crossed, but the correct timing of the timetable with regard to the current range of train traffic can help to avoid the problems.[15]

Conclusion

At present, mobility is a part of everyday activities in cities. The increasing number of passenger and freight vehicles on the widening transport infrastructure causes a number of problems that can not be solved alone, but also in terms of impacts on the whole area. The simiral problems are in the city of Trenčín. The measures, that have been taken so far, have not been able to disburden the city from frequent congestion and traffic collapse. These problems preclude to improve the transport service in the city.

Due to this persistent unfavorable situation in the city of Trenčín, the purpose of the contribution was to propose measures for the use of regional passenger railway transport for the city transport service. To meet this goal, specific measures have been proposed for the deployment of the transport system and technology together with a regional rail transport. For Trenčín, the proposal brings new alternatives to solve the transport problems and to reduce the intensity of individual car traffic, thus relieving the city center from a large amount of road traffic. At the same time it supports public transport systems and also the use of regional rail transport, thereby improving the ecological and environmental aspects of the city.[12]

The proposed concept that is being considered in Trenčín, should become a key global technology in urban logistics. It should help optimize passenger transport in cities and tackle global transport problems.

The paper is supported by the VEGA Agencyunder Project 1/0095/16, 'Assessment of the quality of connections on the transport network as a tool to enhance the competitiveness of public passenger transport system', that is solved at Faculty of Operations and Economics of Transport and Communication, Universityof Žilina.

References

1. D. Vidriková. Crisis Management, **9. 2**, (2010), online available: http://fsi.uniza.sk/kkm/files/admincasopis/KM%202%202010/ODBORNE/Vidrikova.pdf
2. M. Kendra, J. Mašek. Perner´s Contacts, **3**, no. **3** (2008)
3. L. Pečený, J. Gašparík, V. Gáborová, *ICTTE Belegrad 2016: Proceedings of the 3rd International Conference on Traffic and Transport Engineering* 462-466, (2016)
4. T. Skrúcaný M. Kendra, M. Škorupa M. *Conference: 12th International Scientific Conference of Young Scientists on Sustainable, Modern and Safe Transport*, 806–81, (2017)
5. H. Cisarova *"Modelling of transport services in the selected area using city logistics."* in Czech, Dissertation thesis, University of Pardubice, (2012)
6. C. Ambrosini, & J. L. Routhier, Transport Reviews, **24(1)**, 57-77. (2004)
7. M. Dedík, J. Gašparík, Z. Záhumenská, *Logi 2017, Conference Proceedings, 18th International Scientific Conference*, **134**, (2017), DOI: 10.1051/matecconf/201713400009.
8. .J. Blašković, B. Abramović, D. Šipuš,. International Journal for Traffic and Transport Engineering (IJTTE) **7(4)**, 430-442 (2017)
9. T. Čechovič *"Advisement of the use of a regional railway transport on the city Trenčín"* in Slovak, Diploma thesis, Žilinská univerzita v Žiline, Slovakia, 2018
10. M. Kvizda, Rewiew Of Economic Perspectives. **17**, **3** 331-334. (2017)
11. TRENCIN-INAK.SK, 2018. Online available: http://www.trencin-inak.sk/actuality.php
12. L. Bartuska, V. Biba, R. Kampf. *Proceedings of the Third International Conference on Traffic and Transport Engineering.* (ICTTE), 900-904 (2016)
13. Š. Kudláč, J. Majerčák, P. Majerčák, Transportation Research Procedia, **28**, 45-50..(2017)
14. TRENCIN.SK, 2018. Online available: https://trencin.sk/pre-obcanov/o-meste/mapy-mesta/

Computer simulation of railway wagon brake disc casting

Peter Futáš[1] , Alena Pribulová[1], Gabriel Fedorko[2], Vieroslav Molnár[3] and Nikoleta Mikušová[2]

[1]Technical University of Kosice, Department of Metallurgy, Park Komenského 14, 040 01 Košice, Slovakia
[2]Technical University of Košice, F BERG, Logistics Department, Park Komenského 14, 040 01 Košice, Slovakia
[3]Technical University of Košice, FVT, Department of Computer Aided Manufacturing Technologies, Bayerova 1, 080 01 Prešov, Slovakia

Abstract. CAD systems serving as the model for real industrial processes are almost an inseparable part of manufacturing in many industry branches. The use of simulation programs in the foundry industry is perceived as a tool for the visualization of the processes involved in the production of castings. They are also used to resolve possible issues and imperfections before the production begins. Consequently, it enhances the quality and reduces product-manufacturing-related financial costs. The paper tackles the issue of how to resolve the occurrence of defects – shrinkages in the railway wagon disc casting while using a simulation program. The 3D model of the casting and of the system of gates was constructed in the CATIA CAD system. The mere simulation of the casting and solidification process was done in the NovaFlow&Solid program. After evaluating the results of the simulation, the correction of the casting and gates was made in order at once to enhance the quality of the casting and reduce the production costs in the manufacturing of this casting. In the production of the brake disk casting, the main issues are little usage of molten metal during casting and the considerable occurrence of shrinkages, which causes the increased economic costs of its production.

1 Introduction

The simulation programs and simulations performed in them are currently common in the foundry and other industrial enterprises. They have a wide scope of usage since during the development of a particular product or process computer simulations can demonstrate a very detailed course of component production or of a process that is not actually observable under real conditions. It allows designers to look into the ongoing processes, to identify with high precision deficiencies and errors that could occur in the real production, and to modify them to ensure quality enhancement and the reliability of simulated processes.

Rail transport plays an important role in different types of logistics systems and in a wide range of industrial branches. Its operation is closely linked to high demands on safety and reliability. In order to meet the requirements, more attention needs to be paid not only to the operation of individual components and equipment but also to their actual manufacturing. The most important elements that are subjected to extreme stress during operation are railway wagon brake discs. They are a key factor in ensuring safe operation and are produced using gravity casting technology. One of the main factors for the proper functioning of a brake disc is the quality of the original blank casting.

The resultant quality of the casting is influenced by many factors and foundry principles, such as non-directional solidification, risering, and cooling. Another important accompanying phenomenon is the shrinkage of the alloy during the casting process. This starts when the cast alloy is in the liquid state and lasts till the ambient temperature is reached. The reduction or change in volume that occurs in the liquid state is called shrinking, a distinctive foundry property of the material [1], [2]. The changes in the casting volume during the solidification process, resulting in the creation of shrinkages, are a natural phenomenon that cannot be avoided. Shrinking is a phenomenon that involves the reduction of the casting size during the transition from the liquid to the solid state. This phenomenon also causes that cavity shrinkage and casting porosity occurs. [3].

Therefore, it is important that the coarser and more massive parts of the casting rise upwards so that the coarser parts could be equipped with risers that would more easily ensure directional solidification [4], [5], [6].

2 Modeling and simulation of casting

In the production of high-quality castings, it is now necessary to use modern computer systems that can evaluate all the influences affecting the quality of the casting produced before its introduction into production.

Using computer simulation, we can find out how the mold cavity will fill and how the molten metal will solidify. From the filling and solidification simulation results, we can evaluate whether the system of gates and risers provides for directional solidification and also

* Corresponding author: gabriel.fedorko@tuke.sk

whether it ensures the production of a defect-free casting. With computer simulation, it is easy to change the system of gates and risers without the need to produce a new one. Simulation also allows us to detect thermal stresses in castings [7].
Try to ensure that lines are no thinner than 0.25 point.

Fig. 1. Casting of the railway wagon brake disk.

The use of computer simulation facilitates rapid resolution of technological problems already in the pre-fabrication phase of the casting process, methodical elimination of technological errors, shortening of the preparatory time, and reduction of production costs.

To create a 3D model of a railway wagon brake disc casting (Fig. 1) as well as that of the system of gates and risers, CATIA V5R20 program was used (Fig. 2).

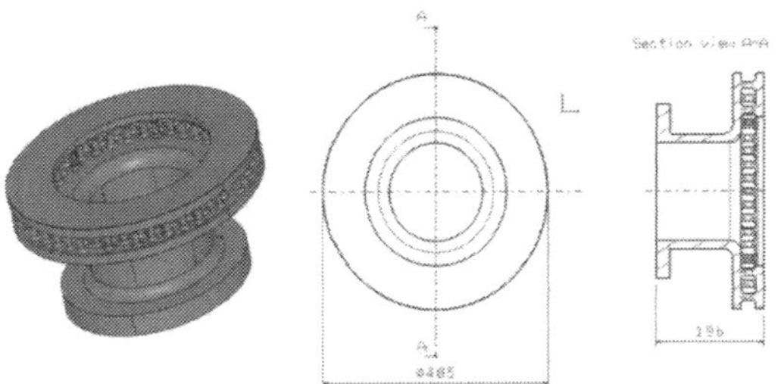

Fig. 2. 3D model of brake disc.

Fig. 3 shows a complete assembly of molded casting and a gate system with a single gate.

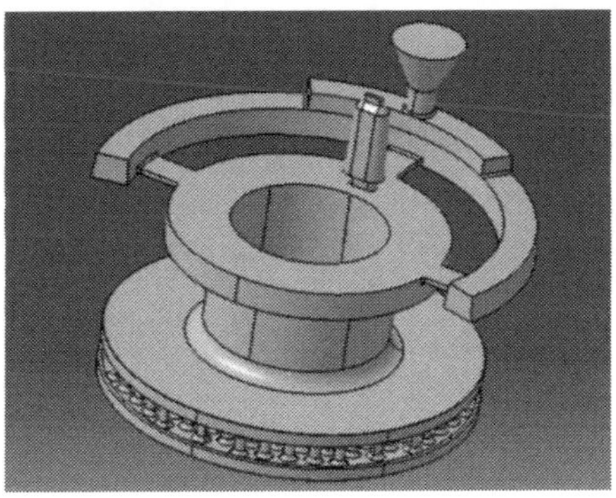

Fig. 3. 3D model of brake disc with a gate system.

The created assembly of the casting and the gate system was subjected to the computer simulation of casting and solidification processes using the NovaFlow & Solid program. This simulation program from Novacast, Sweden, is being used, thanks to its simplicity and speed, to determine the course of filling and solidification of the molten metal in the foundry form, to find out where the foundry defects such as shrinkages and porosity can occur. This program is based on the Control Volume Method, which ensures accurate and fast creation of the simulation network. When creating a simulation network, the basic rule is that at least two cells need to be present in the narrowest cross-section. The finer a simulation network is, the more precise mere simulation is; however, the time needed to carry out the simulation is prolonged. In Fig. 4 is a close-up of a brake disk simulation network.

Fig. 4. A close-up of brake disk simulation network.

The input values that were used to simulate casting and solidification processes are given in Table 1.

Table 1. Input values to simulate casting and solidification processes.

Casting material	EN-GJS-500-7
Mold material	Silica sand
Casting process temperature [° C]	1390
Mold temperature [° C]	25
Air temperature in the mold	25
Filling rate [kg.s^{-1}]	0.5
Moisture of sand mixture [%]	3

3 Analysis of casting and solidification simulation results

Fig. 5 shows the fill rate at 20%, 40%, and 95%.

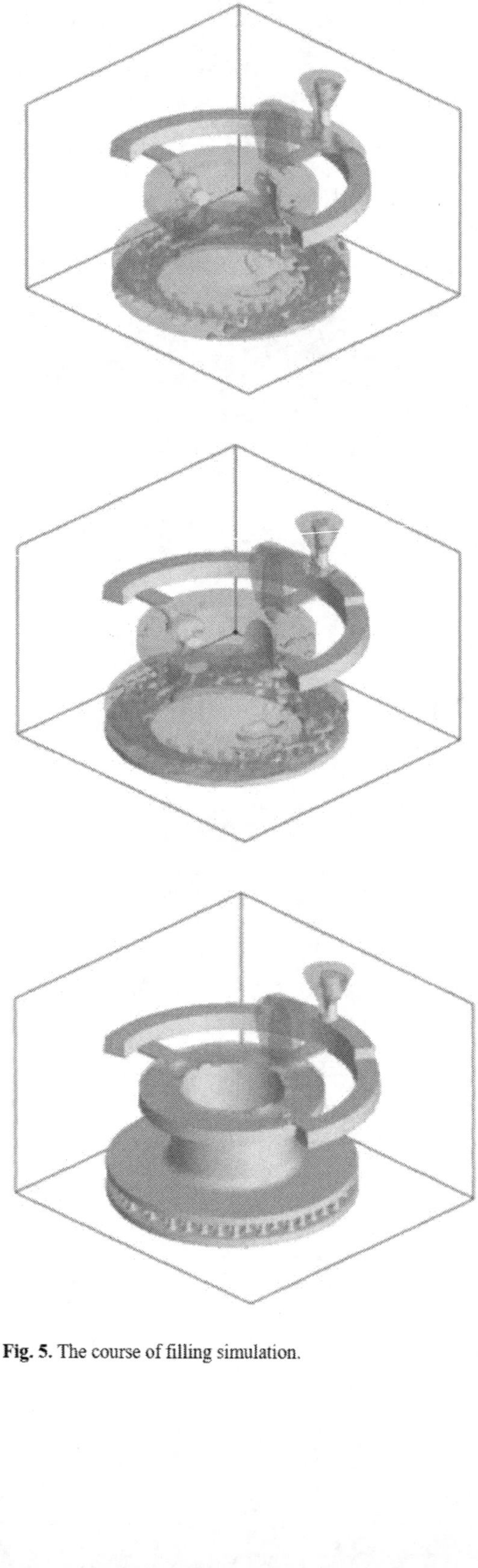

Fig. 5. The course of filling simulation.

From the filling simulation, it can be stated that the filling was running smoothly without spraying, which is a positive factor in filling the mold cavity. After the lower part of the casting was filled, the casting began to solidify from the distant places towards the gate system. The solidification process is shown in Fig. 6. Figures show the course of solidification at 5%, 20%, 40%, and 80% moltenness. The casting solidifies as the last one and the shrinkage present does not interfere with the casting, but is formed in the riser, Fig.7. In case the shrinkage interferes with the casting, it would be necessary to adjust the size of the riser to eliminate it.

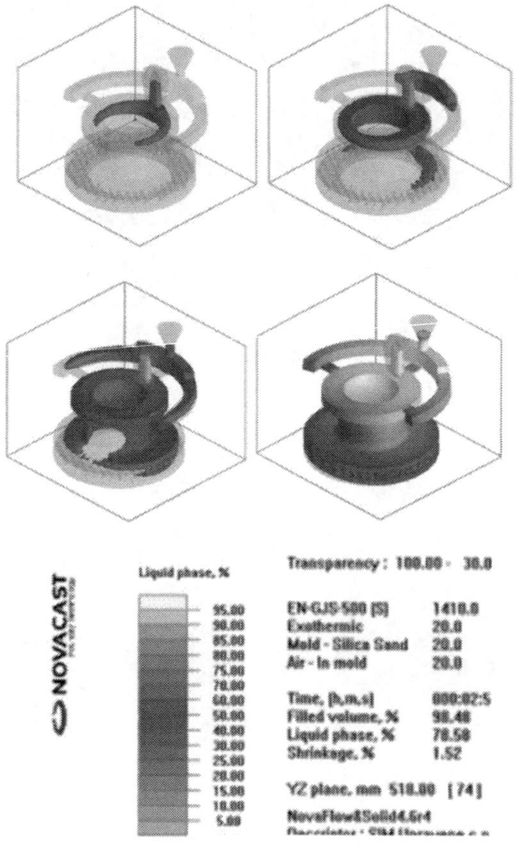

Fig. 6. The course of the simulation process.

4 Conclusion

The computer simulation of foundry processes brings countless benefits to casting production, and consequently cost savings when introducing a new type of casting into production. It offers the possibility of verifying the proposed gate system and the technology of the casting construction and its suitability for the casting process.

As the simulation of the casting and solidification of the railway wagon brake disc casting shows, the designed gate and riser system is sufficient and the casting solidified without adverse shrinkages and pores that would interfere with the casting. Also filling the mold cavity was smooth and would not cause erosion of the mold in the casting process.

The results obtained served to actually produce the casting and coincided with the results obtained in computer simulation.

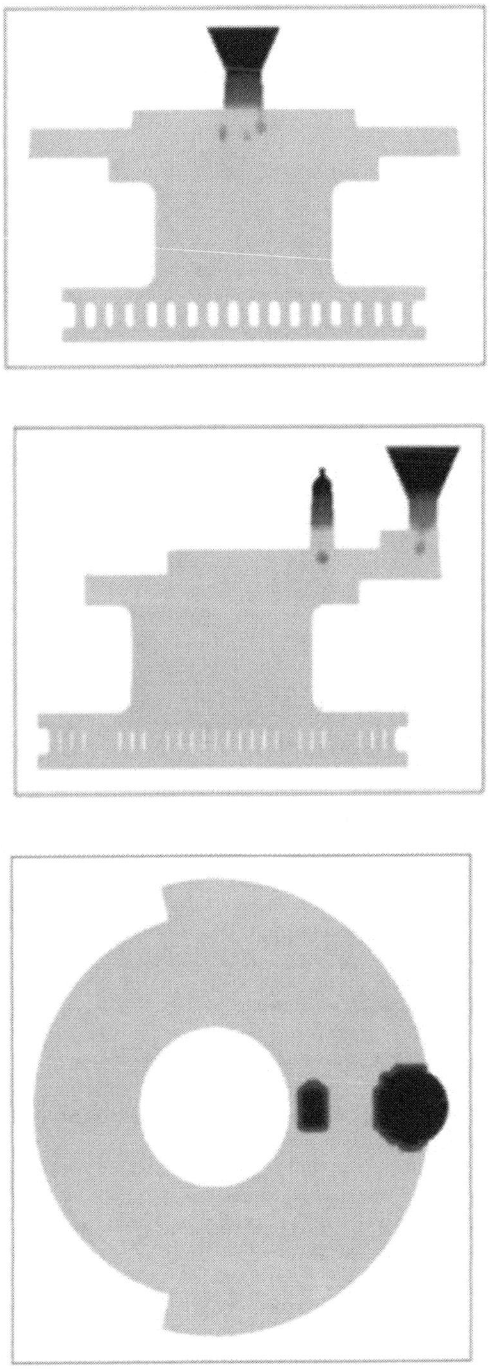

Fig. 7a. 3D model of brake disc with a gate system.

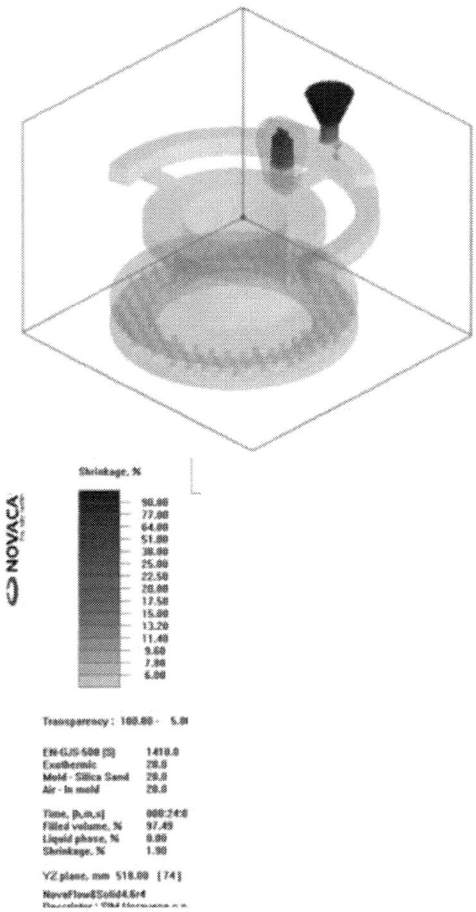

Fig. 7b. 3D model of brake disc with a gate system.

The present research is part of the following research grant projects: VEGA 1/0073/17, VEGA 1/0063/16, VEGA 1/0403/18, VEGA 1/059/19, KEGA 018TUKE-4/2016.

References

1. M. Holtzer, R. Danko, M. Gorny, Int. J. Cast.Wat. **29**, 9 (2016)
2. C. Labrecque, M. Gagne, Can. M. Q. **5**, 36 (1998)
3. C. Yeung, H. Zhao, W. Lee, Mat. Char. **40**, 8 (1998)
4. T. Skrzypczak, W. Węgrzyn-Skrzypczak, Arch. Found. Eng. **10**, 6 (2010)
5. K.M. Pedersen, N. Tiedje, Meas. **41**, 10 (2008)
6. P.K. Basuktar, C.R. Loper Jr., C.L. Babu, AFS. Trans. **78**, 8 (1970)
7. P. Futas, J. Imrich, I. Vaskova., G. Fedorko, V. Molnar, P. Kacmary, J. Sc. R. Prod. **2**, 6 (2013)

Process of modelling the freight wagon multibody system and analysing its dynamic properties by means of simulation computations

Ján Dižo[1,*], *Miroslav* Blatnický[1] and *Alfréd* Pavlík[1]

[1]University of Žilina, Faculty of Mechanical Engineering, Department of Transport and Handling Machines, Univerzitná 8215/1, 010 26, Žilina, Slovak Republic

Abstract. Computational modelling and simulations are nowadays widely used for design of railway vehicles, analysing and assessment of their dynamic properties and also for detection and prediction of the mechanism of deterioration and sources of damages. Therefore it is necessary to develop the adequate virtual model of an analysed railway vehicle as well as a track. For these purposes many simulations software packages are developed. In these computational programs a railway vehicle is modelled as a mechanical system usually consists of many bodies called a multibody system. In this paper the process of computational modelling of a freight wagon multibody system is outlined. Moreover, there are presented also results of dynamic simulation of a freight wagon running on a real track section. The analysed intermodal freight wagon was equipped with Y25 bogie. The freight wagon model, simulation analyses and ride properties evaluation were performed by means of a commercial simulation program.

1 Introduction

The design of a railway vehicle has to meet strict requirements for its commissioning in real operation conditions. Production and operation of railway vehicles is closely linked to economic factors. Results of simulation calculations significantly contribute to the estimation and prediction of railway vehicles behaviour. Implementation of virtual reality tools requires creating a representative virtual model of a railway vehicle, which sufficiently and accurately reflects its essential mechanical properties. Simulations of railway vehicles can be performed using some commercial software working with a railway vehicle model as a system of mass points, which represent individual railway vehicle body interconnected by joints, couplings, constraints etc.

The issue of investigation of railway vehicles dynamic behaviours is still current especially in terms of development of new railway vehicles and their approval in operation, analysis of the interaction of a railway vehicle and track as well as prediction of railway vehicles behaviour running on a track according to the changes of a railway vehicle design or a track.

2 Modelling the freight wagon multibody system

The production of a new railway vehicle, i. e. also a freight wagon, is consists of: the design of a prototype, development phase, optimization of a freight wagon model, construction of a freight wagon, verification and validation of a freight wagon. The development phase of a new railway vehicle implies not only static analysis but also dynamic analysis in order to obtain data about its dynamic properties.

Phases of development and optimization of railway vehicles use different ways of analysing the dynamic and ride properties of a railway vehicle. Computational simulations of a railway vehicle running deal with dynamic of railway vehicle multibody system (its reaction on a track excitation) [1-5].

After the process of preparing and creating the railway vehicle multibody system can be performed with it a multibody simulation. What is the multibody simulation?

The multibody simulation, also known as multibody system simulation, is used for prediction and optimisation of dynamic behaviour of any type of multibody system by solving the equations of motion [6, 7]. The name multibody can refer to a wide range of systems, thus of course, a railway vehicle mechanical system.

The bodies of a railway vehicle multibody system are linked by means of joints and other kinematic constraints, which allow certain motions and restrict others. Bodies can be rigid or flexible. The system is completed by excitation elements and sensors for measuring the desired outputs [8]. Dynamic properties are described by motions of a railway vehicle system due to the applied forces and the inertia characteristics of bodies, i.e. their mass, moments of inertia and position of the centre of gravity.

* Corresponding author: jan.dizo@fstroj.uniza.sk

The railway vehicle multibody system approach, with the development of computer technology, is clearly an added value in the analysis and design of railway vehicle [9].

Within a railway vehicle simulation the contact mechanics entries into calculation as well [10]. Analysis of the contact mechanics is used for the investigation of the wheel/rail phenomenon conditions [11, 12], such as creep forces, stress distribution, rolling contact fatigue, wear etc. [13, 14].

Computer modelling the railway vehicles allows for a user to test out various situations without spending the time, money and use of equipment to test them on a track [15, 16]. Modelling can predict conditions of derailment situation, i. e. what speed derailment of wheel/rail can occur at or under which conditions it may be prevented [17-19].

In the field of modelling a railway vehicle, during last four decades, computer hardware and software are highly developed and used to simulate the dynamic relations. This simulation programs help universities, railway organisation and train manufacturing to deal with own related issues and project such as reducing the number of (expensive) field measurements [20, 21], parameter studies on the dynamic behaviour of the railway vehicle, predicting and reducing the time and cost of increasing the railway vehicle performance like increasing speed of the vehicle, increasing axle load.

Using modern computer software there is possible to carry out realistic simulations of dynamic behaviours of railway vehicles. The theoretical basis of the mathematical modelling used are mature and reliable, therefore current simulations tools are often originally written by research institutes to powerful, validated and user-friendly software.

There have been developed several computer software to help engineers simulating the dynamic properties of railway vehicles. These multibody tools have been created to have more user friendly environments with more graphical features, animations and simpler coding. All railway simulation tools, more or less, have the same procedure. Simpack, VI-Rail (based on Adams/Rail), Universal Mechanism and also Vampire, Nucars and Gensys are recently most used software in the field of railway vehicle multibody simulations [22-24].

The setting up a mechanical model of a railway vehicle is the first step in every railway vehicle multibody simulation. Such a model constitutes the base for its mathematical description using the equations of motion, obtained with the aid of physical principles and laws (Newton's law, etc.). The multibody system approach is the powerful and the most widely used method for this procedure, especially if the vehicle's running behaviours have to be analysed.

As it was mentioned above, a railway vehicle multibody system consists of rigid bodies (eventually flexible body) interconnected via massless force elements, joints or similar constraints. Due to relative motions of the system's bodies, force elements generate additional forces and torques. Springs, dampers or actuators combined in primary and secondary suspension of a railway vehicle are typical examples of such force elements.

To take the flexibility of lightweight structures into account the interface between the multibody software and Finite Element software has been implemented in principle into every multibody software.

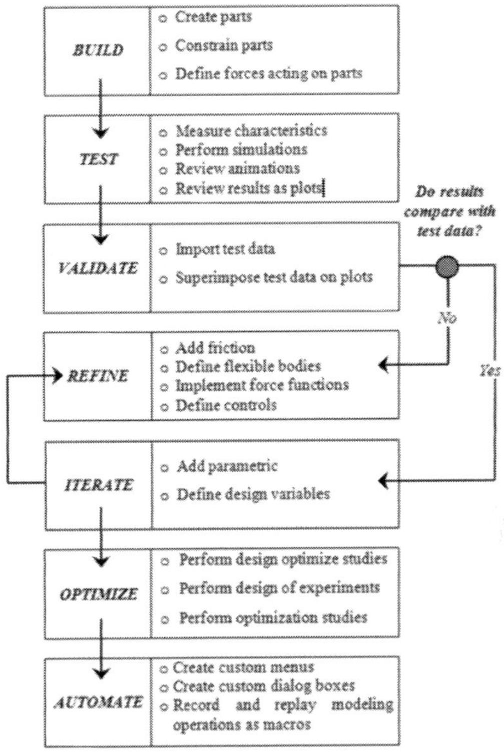

Fig. 1. Process of modelling and simulation of a railway vehicle

Modelling of an entire railway vehicle and a railway track system is divided into three main parts:

- vehicle model,
- track model
- wheel/rail contact.

In view of a railway vehicle system and track subsystem, forces generated by wheel/rail contact can be regarded as external forces. The excitation of vehicle and track leads to motions. The corresponding kinematic quantities are fed back to the wheel/rail contact.

With the advent of powerful computers simulation of complex mechanical system has become a real possibility. The railway vehicle computational model can be constructed and run on typical or measured track in a virtual environment and a wide range of possible designs or parameter changes can be investigated.

Inputs to a railway vehicle multibody system are usually considered in each wheelset. Typical inputs are vertical and lateral track irregularities and deviations in gauge and cross level.

Outputs from the model can be set up to provide accurate predictions of the dynamic behaviour of a railway vehicle and its interaction with a track.

The rail transport remains the most cost effective method for moving freight between two locations connected by land. This is due to the low energy loss of metal on metal contact between wheels and rail.

2.1 The multibody model of a freight wagon

The typical freight wagon model consists of three main subsystems: the body of wagon, the front bogie and the rear bogie. The body of a freight wagon, bogie frames and wheelsets are dominating body components. They represent main vehicle mass. When the body of wagon, bogie frames, as well as wheelsets are treated as rigid bodies they are defined by their mass-inertia characteristics (mass, moments of inertia and the position of the centre of gravity) and by the relative position of the bogies with respect to the body of wagon itself [25]. Each rigid body has six degrees of freedom (three translations and three rotations). In some cases of a freight wagon running simulation the structural flexibility of a body is necessary to consider.

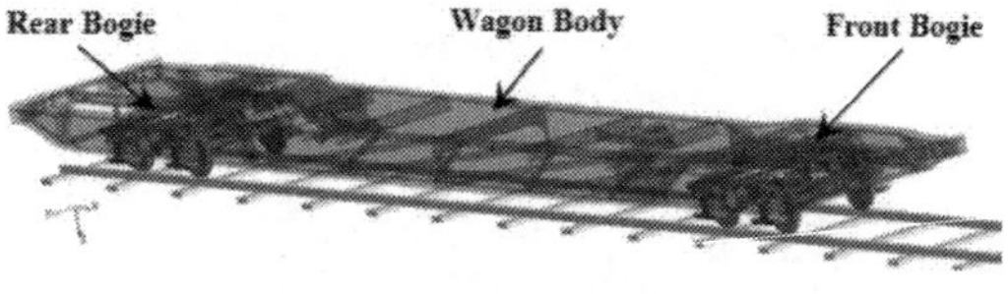

Fig. 2. The virtual model of an intermodal freight wagon

All bodies of a wagon are connected by springs and dampers which form primary and secondary suspension. Primary suspension is used to the connection between the wheelset and the bogie frame. The secondary suspension is used to the connection between the bogie and the body of wagon. In case of the modelled Y25 bogie the secondary suspension is consists of the centre pivot and two sidebearers.

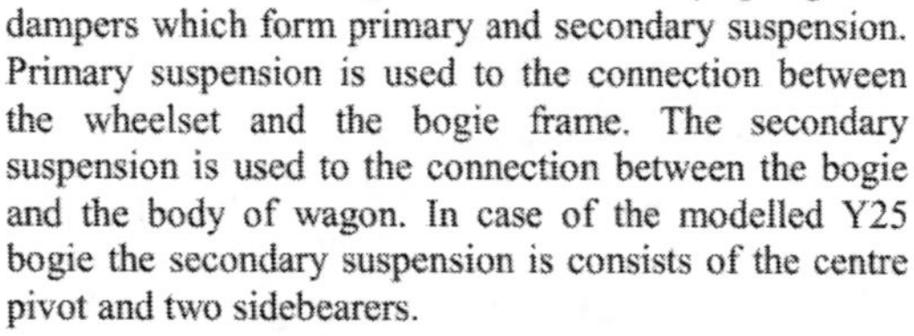

Fig. 3. The virtual model of the Y25 freight wagon bogie

2.1.1 Wagon body

The body of wagon usually holds main mass of a freight wagon. The mass properties of the body of wagon structure are not difficult to calculate: CAD model or finite (FE) analysis can give the mass, centre of gravity position and inertia moments quite easily.

In the most applications there is important to consider the structural flexibility of wagon body. In railway application dynamics simulation it is desirable to represent the structural flexibility of wagon body by only a limited number of degrees of freedom in addition to the six rigid body degrees of freedom (longitudinal, lateral, vertical, roll, pitch and yaw motions).

2.1.2 Wagon bogie

When it comes to the bogie frame modelling we should first keep in mind that the most common bogie type worldwide, the three-piece bogie, consist of three bodies and not only of a single bogie frame body. For a bogie frame, seen as a one-piece metal structure, the assumption of essentially rigid body behaviour is fair in most applications. Sometimes the torsional flexibility about a longitudinal axis, is considered representing an improved negotiation of twisted tracks. The bogie flexibility can be determined through finite element analysis [26] and may be represented in the freight wagon multibody simulation by a limited set of eigenmodes of free frame body. Such analysis can also give the mass and inertia moments of the bogie frame [27, 28].

2.1.3 Wheelset

A plain wheelset normally consists of one solid unit with two wheels on a common wheel axle. The wheel diameter is often 0.7 m to 1.0 m. The axle diameter is usually 0.15 m to 0.20 m. A typical mass for a plain wheelset is 1,000 kg to 1,500 kg. Additional mass can be introduced through brake discs, mounted on the axle or wheels and traction gear. The axle boxes also add some mass but they should not be considered in calculating the wheelset pitch inertia moment. In fact, they are created as separate bodies in the modelling. Wheelsets are often modelled as rigid bodies, but wheel axle flexibility may affect the vehicle/track interaction.

2.1.4 Suspension elements

In our case, suspension components in freight wagons are coil spring in combination with friction dampers. They play important roles in reducing bogie frame and wagon body accelerations as well as dynamic wheel/rail forces. Bump stop and antiroll bars may therefore be introduced to mitigate such problems, the bars also reducing quasistatic lateral acceleration components although their man task is to transfer longitudinal forces between bogie and wagon body during acceleration or retardation. Some freight wagons are equipped with the friction damper. The most common friction model is that suggested by Coulomb's law. For a one-dimensional case such a model in series with a linear spring, which together has a linear spring in parallel, produces a resulting force – displacement.

2.2 The multibody model of a track

Nominal geometry (track layout), track flexibility as well as track irregularities must be modelled and described. A simple moving track model for lateral, vertical and roll flexibilities of the track is shown on Fig. 6.

 https://doi.org/10.1051/matecconf/201823500027

Such a simple model may be sufficient for analysis of the freight wagon interaction with the track. Stiffness and damping may have linear or nonlinear characteristics. Actual numerical data reflect the track construction, i.e. type of rails, fastenings, rail pads, sleepers and ballast, as well as freight wagon axle load. The track flexibility will, above all, influence track forces, but under certain circumstances lateral dynamic stability will also be affected.

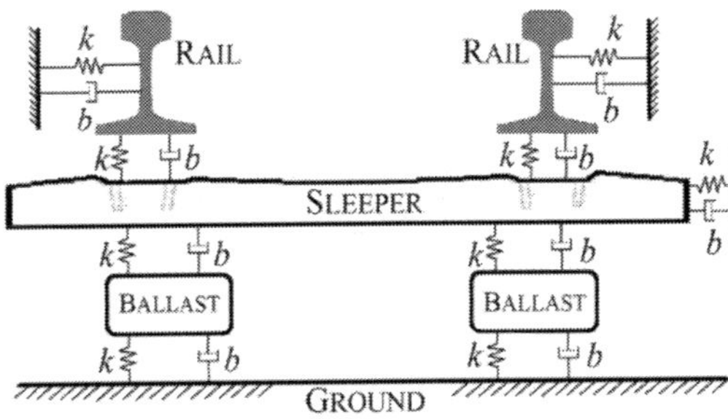

Fig. 4. Example of a track model

The nominal track geometry (layout) is defined by circular curve radii and lengths, lengths and types of transition curves, track cants, etc.

Track irregularities are normally given as lateral and vertical deviations of the track centreline (from nominal geometry) and by deviations in track cant and track gauge. It is important, track irregularities are representative for a longer section of track and they are also representative of the worst condition to be considered. A statistical analysis of track data is, therefore, very often needed.

2.2 The wheel/rail contact model

The crucial importance for the dynamic interaction between freight wagon and track has the characteristics of wheel/rail contact.

The contact functions are usually calculated in the PreProcessor program. The wheel and rail geometry, rail inclination, track gauge and wheelset flange spacing are the main inputs parameters.

The wheel-rail element always describes kinematics of one wheel relative to one rail. The wheel/rail contact is actually a complex and imperfect link. There are several approaches for the wheel/rail contact calculation used in commercial software.

The Wheel/rail contact defines, how the tangential forces and tangential torque for the contact patches are calculated. The wheel/rail contact search can be separated into three steps:

- identification of the contact patch locations, sizes and shapes, then for each contact patch that has been found:
- determination of the normal contact force,
- determination of the tangential contact forces and the tangential torque.

These steps may be executed separately or combined. The main input to most wheel/rail tangential force and torque calculation methods are the longitudinal (x), lateral (y) and spin (about z) creepages. The outputs ate the longitudinal and lateral tangential forces and sometimes the tangential torque [29-31].

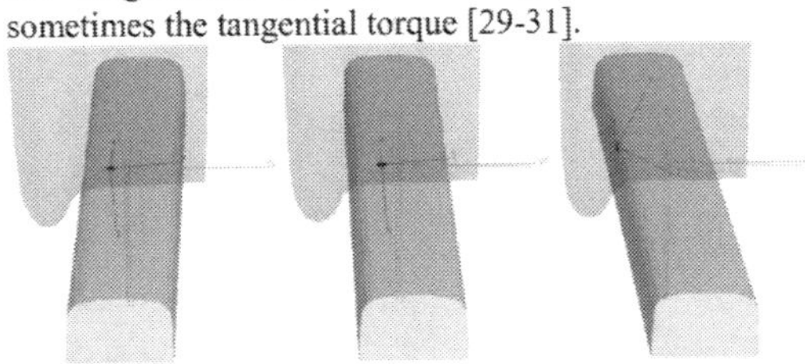

Fig. 5. Contact reference point location and coordinate system orientation for different yaw angle

3 Analysis of ride properties of a freight wagon running on a track

There are certain criteria for the railway vehicle assessment. One of the criteria for the assessment of the curving safety is the derailment quotient assessment. The derailment quotient is the guiding force and vertical force ratio. The derailment quotient limit value depends on a curve radius. The limit value is valid 0.8 for the curves of R > 250 m on the length of 2 m and for smaller radii and for superelevation ramp 1.2. All requirements that railway vehicles must meet for the operation inclusion are referred in the relevant standard and regulations [32].

For purposes of dynamical analysis we have created the multibody model of a freight wagon, which is designed for intermodal transport representing current efficient solution of transport system [33, 34]. In our research we have chosen real track section in Slovakia for a freight wagon simulation analysis considering available important parameters. There are sections with different curve radii and the resulting section of the transmission curve and superelevation ramp on this track. It has been suitable for performing dynamic analysis and the derailment quotient assessment.

The total track length almost 6000 m. There are three right curves and four left curves on this track (seven curves total). The horizontal profile of a track is shown in Fig. 6.

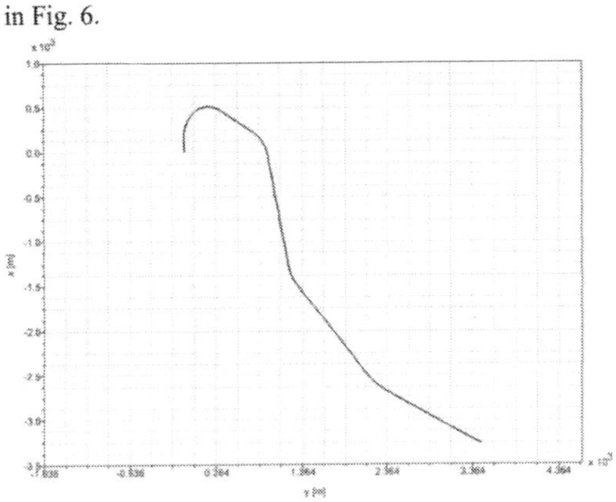

Fig. 6. The horizontal profile of a real track section

For wheels of wagon the S1002 rail profile was chosen. The track model was created by two rails with the UIC60 profile and with inclination of 1:40. There were considered measured track irregularities prescribed by discrete form.

There were used following input parameters:

- the tare wagon model and the loaded wagon model,
- the general wheel/rail contact model,
- the real track section track model.

During a freight wagon testing the most attention was focused on the assessment of:

- lateral wheel forces *(Y)*,
- vertical wheel forces (*Q*),
- derailment quotient (*Y/Q*).

The tested freight wagon for both loading conditions has run on the track at the constant speed of 60 km/h. It means, the wagon has travelled the determined distance for 352 s. For the evaluated ride properties the sample frequency of 160 Hz was chosen.

Output signals of the evaluated parameters in the wheel/rail contact area are marked according graphics in the Fig. 7.

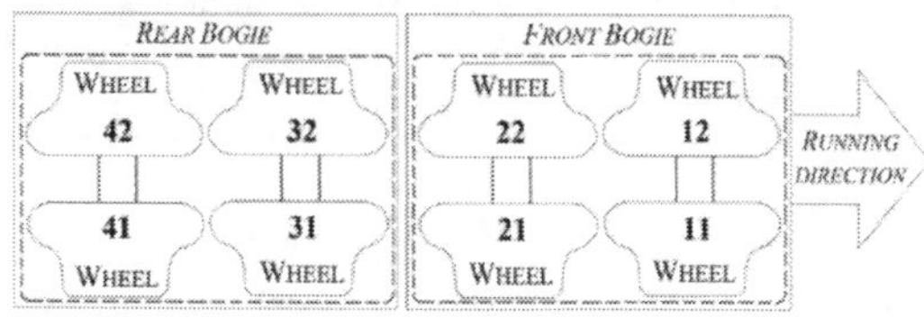

Fig. 7. Positions of output signals on wagon bogie

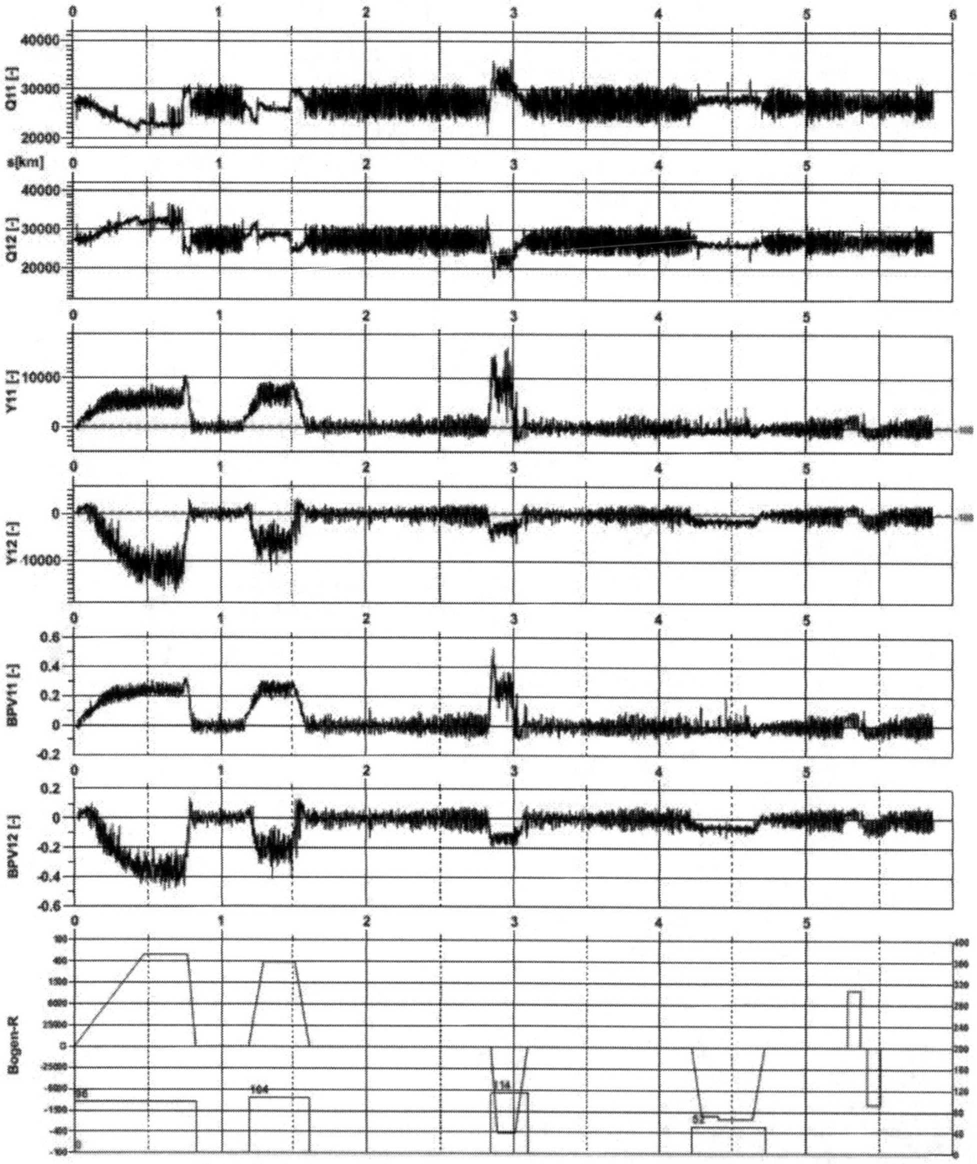

Fig. 8. Evaluation of selected outputs quantities for the tare freight wagon

https://doi.org/10.1051/matecconf/201823500027

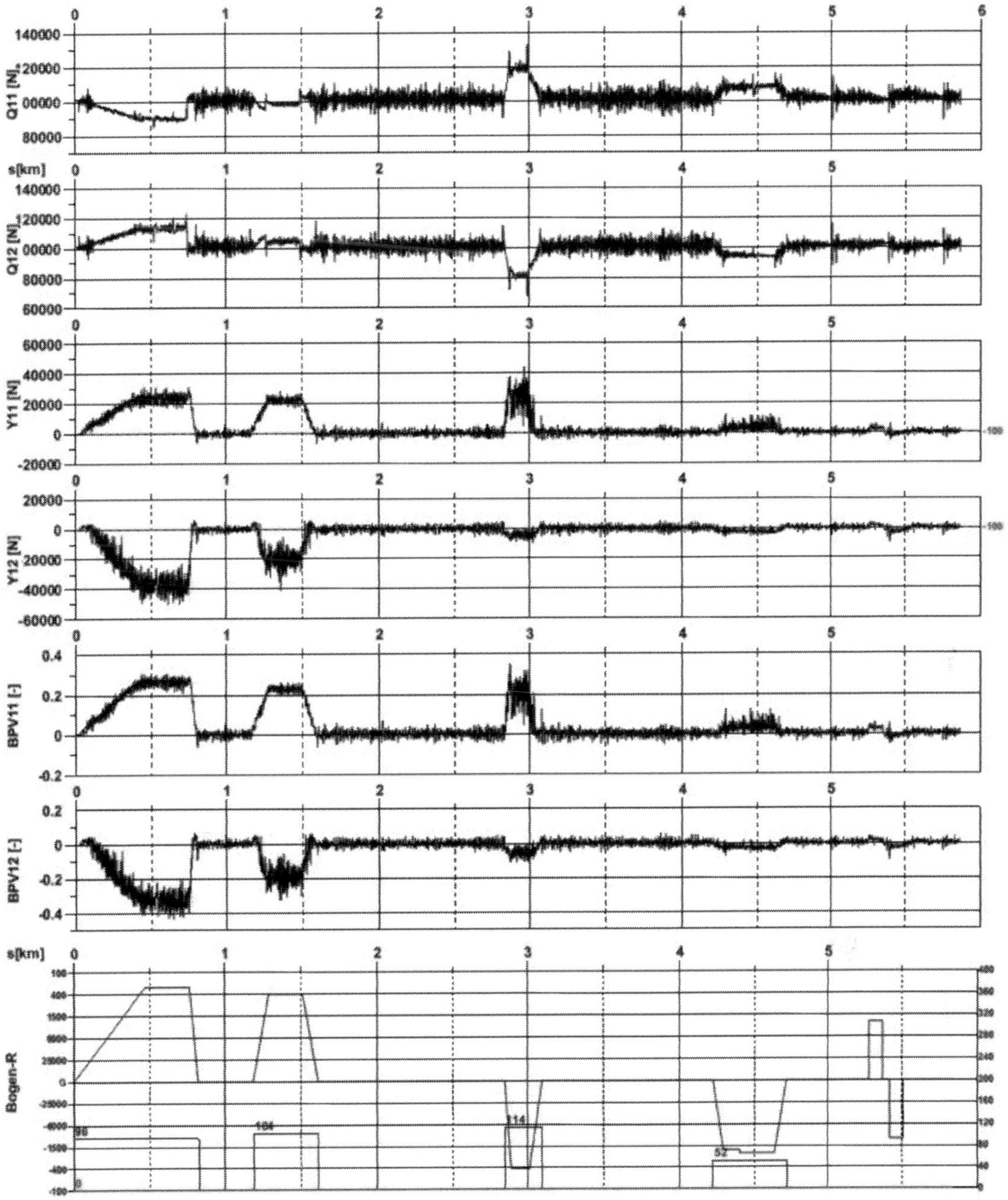

Fig. 9. Evaluation of selected outputs quantities for the loaded freight wagon

Results of dynamic analyses are shown in Fig. 8 and Fig. 9. In these cumulative graphs the comparison of comprehensive values of results are shown. These graphs include seven parts. The vertical forces are in the first and second parts, lateral forces in the third and fourth parts. Waveforms of the derailment quotients are outlined in the fifth and sixth parts and the track curvature is in the last (seventh) part. The larges values of lateral forces were achieved on the first axle of the front bogie, on the guiding wheel.

From waveforms of the vertical forces of the tare wagon (Fig 8) and the loaded wagon (Fig. 9) we can see, when the wagon is entering a curve, the vertical forces and the lateral forces are significantly increasing. The force values depend on the wagon weight.

The irregularities cause excitation of a freight wagon mechanical system. The irregularities were created on both rails in the *y* (lateral) direction and in the *z* (vertical) directions. When comparing vertical forces for tare and loaded wagon can we see the larger damping of these forces occur in the case of the loaded wagon running.

This is because of the construction friction dampers used in the freight Y25 bogies, which are characterized by the friction force dependent on the load of a wagon.

For the derailment quotient assessment we came out again from the waveforms in Fig. 8 and Fig. 9. The greatest values of the derailment quotient reach the guiding wheel for the tare state, namely $Y/Q = 0.485$. In case of the loaded state of the wagon the greatest value $Y/Q = 0.405$ was identified. Observed values indicate, that the greatest Y/Q ratio (the worst ride properties) was achieved on the guiding wheel of the wagon. This ratio is changing inversely with the wagon load (or axleload). It follows the worst risk of overriding the flange occurs, when the tare wagon is running in a curve.

4 Conclusion

At present, railway vehicle dynamics constitutes a significant and indispensable part of railway vehicle engineering. During the design period a prognosis of the running characteristics can be carried out with the aid of computer simulations. This enables reduction in the extent of the test and shortening of the vehicle design period.

The development of software tools for simulation of dynamic behaviour of railway vehicle on track has grown with the development of computing power. Software are now combined into a number of powerful and reliable computer simulation packages and improvements have also been made in modelling the components of vehicle/track systems.

The vehicle dynamic analysis represents a complex issue utilizing methods of mechanics, material engineering, engineering technology, mathematic applied into the railway vehicles area. Simulations represent a robust way, through which we can get adequate information resulting from increasing railway transport requirements.

In this paper ride properties of a long freight wagon equipped by Y25 bogies were presented. As evaluated parameters the vertical and the lateral forces and the derailment quotient were assessed. These parameters were evaluated for the tare state and also for the loaded state (max. axle load of 22.5 t) of the freight wagon. The wagon has run on the real track section model. The measured irregularities were applied into the track model. From the calculated values the highest value of derailment quotient was achieved on the guiding wheel of the tare wagon in the first curve. This value is less than the limit value of 1.2 for this track section.

This work was supported by the Cultural and Educational Grant Agency of the Ministry of Education of the Slo-vak Republic in project No. KEGA 077ŽU-4/2017: Modernization of the Vehicles and engines study program.

The work was also supported by the project No. APVV-0842-11: Equivalent railway operation load simulator on the roller rig and VEGA No. 1/0558/18: Research of the interaction of a braked railway wheelset and track in simulated operational conditions of a vehicle running in a track on the test stand.

Research-Educational Center of Rail Vehicles (VVCKV)

References

1. J. Soukup, J. Skocilas, B. Skocilasova, *Assessment of railway wagon suspension characteristics*. Mechanical Systems and Signal Processing **89**, 67-77 (2017)
2. O. Lunys, S. Dailydka, S. Steisunas, G. Bureika, *Analysis of freight wagon wheel failure detection in Lithuanian Railways*. Procedia Engineering **134**, 64-71 (2016)
3. M. Saga, L. Jakubovicova, *Simulation of vertical vehicle non-stationary random vibrations considering various speeds*. Scientific journal of Silesian University of Technology - Series Transport 84, 113-118 (2014)
4. V. Hauser, O. Nozhenko, K. Kravchenko, M. Loulova, J. Gerlici, T. Lack, *Proposal of a steering mechanism for tram bogie with three axle boxes*. Procedia Engineering **192**, 289-294 (2017)
5. V. Hauser, O. Nozhenko, K. Kravchenko, M. Loulova, J. Gerlici, T. Lack, *Impact of three axle boxes bogie to the tram behavior when passing curved track*. Procedia Engineering **192**, 295-300 (2017)
6. P. Hejma, M. Svoboda, J. Kampo, J. Soukup, *Analytic analysis of a cam mechanism*. Procedia Engineering **177**, 3-10 (2017)
7. P. Baran, P. Šťastniak, P. Kukuča, M. Brezáni, *Investigation of kinematic parameters of two nonconventional piston machines with wobble board*. MATEC Web of Conferences **157** (2018)
8. J. Soukup, B. Skocilasova, J. Skocilas, *Vibration of mechanical system with higher degrees of freedom: solution of the frequency equations*. Procedia Engineering **177**, 17-24 (2017)
9. A. Falendysh, M. Volodarets, V. Hatchenko, I. Vykhopen, *Software analysis for modeling the parameters of shunting locomotive chassis*. MATEC Web of Conferences **116** (2017)
10. A. Golubenko, S. Sapronova, V. Tkachenko, *Kinematics of point-to-point contact of wheels with a rails*. Transport problems **2**(3), 55-59 (2007)
11. J. Gerlici, T. Lack, *Railway wheel and rail head profiles development based on the geometric characteristics shapes*. Wear **271**(1-2), 246-258 (2011)
12. J. Gerlici, T. Lack, *Contact geometry influence on the rail / wheel surface stress distribution*. Procedia Engineering **2**(1), 2249-2257 (2010)
13. L. Smetanka, P. Šťastniak, J. Harušinec, *Wear research of railway wheelset profile by using*

computer simulation, MATEC Web of Conferences **157** (2018)

14. G. Vaiciunas, G. Bureika, S. Steisunas, *Research on metal fatigue of rail vehicle wheel considering the wear intensity of rolling surface*. Eksploatacja i Niezawodnosc – Maintenance and Reliability **20**(1), 24-29 (2018)
15. V. Tkachenko, S. Sapronova, *Steerability of railway vehicles*. Transport problems **2**(4), 9-16 (2007)
16. A. Falendysh, P. Kharlamov, O. Kletska, N. Volodarets, *Calculation of the parameters of hybrid shunting locomotive*. Transport Research Procedia **14**, 665-671 (2016)
17. J. Gerlici, T. Lack, *Rail vehicles brake components test bench utilisation*. Applied mechanics and materials. ISSN 1660-9336. Vol. **486**, pp. 379-386 (2014).
18. J. Gerlici, T. Lack, J. Harušinec, *Realistic simulation of railway operation on the RAILBCOT test stand*. Applied mechanics and materials **486**, 387-395 (2014)
19. J. Gerlici, T. Lack, J. Harušinec, *Development of test stand prototype for rail vehicles brake components testing*. Communications - Scientific Letters of the University of Žilina **16**(3a), 27-32 (2014)
20. F. Klimenda, J. Kampo, P. Hejma, *Vibration measurement of conveyor rollers*. Procedia Engineering **136**, 198-203 (2016)
21. J. Harušinec, A. Suchánek, M. Loulová, P. Strážovec, Design of a device to simulate environmental influences on the course of brake tests. MATEC Web of Conferences **157** (2018)
22. A. Sapietova, V. Dekys, *Dynamic analysis of rotating machines in MSC:ADAMS*. Procedia Engineering **136**, 143-149 (2016)
23. M. Chalupa, L. Severa, R. Vlach, *Vehicle track loading simulation*. Proceedings of SPIE - The International Society for Optical Engineering (2012)
24. M. Chalupa, J. Veverka, R. Vlach, *Dynamic loading of vehicle track simulation*. Engineering Mechanics (2011)
25. J. Gerlici, T. Lack, *Modified HHT method for vehicle vibration analysis in time domain utilisation*. Applied mechanics and materials **486**, 396-405 (2014)
26. A. Suchánek, J. Harušinec, M. Loulová, P. Strážovec, Analysis of the distribution of temperature fields in the braked railway wheel. MATEC Web of Conferences **157** (2018)
27. M. Gorbunov, J. Gerlici, S. Kara, O. Nozhenko, G. Chernyak, K. Kravchenko, T. Lack, *New principle schemes of freight cars bogies*. Manufacturing Technology **18**(2), 233-238 (2018)
28. J. Gerlici, M. Gorbunov, O. Nozhenko, V. Pistek, S. Kara, T. Lack, K. Kravchenko, *About creation of bogie of the freight car*. Communication - Scientific Letters of the University of Zilina 19(2), 29-35 (2017)
29. T. Lack, J. Gerlici, *A modified strip method to speed up the calculation of normal stress between wheel and rail*. Applied mechanics and materials **486**, 359-370 (2014)
30. T. Lack, J. Gerlici, *A modified strip method to speed up the tangential stress between wheel and rail calculation*. Applied mechanics and materials **486**, 371-378 (2014)
31. S. Sapronova, V. Tkachenko, N. Kramar, A. Voron'ko, *Regularities of shaping of a wheel profile as a result of deterioration of the rolling surface in exploitation*. Transport Problems **3**(4), 47-54 (2008)
32. UIC CODE (2009): 518 OR *Testing and approval of railway vehicles from the point of view of their dynamic behaviour – Safety – Track fatigue – Running behaviour*. 4th edition, September 2009
33. T. Skrucany, J. Ponicky, M. Kendra, J. Gnap, *Comparison of railway and road passenger transport in energy consumption and CHG production*. Proceedings of the third international conference of traffic and transport engineering, 744-749 (2016)
34. I. Rybicka, J. Caban, J. Vrabel, B. Sarkan, O. Stopka, W. Misztal, *Analysis of the safety systems damage on the example of a suburban transport enterprise*. Proceedings of XI International Science-Technical Conference Automotive Safety (2018)
35. A. Suchánek, J. Harušinec, M. Loulová, P. Strážovec, Analysis of the distribution of temperature fields in the braked railway wheel. MATEC Web of Conferences **157** (2018)
36. J. Harušinec, A. Suchánek, M. Loulová, P. Strážovec, Design of a device to simulate environmental influences on the course of brake tests. MATEC Web of Conferences **157** (2018)

Brake actuator optimization of the brake test stand as a tool for improvement railway safety

Jozef Harušinec[1,*], *Andrej* Suchánek[1], *Pavol* Šťastniak[1] and *Peter* Strážovec[1]

[1]University of Žilina, Faculty of Mechanical Engineering, Department of Transport and Handling Machines, Univerzitná 8215/1, 010 26 Žilina, Slovak Republic

Abstract. In recent years have been the speed, comfort and safety of rail transport significantly increased. This asks for higher demands on reliability and safety of rail vehicles parts meeting. The important is research the properties of brake pair, therefore build specialized test equipment, where it is possible to programmatically load wheels, wheelset, bogie or complete vehicles. Tests on the friction test stand brake represent a special group, which replace the expensive tests of complete vehicles, that are more costly. The test stand brake allows the implementation of experimental tests of the actual brake units and components under simulated conditions.

1 Braking of rail vehicles

Braking is an important phase of the drive rail vehicle, which change the kinetic and potential energy obtained by driving to another kind energy. Brakes rail vehicles are the most important safety device.

Braking is almost always the increasing of level road resistance in order to keep required speed, stopping on a certain place or to ensure vehicle against spontaneous drive-away on downhill grade. Every powered vehicle with any kind of train has to stop in specified braking distance to avoid threatening the safety of rail traffic.

2 Properties affecting the braking of rail vehicles

2.1 Thermal stress of the braked railway wheel

Property monitored in detail in railway vehicles braking is the heat generated by friction of brake block on the wheel.

A part of frictional heat comes from friction of individual surfaces of wheels and a part of frictional heat is from brake block. Temperature reached by the steel wheel friction and non-metallic block is as high, however, due to the lower thermal conductivity of the block the bigger strain is put on wheel.

The Fig. 1 shows the progress of non-stationary temperature fields in braked railway wheel. When braking, the most heated surface is the wheel track. Temperature distribution along wheel cross section depends on:

- size of the contact force,
- speed and braking time,
- type of braking block,
- wheel material.

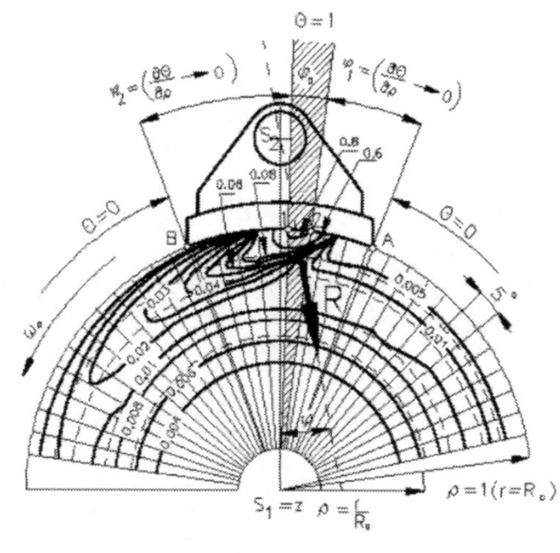

Fig. 1. The process of non-stationary temperature fields in braked railway wheel.

Fact that can be considered is that during braking 66% of created heat passes to the wheel for cast iron blocks and up to 90% to non-metallic blocks.

Another factor that significantly affects the thermal stress is placing of the brake shoe on the wheel. The worst option is overhanging brake block over flange of the wheel. Very unfavourable position is also overlapping of brake block on the outside of the wheel, where the brake block can even rub the external body rim. In both cases there is high heat flux, stress, and thus increases the possibility of damage to the wheel.

* Corresponding author: jozef.harusinec@fstroj.uniza.sk

 https://doi.org/10.1051/matecconf/201823500028

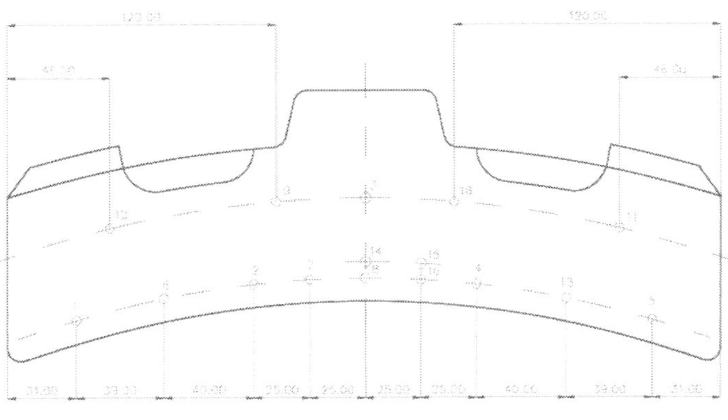

Fig. 2. Location of thermocouples in the test of cast iron.

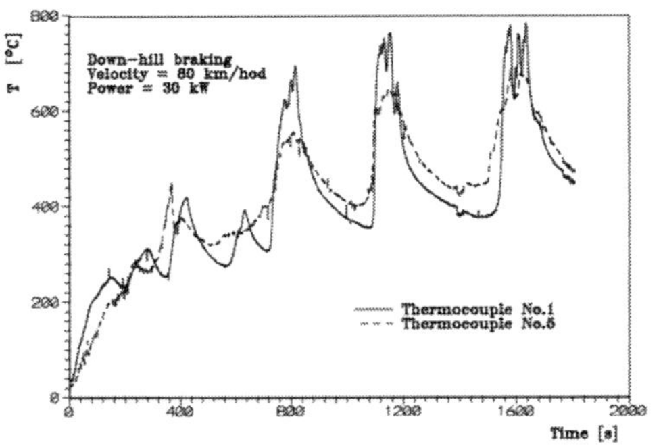

Fig. 3. Temperatures measured in a distance of 1.5 mm from the wheel tread, at a max. distance from the middleblockaxis.

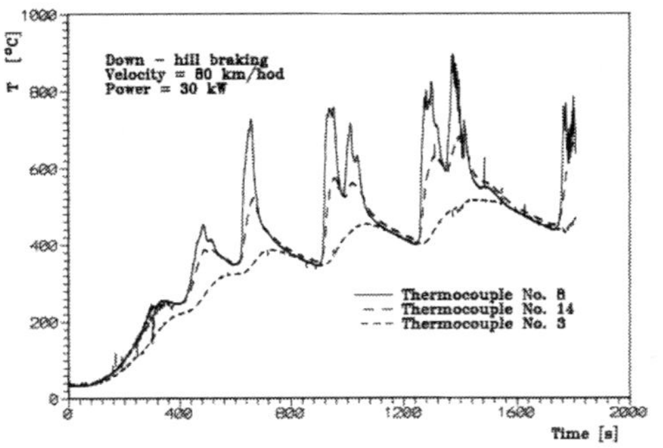

Fig. 4. Temperatures measured in a distance of 1.5, 8 and 45 mm from the wheel tread, in the middleblockaxis.

When braking on an incline at a speed of 80 km/h and braking power 30 kW on test brake bench UIC on Department of transport and handling machines was found in the research the distribution of temperatures in thermocouples. The distribution of thermocouples is shown in Fig. 3 and Fig. 4.

Coatings from rails, which are caused by heating of the wheel, may occur in traffic to tread wheels – if the adhesive forces are only slightly larger than the friction forces. Hardness of coatings is very high and in extreme cases can reach up to 11 mm thickness. Equally there may be a coating from material of the brake block, when the wheel is braked more force and it develops the large amount of heat.

An important factor that affects the production of heat in the braking is friction coefficient between the brake block and wheel tread. Its size decreases with warming of brake block.

Further warming of brake blocks is related to braking time. The practice is known, that mild and long slopes are more dangerous than sharp and short slopes. Brake block is heated by prolonged braking and changes its shape due to temperature changes, and thus support surface is a deformed while keeping the brake force, resulting in an excessive increase in pressure and thereby decrease the friction coefficient. There will be more heat per certain time unit, how can penetrate deep into the material of the wheel and brake block and then into the air. This leads to the melting of surface layer in the contact wheel and brake blocks. Molten metal and impurities lubricate the contact surfaces and thereby reducing braking effect, that is, begins to decrease friction and temperature also. Molten material solidifies due to the reduction temperature. Dry friction between wheel and brake blocks increases and this leads again to an increase in temperature.

2.2 The friction in rail vehicles

The basic element of the problem of rail vehicles during braking can be classified adhesion and friction. Adhesion is the ability to transfer tangential forces by contact area between wheel and rail. Friction has the ability to ensure the creation of tangential forces between wheel and brake blocks.

The basic consideration of effective braking based on the conditions of adhesion, which is the limiting factor for rail transport.

If frictional forces acting on the wheel will be greater than the adhesion, wheel is blocked and begins to slide, there is theoretically elliptical area with a depth into the wheel, the effect of friction wheel - rail is significantly lower than in rolling the wheel. At the same time the place area leads to high overheating and thus a change in the structure of the material. Deformed wheel loses circularity. Wheel becomes exciter dynamic oscillations at the next ride and it deteriorates the drive properties of vehicle. When the wheel rotates during braking just a few degrees short groove will be created, which leads to influence of the structure of the material and fast damage of wheel.

Friction coefficient between the wheel tread and brake blocks mainly depends on the material of brake block. The values of the friction coefficients are high at low speeds. The friction coefficient of cast iron brake blocks is significantly lower than for non-metallic.

3 Use of composite brake blocks

Cast iron brake blocks in operation causing considerable noise, which is not invoked only by braking but rolling wheels braked of cast iron blocks on rails These wheels have on tread the deposits of cast iron, that create inequalities causing vibrations. One of the measures to reduce railway noise has been a requirement for manufacturers of friction materials, that began to develop brake blocks of composite materials. These composite blocks that still bring the noise reduction effects, but on the other hand, it should be borne in mind that, have a lower heat dissipation and therefore own wheel more damaged, especially in the surface layers of the tread. Materials are usually made up of polymer matrix enriched with various materials such as small particles of pumice, dolomite, aluminium oxide, silicon oxide, zinc oxides, magnesium and copper, graphite powder, iron powder, powder of white cast iron, cast iron powder phosphor, barium sulfate, copper, aluminum, mild steel scrap, metal fibers of brass. Currently, are fillers asbestos-based, lead and zinc environmental reasons completely replaced by other fillers and they use mainly fast hardening powders of various alloys and ceramic particles, which are characterized by high hardness, such as corundum.

Composite brake blocks can be divided into organic and metal-ceramic. The main difference between organic and metal-ceramic brake blocks is in materials, in the production and subsequent heat treatment.

The problem in the introduction of composite brake blocks is the friction coefficient, which is much higher than cast iron block. Replacement of these blocks is not possible. Therefore come the requirement from a railway companies to produce of composite brake blocks, which should have the friction coefficient same as cast iron brake blocks and may be freely exchangeable in existing rail vehicles. These brake blocks are marked LL. Blocks, which have a slightly higher the friction coefficient are marked L and blocks with the highest friction coefficient are marked K.

The disadvantage of composite blocks is their price, which is 5 times higher than that of cast iron blocks. Various literary sources but ultimately argue that operating costs of K blocks are lower, but their life is 4-5 times higher.

4 Rail vehicles brake components test stand RAILBCOT

Purpose of test stand RAILBCOT is the analysis of changes geometry tread profiles of railway wheelset. Change is induced by simulated working load under laboratory conditions, that mimic real traffic.

Activity test bench consists of active loading the test wheelset different forces, which will be simultaneously simulated in combination with change of engine torque, engine speed, change wheel load, resizing of free channel gauge, the angle of attack wheelset change and variable braking mode, which is based on the independent activity of two and two brake units with changes in their organization, shape and material composition brake blocks.

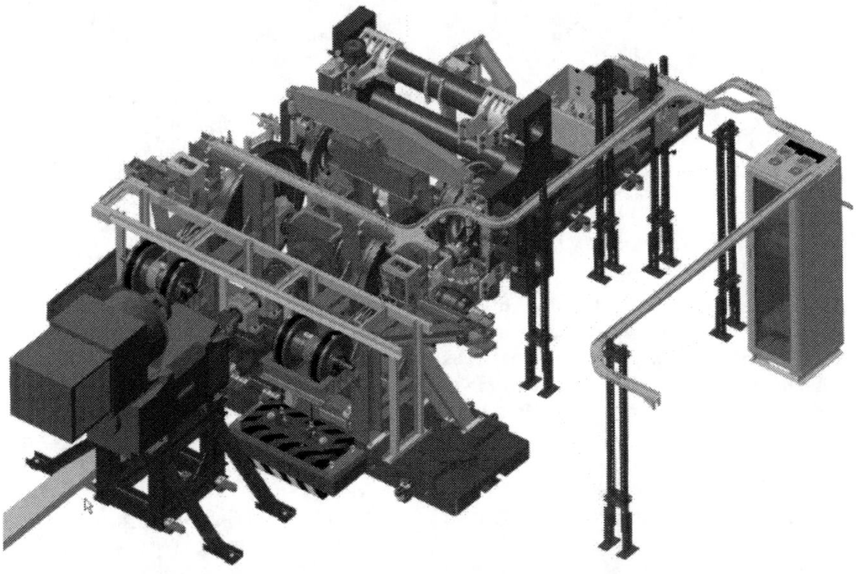

Fig. 5. Model of the test stand RAILBCOT.

 https://doi.org/10.1051/matecconf/201823500028

4.1 Description of the test stand RAILBCOT

Test stand RAILBCOT (Fig. 5) is built at the Department of Transport and Handling Machines. Will be used to detect changes of profiles of wheels and rails, and hence changes in running properties wheelset. The test consists of two processes. The first is rolling the wheel on the rotating rail. The second is the process of braking wheelset with integrated brake units. Each wheelset is braked by two brake units. Each unit has a separate brake control pressure in the cylinder and it is possible to simulate the bilateral and unilateral action of brake blocks.

The principle of making the wheel forces and braking forces are shown in (Fig. 6). The wheelset (3) is pressed to powered rotating rail (4) by mechanism (5). This mechanism consists of a set of pulleys and size of the workforce is set by the appropriate number of segments weights. On the test bench can simulate wheel loads up to 25 tons per axle. Braking force is inferring with the brake unit (2), which is attached in the bracket on the frame. Brake unit is controlled with compressed air, which is supplied by flexible hose (1).

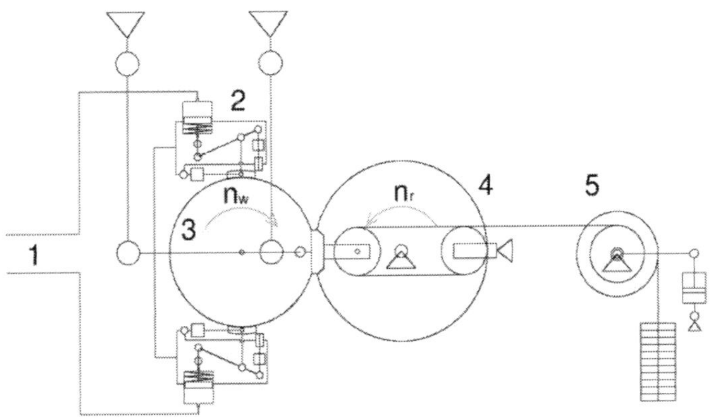

Fig. 6. Scheme arrangement of the test stand RAILBCOT

Production of compressed air and regulating pressure provides air distribution (Fig. 7). Compressed air at a pressure of 1 MPa is formed by compressor 3DSK 75 (1), which is powered by engine 1AL9 130 2LA (3). Power transfer is secured by belt drive. Compressed air is stored in the primary reservoir (4). Air passes through a pressure valve (5) to the secondary reservoir (6). The pressure in the primary reservoir is in the range 0,8-1 MPa (depending on emptying the primary reservoir) and in the secondary reservoir is 0,8 MPa. This arrangement was chosen in terms of smaller impacts in the pneumatic system when refilling pressure air by compressor (1) to the pneumatic system and therefore provides constant pressure 0,8 MPa at the input of the proportional pressure valve MPPE3 126010 B2 (7). Compressed air is supplied by input branch (a) to the valve, where it is on the principle of abandoning air pressure in the flow line (b) into a waste branch (c). The pressure is set by the pre-stress membrane valve, which is controlled electronically.

In order to record the course of braking power when braking process, it was necessary to make adjustments on the used locomotive brake unit.

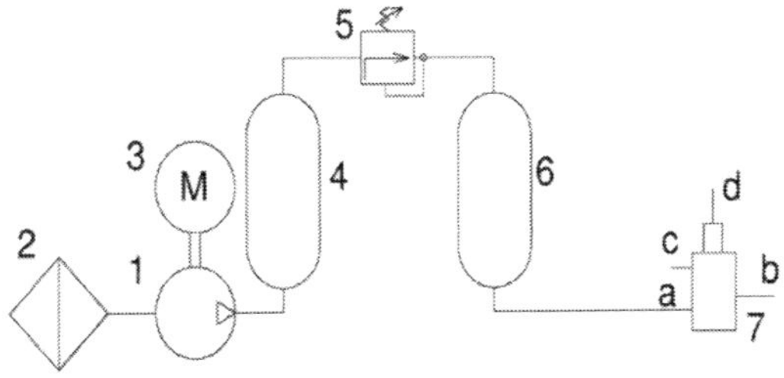

Fig. 7. Air-conditioning distribution system scheme

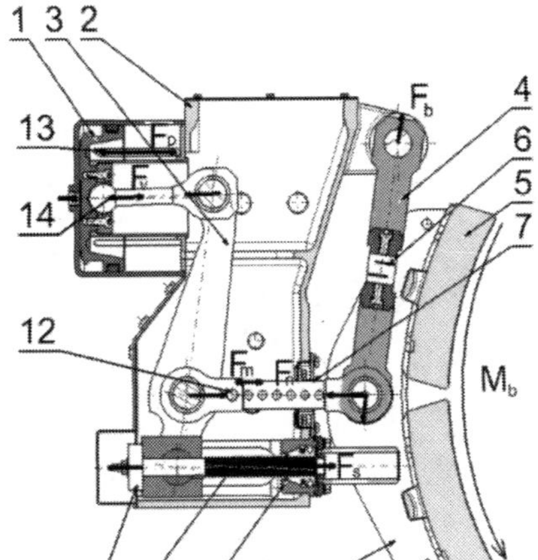

Fig. 8. Forces in the brake unit during braking

 https://doi.org/10.1051/matecconf/201823500028

In Fig. 8 shows a function diagram of a modified brake unit: (1) brake cylinder, (2) frame of brake unit, (3) brake handle transmission lever, (4) suspensions for transmitting tangential forces, (5) brake block, (6) tangential force sensor, (7) rod for inferring normal force, (8) normal force sensor, (9) screw of slack adjuster, (10) frame of slack adjuster, (11) holder of brake blocks, (13) spring of brake cylinder, (14) connecting rod of piston. Force is transmitted through the connecting rod (14) on the brake rigging (3). Through the transfer lever leads to an increase value of braking force, but reduce the distance made by brake blocks. The unit is equipped with one-sided slack adjuster (10, 11). This ensures a constant remoteness between wheel and brake release blocks in the un-braked state.

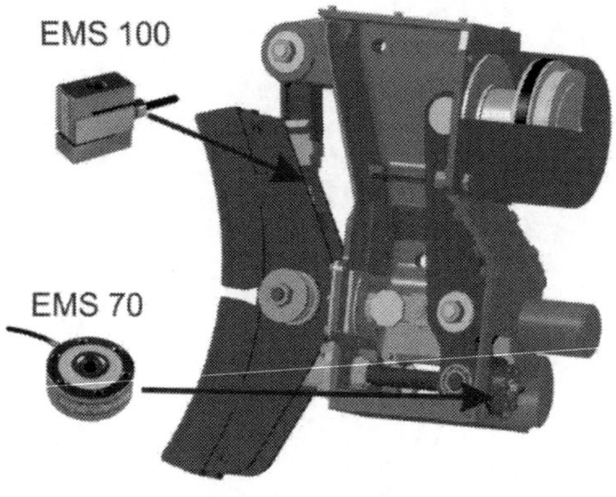

Fig. 9. Location of sensors in the brake body units.

On modified body of slack adjuster is placed force sensor. Direct measurement of the braking force was problematic because the actuator is stressed by buckling. (7). This would adversely affect the accuracy and also the additional stress.

For this reason, is the measurement point on body of slack adjuster. The body is stressed by tensile and it has a good approach when replacing the sensor, in case of its failure. Tangential force sensor (6) is placed between the two adapters (4). These adapters replace the original suspensions. Currently, the test stand RAILBCOT is ready for test operation.

5 Conclusion

For detection and deducing the normal and tangential braking force were two possible options for implementation. The first was to create a new braking system on the brake rigging used in vehicles with installed brake block. The second option is a braking system integrated into a compact unit, so use the brake unit. There was chosen the variant to use the brake unit. Unit works as a compact braking system comprising cylinder with a diameter of 8 inches, brake rigging, one-sided slack adjuster and holder of brake blocks. Brake unit is additionally equipped with force transducers for measurement.

Fig. 10. Location of brake unit in test frame

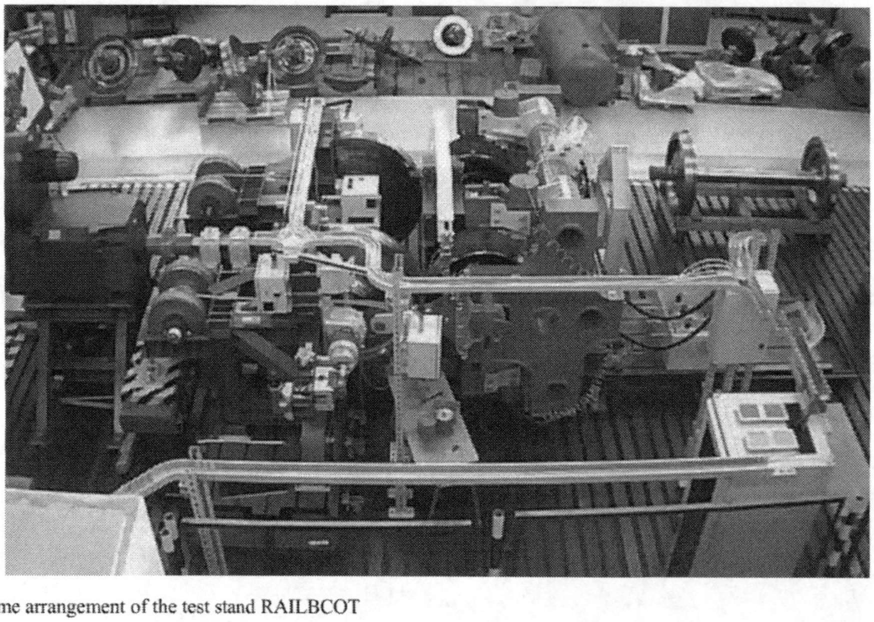

Fig. 6. Scheme arrangement of the test stand RAILBCOT

This work was supported by the Cultural and Educational Grant Agency of the Ministry of Education of the Slo-vak Republic in project No. KEGA 077ŽU-4/2017: Modernization of the Vehicles and engines study program. The work was also supported by the project No. APVV-0842-11: Equivalent railway operation load simulator on the roller rig and VEGA No. 1/0558/18: Research of the interaction of a braked railway wheelset and track in simulated operational conditions of a vehicle running in a track on the test stand.

Research-Educational Center of Rail Vehicles (VVCKV)

References

1. J. Dižo, J. Harušinec, M. Blatnický, Structural analysis of a modified freight wagon bogie frame. MATEC Web of Conferences. ISSN 2261-236X. Vol. 134 (2017).
2. J. Dižo, S. Steišunas, M. Blatnický, *Vibration analysis of a coach with the wheel-flat due to suspension parameters changes*. Procedia Engineering, ISSN 1877-7058. Vol. **192**, pp. 107-112 (2017).
3. J. Gerlici, T. Lack, *Modified HHT method for vehicle vibration analysis in time domain utilisation*. Applied mechanics and materials. ISSN 1660-9336. Vol. **486**, pp. 396-405 (2014).
4. J. Gerlici, T. Lack, *Rail vehicles brake components test bench utilisation*. Applied mechanics and materials. ISSN 1660-9336. Vol. **486**, pp. 379-386 (2014).
5. J. Gerlici, T. Lack, J. Harušinec, *Realistic simulation of railway operation on the RAILBCOT test stand*. Applied mechanics and materials. ISSN 1660-9336. Vol. **486**, pp. 387-395 (2014).
6. T. Lack, J. Gerlici, *A modified strip method to speed up the calculation of normal stress between wheel and rail*. In Applied mechanics and materials. ISSN 1660-9336. Vol. **486**, pp. 359-370 (2014).
7. T. Lack, J. Gerlici, *A modified strip method to speed up the tangential stress between wheel and rail calculation*. Applied mechanics and materials. ISSN 1660-9336. Vol. **486**, pp. 371-378 (2014).
8. J. Gerlici, T. Lack, *Railway wheel and rail head profiles development based on the geometric characteristics shapes*. Wear: An international journal on the science and technology of friction, lubrication and wear. ISSN 0043-1648. Vol. **271**, No. 1-2, Sp. iss., pp. 246-258 (2011).
9. J. Gerlici, T. Lack, *Contact geometry influence on the rail / wheel surface stress distribution*. Procedia Engineering. ISSN 1877-7058. Vol. **2**, Iss. 1, pp. 2249-2257 (2010).
10. T. Lack, J. Gerlici, Railway wheel and rail roughness analysis. Communications: scientific letters of the University of Žilina. ISSN 1335-4205. Vol. 11, No. 2, pp. 41-48 (2009).
11. V. Hauser, et. al., Proposal of a steering mechanism for tram bogie with three axle boxes. Procedia Engineering. ISSN 1877-7058. Vol. **192**, pp. 289-294 (2017).
12. V. Hauser, et. al., *Impact of three axle boxes bogie to the tram behavior when passing curved track*. Procedia Engineering, ISSN 1877-7058. Vol. **192**, pp. 295-300 (2017).
13. T. Lack, J. Gerlici, *Wheel/rail contact stress evaluation by means of the modified strip method*. Communications – Scientific Letters of the University of Žilina. ISSN 1335-4205. Vol. **15**, pp. 126-132 (2013).
14. T. Lack, J. Gerlici, *Contact area and normal stress determination on railway wheel/rail contact*. Communications – Scientific Letters of the University of Žilina. ISSN 1335-4205. Vol. **7**, pp. 38-45 (2005).
15. T. Lack, J. Gerlici, *Wheel/rail tangential contact stress evaluation by means of the modified strip method.*. Communications – Scientific Letters of the University of Žilina. ISSN 1335-4205. Vol. **16**, pp. 33-39 (2014)

Double tread wheel profile – proposal and justification of specific way how to pass along small radius track

Vladimir Hauser[1,*], *Kateryna* Kravchenko[1], *Olena* Nozhenko[2] and *Mária* Loulová[1]

[1]Faculty of Mechanical Engineering, Department of Transport and Handling Machines, University of Žilina, Univerzitná 8215/1, 010 26 Žilina, Slovak Republic.
[2]Volodymyr Dahl East Ukrainian National University, Tscentralny av. 59a, 93400 Severodonetsk, Ukraine

Abstract. In urban railway environments, in many respects, the problem of tramcars passing small radius arches appears to be topical. In this regard, the authors proposed that the wheelsets should be complemented on the outer side of the wheels with a second tread of a lower radius, intended exclusively for ride in arches of the track with a radius which the conventional wheelsets cannot even theoretically pass without creep. The complemented wheel tread enters into operation (or out of action) as a result of passing over a specially shaped rail located in the intersections before and behind the small radius arc. In this process, there is a change in the wheel-rail contact location - similar to what is commonly used when riding on a facing rail located in the immediate vicinity of the point frog. The paper deals with the comparison of these two cases based on a simulation analysis of vehicle ride in the calculation program.

1 Introduction

Rolling stock operated in the urban railways environment is a significant, efficient and environmentally friendly mass transit system that forms an indispensable part of modern cities. One of the demands placed on trams is the transition through arcs of very small radii. In many European cities, on tracks with a gauge of 1000 mm, where curves with a radius of 50 m are common, under arduous conditions, arches of the track with a radius of 18 m and more are built, with a requirement for the vehicle design to be able to pass a track with a curve of 17 m. [1, 2].

The requirement to pass through a strongly curved section of the track is a major technical problem [3, 4], since in such sections of the track an increased vehicle resistance [5, 6], wear of the wheel-rail pair [7 - 9] and noise [10 - 12] of the vehicle appears. The current, in practice widely used solutions include, in particular, intensive lubrication of the curved sections of a track [13, 14] along with the use of the noise-absorbing elements in the fastening of the rails [15, 16] - that is to say, a measure eliminating the adverse consequences without affecting the cause thereof.

In bogie design, independent wheels mounted in the subframe are commonly used [17]. Although this concept may improve arc passage under certain circumstances, it requires a complicated electronic-mechanical steering system [18, 19] to ensure a correct ride, both in the arc and in the straight track.

In order to provide a simple and at the same time a fully-fledged solution for the passage of the vehicle through the track arc, the authors team proposed own solution (Figure 1, Figure 2), expanding the possibilities of rolling a wheelset consisting of a single rigid body. The design solution is to add a new tread to the outside of the wheels. The added tread has a smaller radius, it is intended solely for the ride in the arches of the track with a radius that the conventional wheelsets cannot even theoretically overcome without creep. This solution, which is the subject of the patent application [20], has been described in detail in the paper [21].

When such a wheelset is entering the arc, the wheel-rail contact point location changes - similar to the usual practice used for ride in the facing rails, located in the immediate vicinity of the point frog [22]. The aim of the paper is to provide a comparison of these two cases by simulation analysis of vehicle driving in the calculation program [23, 24].

2 The method of passing a wheelset with additional tread through an arc of a low radius

The proposed solution is based on the relation defining the minimum radius of the track arc R [mm], which can be used by the conventional wheelsets to pass theoretically without creep:

$$R = \frac{2sr}{\Delta r}, \tag{1}$$

where 2s [mm] - distance of the wheelset junction circles; r [mm] - nominal radius of the wheel; Δr [mm] - the difference between the instant rolling radius of the left and right wheels of the wheel when the track channel clearance is exhausted.

* Corresponding author: vladimir.hauser@fstroj.uniza.sk

The shape of the wheel tread is, as a rule, specific to the transport company of each city, taking into account specific traffic requirements. Modern profiles with proven good ride performance include, among other, the KP-1 profile [25], the characteristics of which can be shown by the following considerations: tramline wheelset, radius r = 340 mm, located in a rail with the NT1 profile, reaching the distance of the contact circles 2s = 1061,9 mm and the value Δr = 5,5 mm is according formula (1) theoretically able to pass the arc with a radius R = 65,64 m without creep, which is in in-sufficient in urban conditions, such as the turnover and some crossings.

The minimization of the variable R in relation (1) in the proposed solution is realized by increasing the fraction de-nominator, while not affecting the existing geometry of the wheel profile. The relationship (1) is modificated to the form (2), the solution described is shown in Fig. 1a) and Fig. 1b).

$$R = \frac{(2s + y)r}{\Delta r + z} \qquad (2)$$

where 2s [mm] - distance of the wheelset contact circles; y [mm] - the lateral displacement of the complemented wheel tread r [mm] - nominal radius of the wheel; Δr [mm] - the difference between the instantaneous rolling radius of the left and right wheels of the wheel when the track channel clearance is exhausted; z [mm] - distance of the vertical displa-cement of the complemented wheel tread.

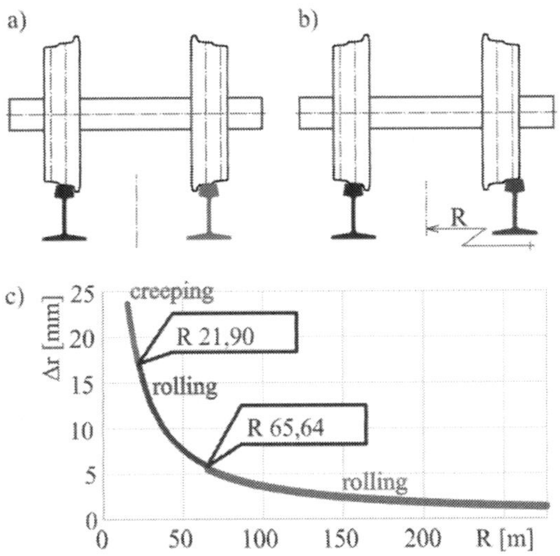

Fig. 1. Wheelset with complemented tread (a) during ride in track according to (1); (b) during ride in track according to (2);c) distribution of the radius of the track arc in terms of the possibility of its overcoming by rolling

The arc of the track of a lesser radius than the relationship (1) defines, is solved specifically - the inner rail is shifted in a vertical and lateral direction about the same values as is the complementary wheel tread shifted from the original one. The complemented wheel tread enters into operation (or acts out of action) as a result of passing over a specially shaped rail located in the track transition curve in places with a radius according to the relationship (1) before (or be-hind) the small radius curve. In this way, it is possible for the wheelset to create sufficient wheel radius difference when passing an arc. The desired, resp. the achievable value Δr depending on the radius of the arc of the track is graphically shown in Fig. 1c).

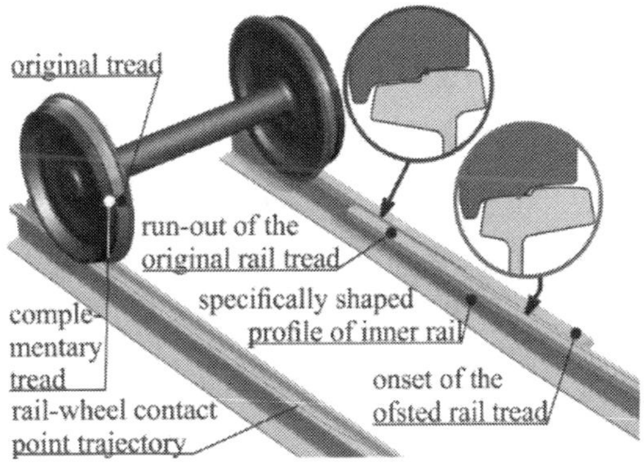

Fig. 2. Visualization of an entry into a small radius curve in a location defined by the relationship (1) with drawing of a trajectory of the contact location on the surface of the rail belts

In the described process, the wheel-rail contact location is changed from the original tread to the complemented part as shown in Fig. 2. The trajectories are drawn on the surface of the track belts where they are in contact with the passing wheelset obtained on the basis of the vehicle simulation analysis described in Chapter 4. The trajectory shows the ride of the first wheelset of the vehicle.

In Fig. 2 is a situation that needs to be evaluated in terms of phenomena occurring in the wheel-rail contact. As an analogous situation we can consider a ride through a facing rail, in common practice located in the immediate vicinity of the point frog of urban railways.

3 The method of passing the wheelset through a frog

One of the specifics of city railways is the way the wheel passes through the frog of the point. This process [22] differs fundamentally from conventional railways [26]. The tramline wheel has a smaller width, so it cannot safely create a two-point contact of the wheel tread with the tip of the frog and the wing rail necessary to overcome the gap created for the flange during the ride in a turning as it is commonly used in the conventional rail environment. The passage of a railway wheel through a point frog together with the comparison of the width of the railway [27] and the tramline wheel [25] is shown in Fig. 3.

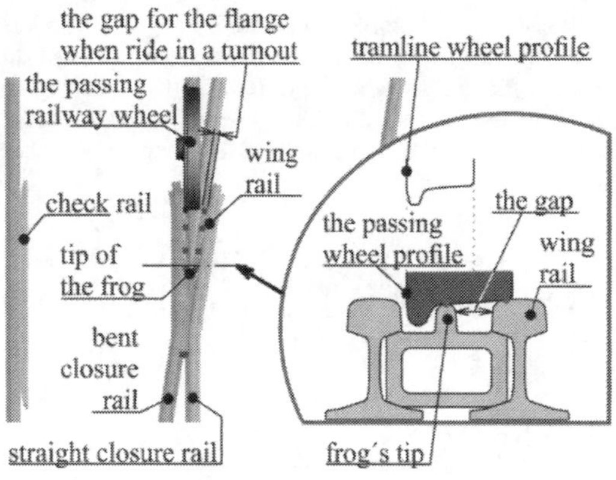

Fig. 3. Visualization of railway wheel passage through a point frog

Before passing a frog, the tramline wheel moves its contact point from the tread to the flange by passing it through the facing rail, then passes through the frog on the flange, allowing the wheel tread to pass over the gap created for the flange for ride to a deviation. Subsequently, a contact point is then transferred from the flange back to the tread thanks to the facing rail. The visualization of the passage of the tramline wheel through a city railway point frog is shown in Fig. 4. The trajectories are drawn on the surface of the track belts where they are in contact with the passing wheelset, obtained on the basis of the simulation analysis of the vehicle's ride described in detail in Chapter 4. The displayed trajectory documents the ride of the first wheelset of the vehicle.

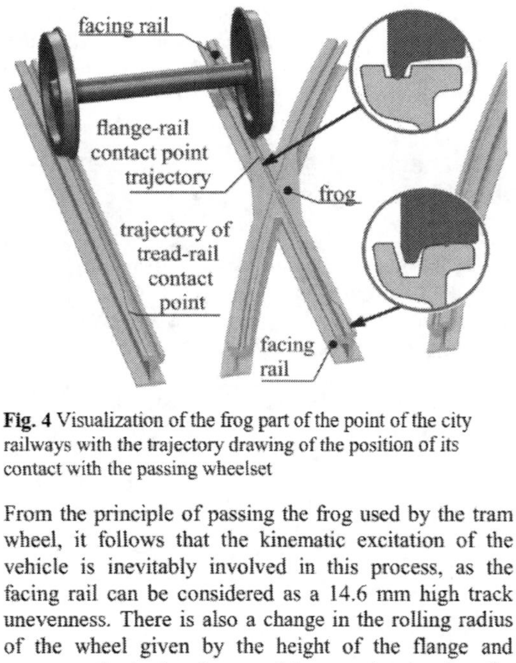

Fig. 4 Visualization of the frog part of the point of the city railways with the trajectory drawing of the position of its contact with the passing wheelset

From the principle of passing the frog used by the tram wheel, it follows that the kinematic excitation of the vehicle is inevitably involved in this process, as the facing rail can be considered as a 14.6 mm high track unevenness. There is also a change in the rolling radius of the wheel given by the height of the flange and consequently to the change of the quantity Δr, thus the wheelset has the tendency to roll over the curved trajectory, the radius of which can be determined from the relation (1) modified to the following form:

$$R = \frac{(2s - y)r}{\Delta r + h} \tag{3}$$

where 2s [mm] - distance of the wheelset's contact circles; y [mm] - the distance of the wheel contact circle during ride on the flange from its nominal position r [mm] - nominal radius of the wheel; Δr [mm] - the difference between the instantaneous rolling radius of the left and right wheels when the track channel clearance is exhausted; h [mm] - height of the flange.
Comparison of the ride of wheelset with the added tread into an arc of small radius and the passage of the wheelset through facing rail and the frog was realized by simulation analysis of the vehicle's ride in a calculation program.

4 The method of passing the wheelset through a frog

In the simulation analysis, a ride of a vehicle similar to T3 tram with its parameters [28] is contemplated, with the exception that the vehicle being considered in the simulation analysis is equipped with bogies with a mechanism for steering the wheelsets to a radial position in accordance to the angle between the bogie frame and the vehicle body. A detailed description of the vehicle simulation model has been published in publications [29].

Passing the investigated track sections with a gauge of 1000 mm is considered at a constant speed of 10 km/h without driving and braking forces. In the case of passing a point frog, the ride on a straight track is considered, the facing rails and the frog are located on the right rail. The facing rails have a length of 800 mm, a longitudinal slope of the bottom of the groove of 1:32, the length of the rolling edge curvature run-out 600 mm [22]. To create the variable geo-metry of each facing rail, 9 profiles of cross section were created in a specific structure necessary to ensure compatibili-ty with the computational program [30]. The following section, representing the frog, has a length of 1000 mm, with a crossing at the centre of this section including a gap for the flange for a ride into deviation. The crossing angle of the rails of 23° represents a sudden change in rail geometry. For a correct processing of this change in the calculation pro-gram, it was necessary to define 101 cross-sections in the section of crossing ± 250 mm. The described track section is shown in Fig. 5.

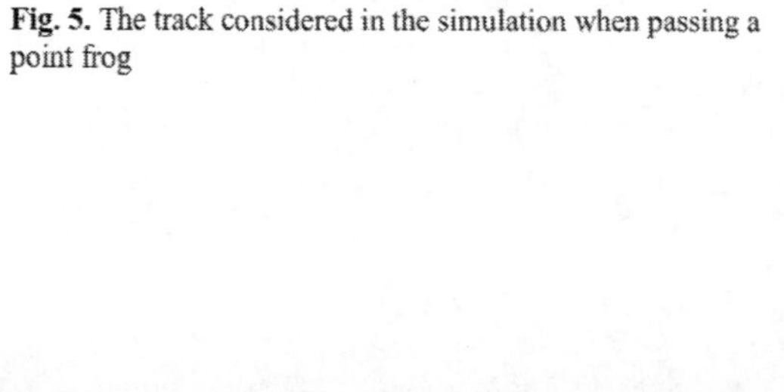

Fig. 5. The track considered in the simulation when passing a point frog

In the case of an arc entry through a specially shaped rail allowing the transfer of the point of contact from the wheel tread to the complemented part, the ride in the clothoidal transition curve connecting the straight track with a right-hand rotated arc of a radius of 25 m is considered. The length of the transition curve is 10 m, it is a tilted arc, where the beginning of the superelevation sine ramp is located in a straight line 5 m before the beginning of the transiti-on curve. The superelevation of the outer rail belt in the arc is 33.6 mm. Basically, this is a geometric arc solution based on the standard [31].

The shaped portion of the rail, i.e. the inner rail of the length of 1.5 m, is designed so that when being passed over, there is no vertical deflection of the passing wheel. The run-out of the original rail tread, as well as the onset of the new tread, have a length of 750 mm and a longitudinal slope of 1: 250. The moment when the contact point is moved in theory occurs when the half-length of the shaped rail is passed. In order to create a variable geometry of the shaped rail, 12 profiles of the cross section were created in the specific structure necessary to ensure compatibility with the calculation program [30]. The described section of the line is shown in Fig. 6.

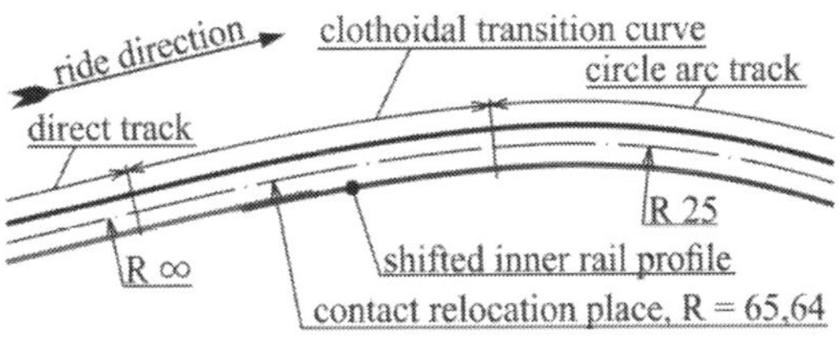

Fig. 6. Track considered in simulation when entering a small radius curve

Vehicle ride described by track sections was simulated in Simpack 9.10.1. [32]. The results obtained in the monito-red quantities are given in Chapter 5.

5 Comparison of calculated parameters courses

In order to compare the course of selected parameters of vehicle effects on the track, it is necessary to harmonize their time track, or another independent variable. In the presented graphs, the independent variable is the distance travelled by the vehicle along the track shown in Fig. 5 and 6, where the zero value of the distance travelled is the place where the first contact of the flange with the bottom of the groove, or the complemented tread and the shifted rail belt on the first wheelset of the vehicle, occurred.

The selected parameters of the wheel-rail contacts of the first bogie of the vehicle passing through the monitored track section are presented. In the following graphs (Figures 7 to 11), the shortened denomination of individual wheels is used: 1L – left wheel of the first wheelset, 1R – right wheel of the first wheelset, 2L – left wheel of the second wheelset, 2R – right wheel of the second wheelset.

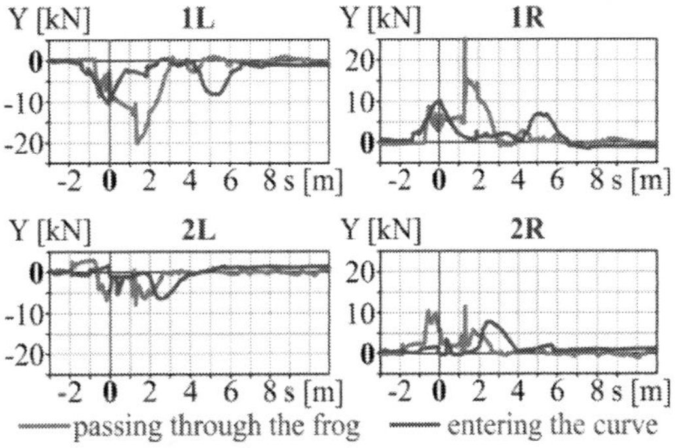

Fig. 7. Lateral forces on the individual wheels of the front bogie

Fig. 7 shows the course of the guiding forces during ride through the monitored track section. In the case of passing through a frog, it is possible to see the moment when the wheel enters the facing rail, which is manifested by the increa-se of the guiding forces at the point s = -0.8 m. At the point s = 1.3 m, it is possible to see a sharp increase in the steering force caused by the re-positioning of the contact location on the tread. Changing the contact point when entering the arc also causes a certain increase in the steering force, but it is milder.

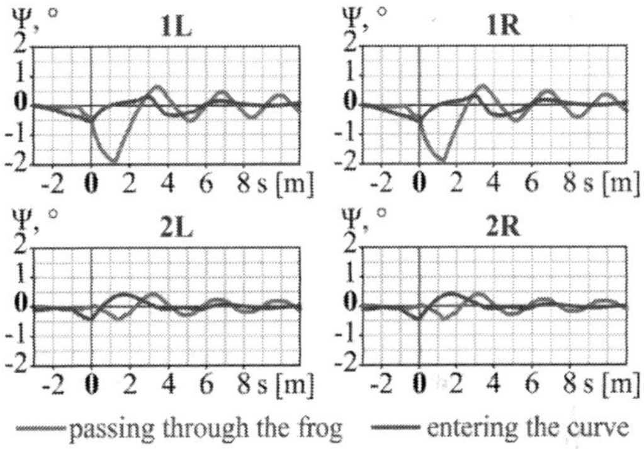

Fig. 8. Courses of angles of attack on the individual wheels of the front bogie

From Fig. 8 it can be deduced that in both cases there is a certain oscillation being induced in the vehicle. More sig-nificant increase in the angles of attack occurs on the first wheelset, reaching almost 2°. Entering the arc causes a larger increase in the angles of attack on the second wheelset, but the values do not reach 0.5°.

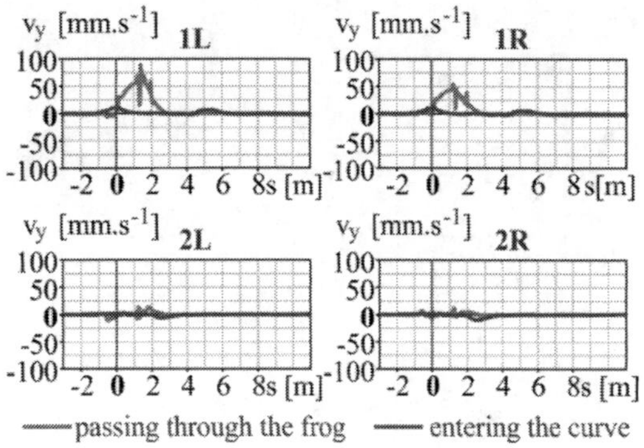

Fig. 9. Course of the lateral creep velocities on the individual wheels of the front bogie

Fig. 9 shows the course of lateral creep speed in the contact point. Extremely high values, approaching 100 mm.s^{-1}, are achieved when passing through the frog. They are caused by the wheelset deflection from the position perpendicular to the track axis as a consequence of its rolling according to relation (3), although it is running in a straight track. A significantly more favourable course of the observed magnitude is reached when entering an arc where it does not exceed 15 mm.s^{-1}.

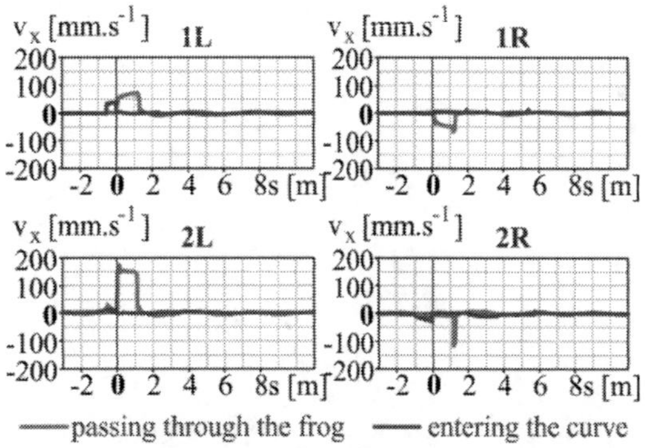

Fig. 10. Course of the longitudinal creep velocities on the individual wheels of the front bogie

The importance of the specially shaped rail geometry as well as the correctness of the proposed solution described by the relation (2) confirms the comparison of the longitudinal creep speeds shown in Fig. 10.

Changing of the contact point location in the process of passing a frog causes extremely high creep velocities valu-es, which on some wheels exceed 100 mm.s-1. The entry into the track arc causes some increase in the longitudinal creep velocity, which for example on the attacking wheel is approaching 40 mm.s-1, but this increase occurs before changing the contact point. At the moment of moving the contact point from one area to the part added, the creep velocity decreases, which in the next part of the monitored section is approaching zero values.

A significant point of view on the described problem is also the intensity of the vehicle's impact on the track, directly affecting the service life of the contact surfaces.

Fig. 11 shows the course of the wear number $W_p[N]$, i.e., the quantity predicting the intensity of the interaction of the contact pair given by:

$$W_p = |T_x.w_x| + |T_y.w_y| + |M_z.\varphi_z| \tag{4}$$

where T_x [N] - the longitudinal creep force; w_x [-] - longitudinal creepage; T_y [N] - lateral creep force; w_y [-] - lateral creepage; M_z [Nm] - creep torque; φ_z [m^{-1}] - spin creepage.

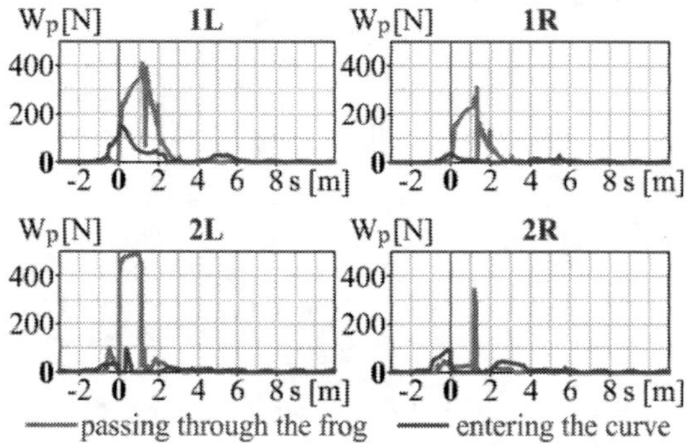

Fig. 11. Course of the sum of the wear number of contact points on the individual wheels of the front bogie

It can be deduced from the courses above that the entrance of the vehicle into the arc using the proposed shaped rail has, as on the vehicle as on the rail, less negative impact than the passage through the frog.

6 Conclusion

The paper analyses the effect of the city railways vehicle on the track when passing a specific track sections, where the wheel-rail contact point moves out of the wheel tread. Using the SIMPACK computational program, two analogous cases were analysed - passing the frog, where the contact point is moved from the tread to the wheel flange; and the entry into the small radius arc by means of a specially designed rail, that allows the contact point to be moved from the tread to the complemented part. On the basis of the performed analysis, it can be evaluated, that changing the contact point when entering the arc by means of the designed, specially shaped rails takes place in a more favourable way, in terms of the steering forces, the angles of attack and the creep velocities. From the wear point of view, it is possible to assume a nearly 4 times longer service life of the proposed shaped rail than the one of a frog point.

This work was supported by the Cultural and Educational Grant Agency of the Ministry of Education of the Slovak Republic in project No. KEGA 077ŽU-4/2017: Modernization of the Vehicles and engines study program. The work was also supported by the project No. APVV-0842-11: Equivalent railway operation load simulator on the roller rig and VEGA No. 1/0558/18: Research of the interaction of a braked railway wheelset and track in simulated operational conditions of a vehicle running in a track on the test stand.

Research - Educational Cetre of Rail Vehicles (VVCKV)

References

1. J. R. Evans, T. K. Y. Lee, C. C. Hon, *Vehicle System Dynamics*. 119-127.(2009)
2. CH. W. Jenks, et. al., *Track Design Handbook for Light Rail Transit*, 686 p. https://www.nap.edu/read/22800/chapter/
3. E. Eken, R. Friberg, *Modelling of dynamic track forces generated by tram vehicles*. 130 p. (2015)
4. P. Šťastniak, L. Smetanka, M. Moravčík, Manufacturing Technology. **17, 2**, 250-256 (2017).
5. V. Tkachenko, S. Sapronova, I. Kulbovskyi, O. Fomin, Eastern European Journal of Enterprise Technologies. **5. 7-89**, 65-72. (2017)
6. J. Gerlici, T. Lack, Applied mechanics and materials. **486**, 379-386 (2014).
7. J. Gerlici, T. Lack, Wear: An international journal on the science and technology of friction, lubrication and wear. **271**, **1-2**, **Sp. iss.**, 246-258 (2011).
8. J. Gerlici, T. Lack, Procedia Engineering. ISSN 1877-7058. **2, 1**, 2249-2257 (2010).
9. P. Baran, M. Brezáni, P. Kukuča, P. Šťastniak, Procedia Engineering. **192**, 34-39 (2017).
10. L. Smetanka, P. Šťastniak, Manufacturing Technology. **17**, **4**, 580-585 (2017).
11. J. Gerlici, T. Lack, J. Harušinec, Applied mechanics and materials. **486**, 387-395 (2014).
12. J. Gerlici, T. Lack, Applied mechanics and materials. **486**, 396-405 (2014).
13. Lincoln (2011) Wayside Rail Lubrication http://www.lincolnindustrial.com/Catalogs/Uploaded/ Cata-log_88/Railroad_W-183-EN-0511.pdf
14. A. Marqueteeken, A. Van Leuven, F. Kopf, *Urban Track. Cost effective track maintenance, re-newal & refurbishment methods. Preventive maintenance of embedded tram tracks Rail wear in curves and special trackwork for trams*. 44p. http://www.urbantrack.eu/images/site/deliverables/ d0208_stib_m24.pdf
15. Vossloh Fastening Systems (2016) Chamber filling elements in covered track. https://www.vossloh.com/01_product_finder/VFS/Kammerfuellelemente/Vos_Produktbroschuere_Kammerfuellelemente_EN.pdf
16. Vossloh Fastening Systems (2016) System W-Tram. https://www.vossloh.com/01_product_ finder/ VFS/System-W-Tram/Vos_Produktbroschuere _Sys-W-Tram_EN.pdf
17. A. Chudzikiewicz, B. Sowinski, *The Dynamics of Vehicles on Roads and Tracks - Proceedings of the **24th** Symposium of the International Association for Vehicle System Dynamics, IAVSD* pp. 1427-1434 (2015).
18. P. Sýkora, T. Lelek, O. Sadílek, Current problems in rail vehicles 2017 – **XXIII**. (2017).
19. T. Lack, J. Gerlici, Applied mechanics and materials. ISSN 1660-9336. Vol. **486**, pp. 371-378 (2014).
20. V. Hauser, J. Gerlici, T. Lack, M. Loulová, O. S. Noženko, K. O. Kravčenko, O. V. Prosvirova, *Railway bogie wheelset and curved track segment*: Patent application no. a201701589 (In Ukrainian) Kyjiv: Ukrajinskyj instytut promyslovoji vlasnosti (UKRPATENT), 2017. 11 s.. (2017).
21. O. Nozhenko, V. Hauser, K. Kravchenko, M. Loulová, Reducing tram car's curve-pasing resistance by double treaded wheel profile. Manufacturing Technology, **18**, **1**, 79-83, (2018).
22. DT-Výhybkárna A Strojírna (2017) *Tram car crossings, components, accessories*. (In Czech) https://www.dtvm.cz/dtvs/cz/produkty/tramvajove-vyhybky/tramvajove-vyhybky-kolejove-konstrukce-pro-tramvajovou-dopravu/
23. J. Dižo, M. Blatnický, Diagnostyka. **17**, **2**, pp. 9-16. (2016)
24. T. Lack, J. Gerlici, Applied mechanics and materials. **486**, 359-370 (2014).
25. J. Zelenka, M. Kohout, *Relationship wheel-rail in the conditions of tramcar operations*. **46**. zasedání odborné skupiny Tramvajové tratě Sdružení Dopravních podniků České republiky University of Pardubice, Jan Perner Faculty Of Transport, 42 p., (2012).
26. A. Hába, J. Zelenka, M. Kohout, Vibroengineering Procedia.. **6**, **10/2015**, 217-222 (2015).
27. UIC Code 510-2 OR (2004) Trailing stock: wheels and wheelsets. Conditions concerning the use of wheels of various diameters. UIC, Paris, 2004.
28. M. D. Ivanov, A. A. Ponomarev, B. K. M. Ieropol'skij, *Tramcars T-3* (In Russian). 240 p. (1977).
29. P. Šťastniak, M. Moravčík, P. Baran, L. Smetanka, MATEC Web of conferences. Vol. **157** (2018).
30. J. Soukup, J. Skočilas, B. Skočilasová, J. Dižo, Procedia Engineering. **177**, 235-241 (2017).
31. TNŽ 73 63 61. Geometrical location and layout of 1000 mm gauge tracks, (In Slovak) ŽSR: Bratislava (2007).
32. Simpack 9.9.2 Documentation. 4556 p. (2016).

Design of a new railway wagon for intermodal transport with the adaptable loading platform

Pavol Šťastniak[1,*], *Pavol* Kurčík and *Alfréd* Pavlík[1]

[1]University of Žilina, Faculty of Mechanical Engineering, Department of Transport and Handling Machines, Univerzitná 8215/1, 010 26 Žilina, Slovak Republic

Abstract. The aim of the article is to present the structural design of freight railway wagon with variable use of loading space with regard to the safe operation and assessment of the properties by the calculation methods of simulation analysis. Virtual model of wagon was created in a computer program PTC/Creo. After creating the construction, the calculation model of the wagon frame was subjected to the static and dynamic analysis in programs ANSYS and ADAMS/Rail. On the basis of computer aided simulation analysis was optimized the chassis frame of the wagon. This wagon will be able to offer even more capacity and utilize less resources and energy than current wagons for intermodal transport.

1 Introduction

Indeed nowadays about 1,5 billion t/km are transported in Europe by lorry at distances farther than 150 km, conversely only 0,4 billion t/km (20%) are transported by train, this entails important costs for fossil fuels. In the nearby future when transportation has to be more sustainable it seems quite clear that freight railways will win the mode choice more often [1].

For this to happen though, it is necessary that freight railways, apart from lowering their prices, significantly improve highly increase the quality of transportation. In that sense, quality standards such as reliability, flexibility, availability, cargo security and safety, punctuality, customisation, marketability, traceability, complementary servicing and time for transport among others have to be improved by railways as well.

Hence, rail freight has the challenge to become excellent and to gain in reputation. There are many actions to increase quality in rail freight transport; one of them is the optimisation of the current wagon fleet to improve availability, flexibility, marketability, commercial speed, cargo security and cost. This optimisation has to respond to the actual trends of transport demand and has to be in consonance with the required and feasible infrastructure upgrades.

The basic idea of new wagon is that in the future, longer loading surfaces without interruptions, as well as more capable platforms with higher axle loads and with lower loading heights will be necessary to increase the capacity of the freight railway transportation.

The time schedule is divided to several subtasks as shown in Fig. 1.

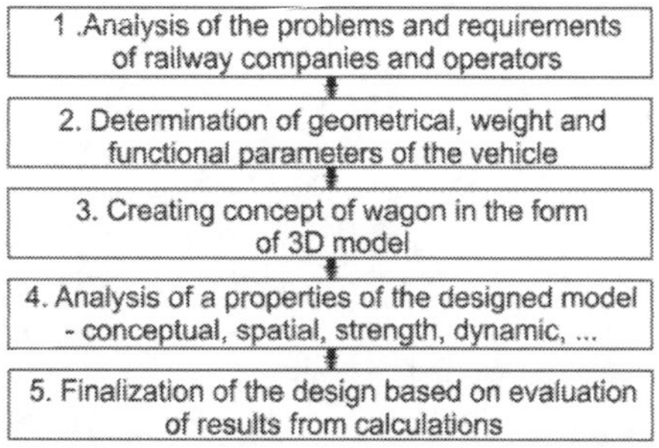

Fig. 1. The time schedule segmented into several subtasks.

2 Analysis of the problems and requirements of railway operators

Intermodal continental transports utilize a large amount of unit types (much more than hinterland (sea) transportation) and that increases the amount of possible loading cases for the trains. In that sense, there would be an optimal wagon for each case but this wagon could be sub-optimally utilized for other cases. This variety of cases makes difficult to know which wagon is the optimal for an averaged situation.

Wagons represent an important investment for companies (c.a. 100.000 € per wagon) and they should be extensively utilized during their whole life cycle (25-30 years) to achieve profitability. For this reason, wagons specialized in one kind of units are usually employed for other unit types even if they are not 100% efficient at it.

From individual wagons were created four trains, whose maximum length were 500 m and have been investigated the parameters listed in Fig. 2.

* Corresponding author: pavol.stastniak@fstroj.uniza.sk

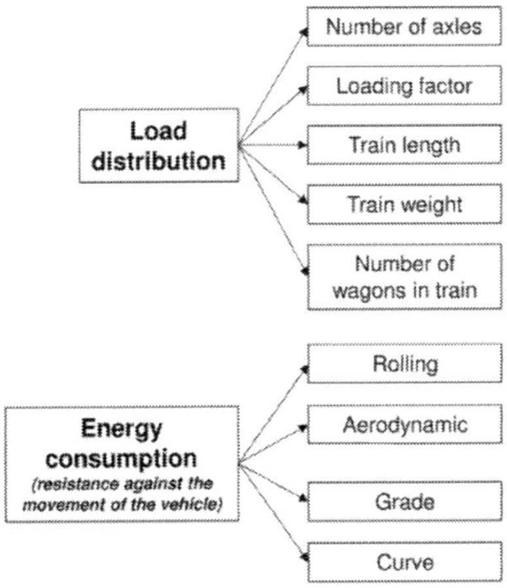

Fig. 2. The time schedule segmented into several subtasks.

Concluding that:

- 106 ft wagon suits optimally the FTL (full truck load, semitrailers) segment and performs quite reasonably in other segments except the pure bulk ones.
- 60 ft wagon is the optimal wagon for the pure bulk segment.
- 80 ft wagon would offer better performance than 106 ft in averaged situations (mainstream shuttles), but it is not recommended for specialized FTL transports, however it performs better than 106 ft wagon in bulk segments.
- 90 ft wagon would offer better performance than 80 ft if in the future the 45 ft unit is widely introduced.

since it would enable better utilization of space (loading length) on trains than existing wagon technologies. The 80 ft wagons would be able to transport same of even more amount of units with the fewer axles and the less deadweight. Furthermore, the aerodynamics would improve (fewer gaps between containers, fewer bogies per meter) and the noise emissions would be reduced due fewer axles per train.

Longer loading lengths 85 ft and 90 ft could have an advantage too, but only if the 45 ft unit is widely introduced and if it dominates in intermodal traffics, which is not the actual case. A revision of this issue has to take place in the approximately 5 years.

The strategic procedure would be to design a 80 ft without pocket and try to make it as cheap as possible. By this it could be very competitive in its market segment.

3 Concept of wagon in the form of 3D model

The model presents a complete structure of the 4 – axle freight wagon (Fig. 3) for transportation of ISO containers and swap bodies which should meet prescriptions TSI-WAG, valid regulations UIC, agreement on the reciprocal use of freight wagons in international traffic AVV (RIV), recommendations ERRI and norms EN.

Chassis frame (Fig. 4) is welded the steel construction consisting of two side longitudinal beams (in the most of "I" profiles various sections) which through the cross beams create a support frame. On

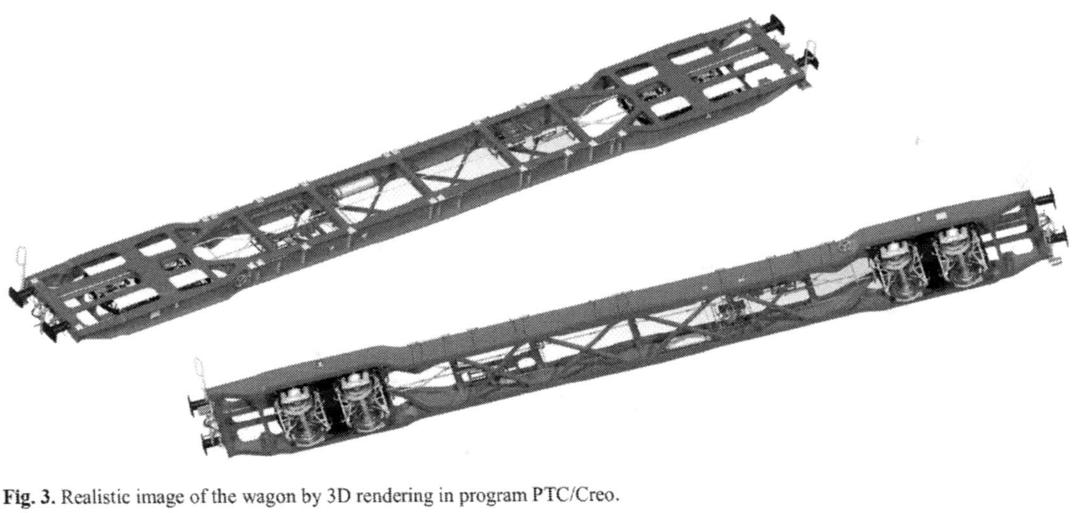

Fig. 3. Realistic image of the wagon by 3D rendering in program PTC/Creo.

The results of the capacity simulations have showed that an 80 ft long wagon could lead to an important advantages in efficiency. These advantages would be amplified by averaged cases with mixture of units, for example in the case of a shuttle between the two important continental terminals with an unknown and varying proportion of units.

It could be concluded that the 80 ft wagon would bring about an important gain for continental transports

lower flange plate of main cross-beam there is rotating lead for upper of rotating pivot standard type and arms for sprung side bearers. The passage from the front to the center part of the wagon is branched because of better layout longitudinal forces. The center section of chassis frame contains the cross brace by reason optimizing the torsional stiffness of the wagon. The material thickness

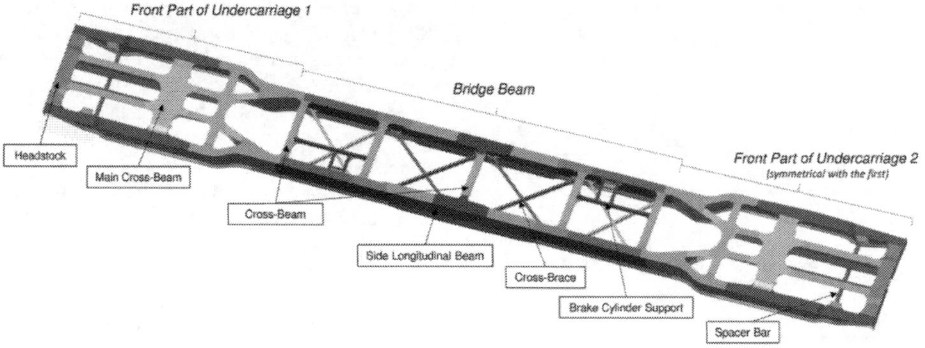

Fig. 4. Main parts of the chassis frame.

of individual structural parts has been optimized based on strength analysis simulations.

4 Computer aided simulation analyzes

Integral part of the development of rail vehicles are ride tests performed on test tracks and static tests on test stands. These tests are expensive and nowadays are using the modern computer technologies that can simulate riding a vehicle/train on the track and detect selected parameters [2, 3].

For the proposed wagon were performed following simulation analyzes:

- Static evaluation:
  - wagon torsional stiffness,
  - max. vertical load forces,
  - wagon rising on one side,
  - all wagon rising,
  - longitudinal forces,
  - load combination.
- Dynamic evaluation:
  - modal analysis,
  - simulation of the ride through the S-curve,
  - geometric characteristics of wheelset and track,
  - acceleration, wheel forces, guiding forces and safety against derailment.

Investigated construction was subjected to static structural analysis (program ANSYS) based on the standards [4, 5]. On the basis of the 3D model of chassis frame was created calculation model which characteristics are shown in Fig. 5.

The task of structural analysis was to simulate behaviours of stresses and deformations in the proposed construction loaded maximum vertical forces.

The simulation analysis shows that the greatest deflection investigated structures (middle of the wagon) will be 29,92 mm. The report ERRI B12/RP17 states that the maximum deflection of the chassis frame must not exceed 3% of the distance between bogie pivots respectively axis of wheelsets. The results of the simulations show that the chassis frame satisfies strength conditions. For further development respectively production of this type of construction it is necessary to verify the results on the real construction on the test stand [6, 7, 8]. Behaviour of displacements and stresses is shown in Fig. 6.

The dynamics analysis was performed in accordance to the technical parameters for two track sections in Cerhenice railway test track circuit. In the first case was taken into account the track with 450 m radius and in

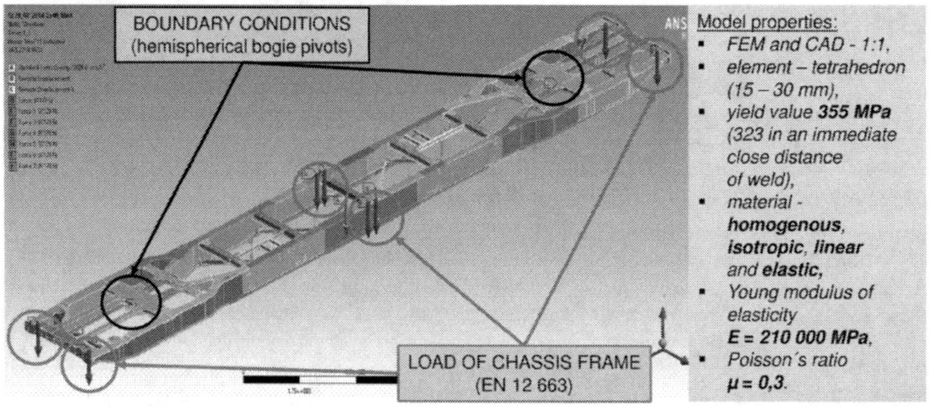

Fig. 5. Computational model – boundary conditions, loads and properties.

https://doi.org/10.1051/matecconf/201823500030

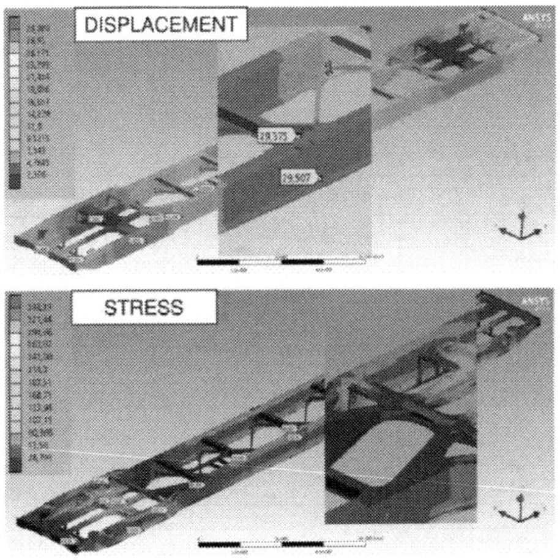

Fig. 6. Results of static simulation analysis – max. displacement and stress.

second case the right track and the velocity of 100 km/h.

The geometric characteristics [9, 10, 11] of railway wheelset and track (Equivalent conicity and Delta-R function) were evaluated for the profiles contact couples S1002 for wheel, UIC60 and R65 rail with inclinations of 1:20 and 1:40 (right track) and for the profiles contact couples S1002 for wheel, S49 rail with inclinations of 1:20 and 1:40 (radius R450 m).

The geometric characteristics analysis for both cases was completed with the Equivalent conicity and Delta R function characteristics (Fig. 7) set for different gauges values (in the interval of 1432-1448 mm).

For both analysis the accelerations, wheel vertical forces and leading lateral forces values, from them the derailment criterion (safety against derailment) was evaluated.

In both cases, the safety against derailment values, were in the interval of safe operation.

With this research and development subjects deals other publications too [12, 13, 14, 15].

5 Conclusion

It is estimated that this wagon will be a sustainable wagon that will be able to offer the same or even more capacity and utilize less resources and energy than current wagons for intermodal transport.

The analysis of the intermodal traffic has enabled to carry a simulation in which wagon capacity performance has been assessed. The primary results speak for an increase of capacity due to wagon use of about 10% in comparison to a realistic reference case. This capacity increase is mainly produced by a better arrangement of containers and by reduction of deadweight of the wagon. When it comes to energy consumption wagon could save up to 18% of the energy necessary to transport a units. This is mostly due to an improvement of the loading factor that enables a better compression of the containers, fewer gaps, a reduced number of axles and a lighter tare per units.

Main advantages over existing design are:

- lower loading plane = transport of containers and swap bodies all dimensions in kinematic gauge G1 (including HiCube containers and swap bodies with a wide of up to 2600 mm),
- distance between pivots to 18 000 mm = no special permissions.

Due to the reduced amount of axles, for same or even more units capacity, wagon will bring about important savings in maintenance and very importantly, it will reduce the noise emissions per train.

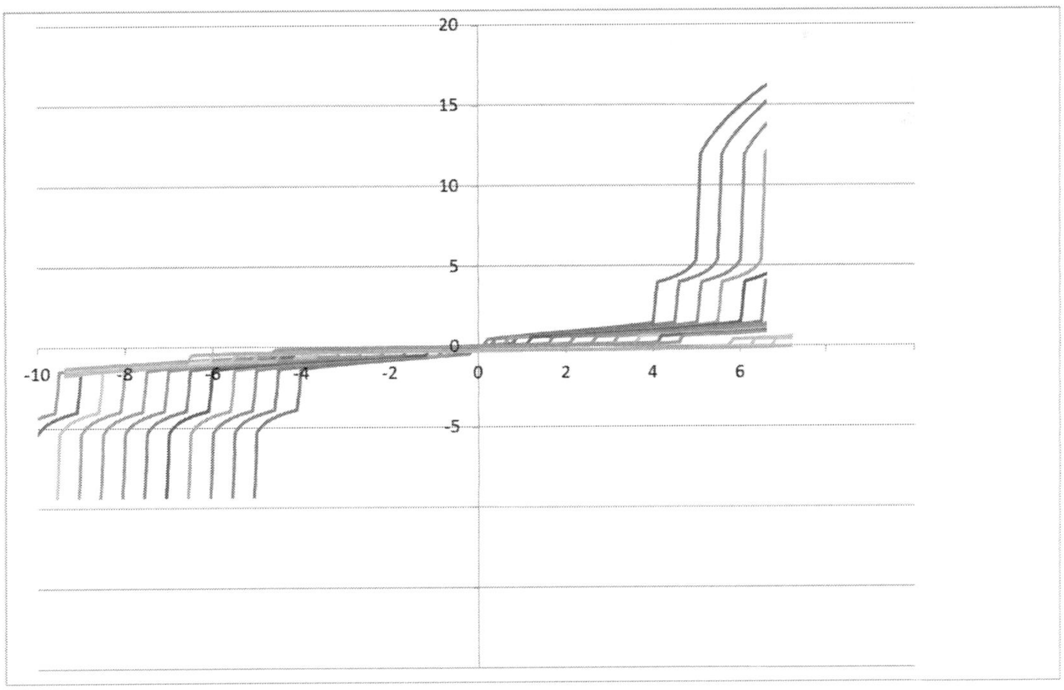

Fig. 7. Delta R function S1002/R65/1:40/gauge 1432 - 1448.

This work was supported by the Cultural and Educational Grant Agency of the Ministry of Education of the Slovak Republic in project No. KEGA 077ŽU-4/2017: Modernization of the Vehicles and engines study program. The work was also supported by the project No. APVV-0842-11: Equivalent railway operation load simulator on the roller rig and VEGA No. 1/0558/18: Research of the interaction of a braked railway wheelset and track in simulated operational conditions of a vehicle running in a track on the test stand.

Research-Educational Center of Rail Vehicles (VVCKV)

References

1. P. Fabián, J. Gerlici, J. Mašek, P. Márton, *Versatile, efficient and long wagon for intermodal transport in Europe*. Communications: scientific letters of the University of Žilina. ISSN 1335-4205. Vol. **15**, No. 2, pp. 118-123 (2013).
2. J. Dižo, J. Harušinec, M. Blatnický, Structural analysis of a modified freight wagon bogie frame. MATEC Web of Conferences. ISSN 2261-236X. Vol. **134** (2017).
3. J. Dižo, S. Steišunas, M. Blatnický, *Vibration analysis of a coach with the wheel-flat due to suspension parameters changes*. Procedia Engineering, ISSN 1877-7058. Vol. **192**, pp. 107-112 (2017).
4. EN 12 663-2: Railway applications. Structural requirements of railway vehicle bodies. Part 2: Freight wagons. European Committee for Standardization (2011).
5. ERRI B12/RP17, No. 8: Programme of Stresses to be carried out on Wagons with Steel Underframe and Body Structure. European Rail Research Institute (1996).
6. J. Gerlici, T. Lack, *Modified HHT method for vehicle vibration analysis in time domain utilisation*. Applied mechanics and materials. ISSN 1660-9336. Vol. **486**, pp. 396-405 (2014).
7. J. Gerlici, T. Lack, *Rail vehicles brake components test bench utilisation*. Applied mechanics and materials. ISSN 1660-9336. Vol. **486**, pp. 379-386 (2014).
8. J. Gerlici, T. Lack, J. Harušinec, *Realistic simulation of railway operation on the RAILBCOT test stand*. Applied mechanics and materials. ISSN 1660-9336. Vol. **486**, pp. 387-395 (2014).
9. T. Lack, J. Gerlici, *A modified strip method to speed up the calculation of normal stress between wheel and rail*. In Applied mechanics and materials. ISSN 1660-9336. Vol. **486**, pp. 359-370 (2014).
10. T. Lack, J. Gerlici, *A modified strip method to speed up the tangential stress between wheel and rail calculation*. Applied mechanics and materials. ISSN 1660-9336. Vol. **486**, pp. 371-378 (2014).
11. J. Gerlici, T. Lack, *Railway wheel and rail head profiles development based on the geometric characteristics shapes*. Wear: An international journal on the science and technology of friction, lubrication and wear. ISSN 0043-1648. Vol. **271**, No. 1-2, Sp. iss., pp. 246-258 (2011).
12. J. Gerlici, T. Lack, *Contact geometry influence on the rail / wheel surface stress distribution*. Procedia Engineering. ISSN 1877-7058. Vol. **2**, Iss. 1, pp. 2249-2257 (2010).
13. T. Lack, J. Gerlici, Railway wheel and rail roughness analysis. Communications: scientific letters of the University of Žilina. ISSN 1335-4205. Vol. **11**, No. 2, pp. 41-48 (2009).
14. V. Hauser, et. al., Proposal of a steering mechanism for tram bogie with three axle boxes. Procedia Engineering. ISSN 1877-7058. Vol. **192**, pp. 289-294 (2017).
15. V. Hauser, et. al., *Impact of three axle boxes bogie to the tram behavior when passing curved track*. Procedia Engineering, ISSN 1877-7058. Vol. **192**, pp. 295-300 (2017).

***** Formatting Issue - Best Available Paper/Graphic *****

Design of new effective train traffic diagram for a railway line

Nikoleta Mikušová, *Miloš* Timák, *Gabriel* Fedorko[1] and *Vieroslav* Molnár[2]

[1]TU Košice F BERG, Logistics Department, Park Komenského 14, 040 01 Košice, Slovakia
[2]TU Košice, Faculty of Manufacturing Technologies, Bayerova 1, 080 01 Prešov, Slovakia

Abstract. The paper presents the result of a case study analyzing the current state of rail transport in Slovakia, namely the current state on the line Humenné – Medzilaborce in the east part of Slovakia and also the current train traffic diagram. On the base of the realized analysis, it was determined the problematic parts of this line, as the insufficient occupation of the transportation line by passengers. This is caused by a lower number of inhabitants of the researched area and also the geographical position of some villages in this area. The aim of this study was therefore to find a solution that would increase interest in the use of rail transport in this region.

1 Introduction

Rail transport, as well as passenger transport and freight transport, has a very important status in the national economy of each country. It is a type of land rail transport along an artificial traffic road which serves to move of material or passengers from the point A to the point B, by the help of rail vehicles. The technical base of rail transport is created by means of transport, i.e. rail driving vehicle for the realization of transport, traffic roads, i.e. railway track and equipment of the rail transport (for example building for transport control, dispatching building, stations, etc [1]. There is about 3624 km of railway line in Slovakia, from this number 3593 km are operated, and electrified are 1576 km. Every year, new proposals are being made for electrification of further kilometers of existing lines, but this innovation is costly and therefore these plans are not always going to be implemented. Electrification also brings additional costs related to the innovation of wagon stocks that must be adapted to this change [2,3]. Table 1 presents the current state of the railway network in Slovakia.

Table 1. Railway network in Slovakia [2]

Length of railways	total: 3624 km	Narrow gauge railway: 50 km
	Operated: 3593 km	Wide-gauge railway: 98 km
	Single-track: 2576 km	Standard gauge: 3444 km
	Double-track: 1016 km	Electrified: 1576 km
Turn-out tracks	Number: 8425	
Bridges	Number: 2317	Length: 52 km
Tunnels	Number: 75	Length: 43 km
Railway crossings	Number: 2205	

2 Train traffic diagram (TTD)

Train traffic diagram (TTD) presents an understanding among all parts of railway transport and organization of train transport. It is a set of measures and tools related to rail transport. TTD is implemented in total railway network, on the base of international railway agreements and conventions. TTD should ensure the speed, safety, and economy of people and goods transport, the required amount of transport and the optimal use of wagon stocks. It should harmonize the activities of stations with adjacent track sections, full use of staff working time and also ensure the highest productivity of work and the lowest costs. Although train movement is often uneven, in the TTD it is presented simplified as a uniform movement between two units. [4-6] It would be unnecessarily complicated to illustrate the actual precise movement of the train and therefore the principles of calculations and graphical depictions of train movements related to the front of the train were adopted. It is possible to classify TTD by indicators, such as the mutual ratio of trains speed in the TTD, number of tracks, ratio of train number for one and reverse direction, an organization of rides for subsequent trains, duration of occupancy of interstation sections, time of operation, level of occupancy and use of track capacity, orderliness. [7] TTDs are classified by the mutual ratio of trains speed as parallel – this type belongs to the simplest type of TTD because all trains have the same driving time in the determined section due to the same weight, length, speed and specific capacity of trains; normal/commercial/out of parallel – train in the same direction are not parallel because they have different driving time, length and transport weight. TTDs are classified by the number of tracks as single-line (bi-directional) – trains in both directions run on one track alternately. Contrary trains can encounter only by

* Corresponding author: gabriel.fedorko@tuke.sk

***** Formatting Issue - Best Available Paper/Graphic *****

crossing in stations or shunts. The quality of TTD can be affected by the suitable layout of the junctions; double-line (one-way) – for trains in the same direction more frequent is the right-hand track [8]. Trains in the opposite direction are not dependent.; multi-track – presented by three-line pars and the outer tracks are used for driving of trains in the same direction and the middle track serves for trains in both directions and it is marked as null. There are four-tracks sections in Slovakia, which include two-track divisions for two tracks for slower trains and two tracks are for faster trains. The ratio of train number for one and reverse directions includes these types of TTDs: the pair – the number of trains is the same in each direction; unpaired – the number of trains is different in each direction. Duration of occupancy of interstation sections include these types of TTDs: identical – duration of occupancy is the same in all interstation sections created by the pair of trains; non-identical – the value of periods for single-track, i.e. times of occupancy and meantime for double-tracks are different. Time of operation presents these types of TTDs basic (for the total time during one year, or in some cases also half-year traffic diagram for summer and winter TTD) and temporary – in the cases for exclusion in tracks, for example, crisis traffic diagram and alternative traffic diagram. By the level of occupancy and use of track capacity TTDs are presented by these types: low manned – the track capacity remains unused; medium manned – are used by normative capacity; high manned – excessive use of capacity. By orderliness, it is possible to classify TTD into non-systematic – because TTD presents a daily repeating model, non-systematic TTD does not exist but it has internal time relations; systematic – characterized by fixed time sequence for driving time in both directions, repeating of rides in the upcoming hours, the same driving time for both directions, among the nodes, the driving time is half or all multiple of the interval. By the periodicity, it is possible to classify TTD into interval – departures of trains are in the interval 1-60 min in the same minute, tact – tracks of trains are in the distance presented by multiple of hours [9]. The base for TTD design is created by these parameters [10,11]:

1. The extent of train traffic – derivation of the number of trains, determination of proprieties and cycle of train sets,
2. Quantitative indicators – data about technical character,
3. Qualitative indicators – a type of speed in the train traffic, other qualitative indicators,
4. Time norms of the TTD – driving time, movement of trains, operational intervals, meantime, electric meantime, arrival meantime, the final value of meantime.

3 The current state of the line Humenné – Medzilaborce

The line Humenné – Medzilaborce (Fig. 1) with the distance 43 km has 18 stations. The crossing of the trains is possible in the stations Humenné, Udavské, Koškovce, Radvaň/Laborec, and Medzilaborce. From the station, Humenné 12 trains are daily dispatched to Medzilaborce. From the station, Medzilaborce has dispatched 11 trains to Humenné. The railway station Humenné is situated on the single-track 191 Michaľany – Medzilaborce – Lupków (Figure 1).

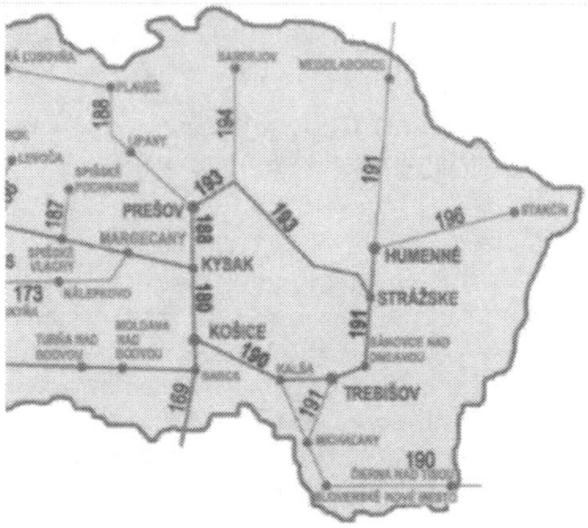

Figure 1. The track 191 presented on the map of Slovak railway network [2]

The track Humenné-Medzilaborce belongs to the least economical route in the Slovakian region. Modernization of track which would bring an increase of operating speed at 160 km/h would allow more efficient and flexible planning of TTD. But due to the geography of this regions, this speed is impossible. Therefore the case study analyses a model that could increase its economy. [12]

4 Modification of TTD and extension of the actual TTD

The aim of the analysis is to add new trains to the current TTD, in addition to the routes from the station Humenné to Košice, Prešov, and Michaľany. The track Humenné-Medzilaborce is considered as closed. By analysis of the current state of TTD on the line Humenné-Medzilaborce, it was found free spaces for completion of the TTD about new trains. The study met driving times and by determination of the times of departure, the study regarded crossing in the stations Medzilaborce, Radvaň/ Laborec, and Udavské. The study also regarded rouds, i.e. if we wanted to add a train, from the station Medzilaborce, the study had to lead out trains by previous or another possible transportation line. [13,14] Figures 1, 2 and 3 present the cycle of driving trail vehicles by the current TTD with new trains by the case study. These are marked as Os 89yy, Os 89xx, Os 89xy and Os 89yx. The Os 89yy has a departure in the time 7:51 from the station Medzilaborce and arrival in the time 8:31 from the station Humenné. Os 89xx has a departure in the time 5:48 from the station Medzilaborce and arrival in the time 7:03 from the station Humenné. Os 89xy has a departure in the time 14:14 from the

https://doi.org/10.1051/matecconf/201823500031

station Humenné and arrival in the time 15:22 from the station Medzilaborce – town. Os 89yx has a departure in the time 21:37 from the station Humenné and arrival in the time 22:37 from the station Medzilaborce. This extension of the current TTD can increase the density of connections.

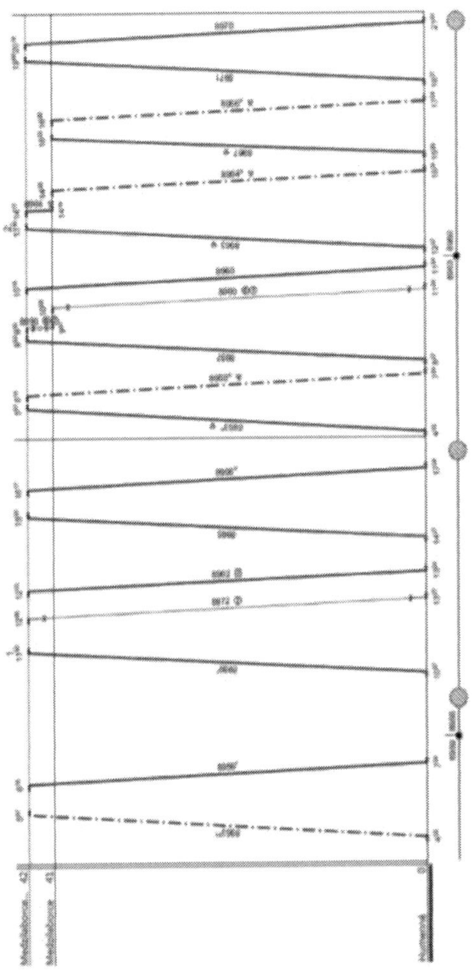

Figure 2 Design of modification of the current cycle of driving trail vehicles by the TTD on the line Humenné-Medzilaborce

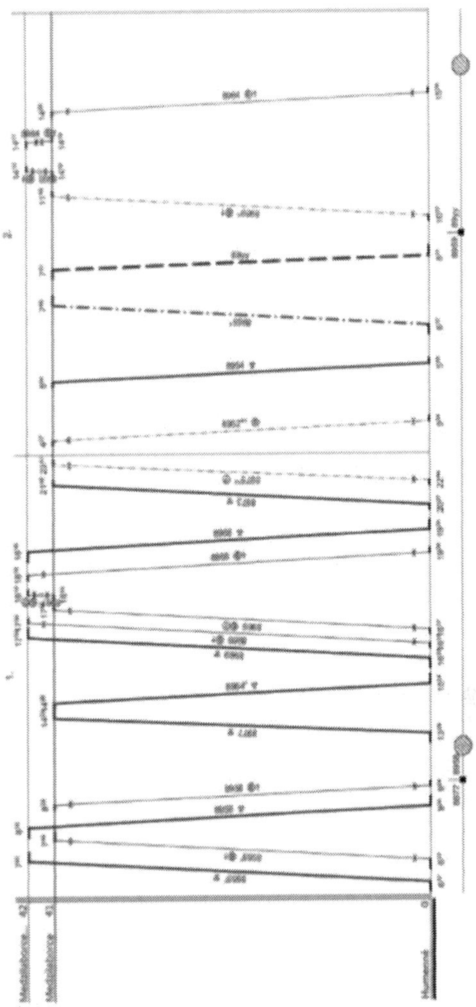

Figure 3 Design of modification of the current cycle of driving trail vehicles by the TTD2 on the line Humenné-Medzilaborce

https://doi.org/10.1051/matecconf/201823500031

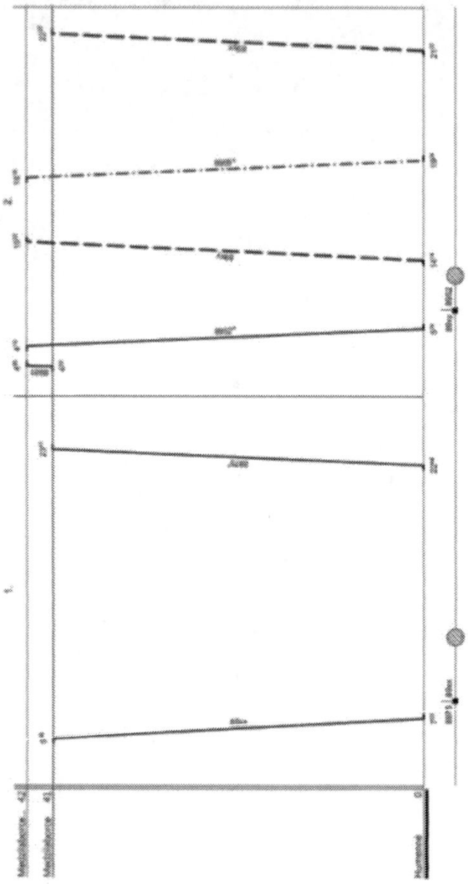

Figure 4 Design of modification of the current cycle of driving trail vehicles by the TTD3 on the line Humenné-Medzilaborce

TTD also brings the turns to need of engine crew and train crew. The turns need of engine crew and train crew is based on the cycles of driving trail vehicles related to the turns of the crew with an emphasis placed on some rules, for example, the maximal duration of the turn, pauses among turns, etc. [15] Turns of engine-drivers and head guards have a different maximal duration of the turns. Fig. 5 presents the design of new turn group for engine drivers. Figures 6, 7, 8 and 9 present design of new turn group for head guards.

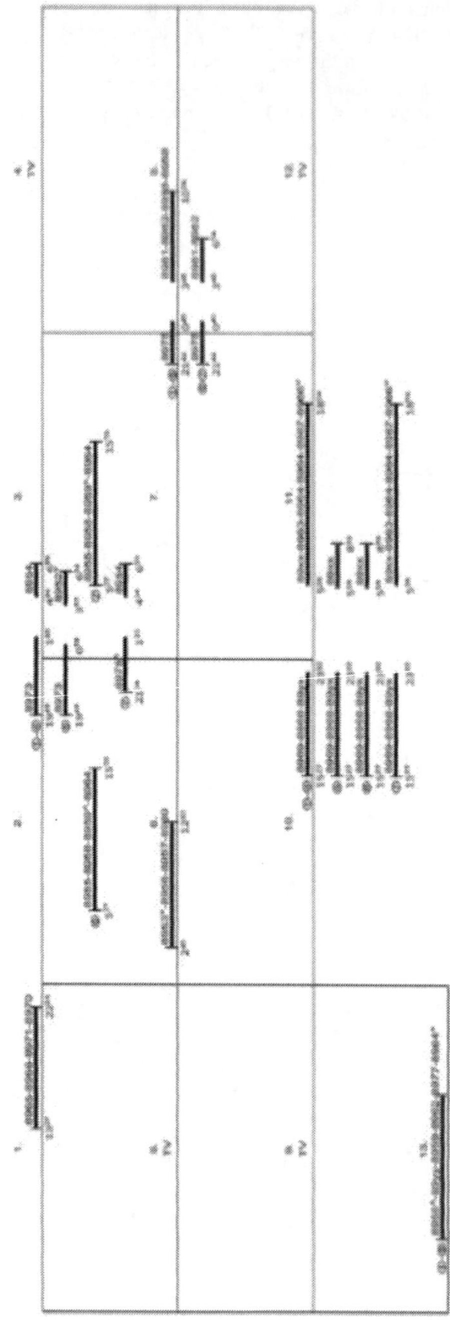

Figure 5 Design of turns group for the engine crew

 https://doi.org/10.1051/matecconf/201823500031

***** Formatting Issue - Best Available Paper/Graphic *****

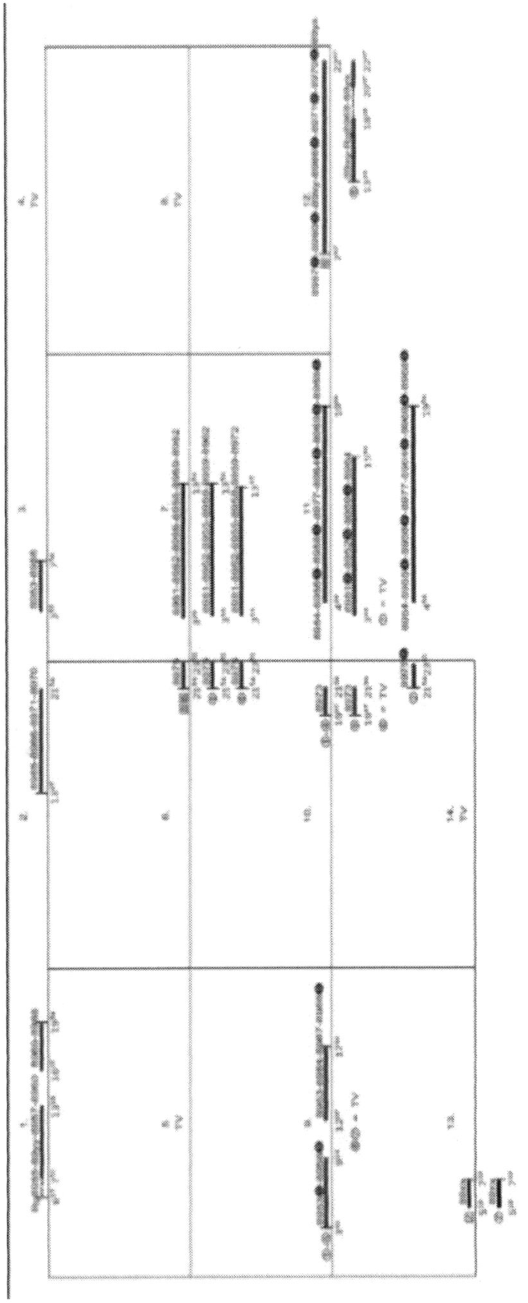

Figure 6 Design of turn group for the train crew

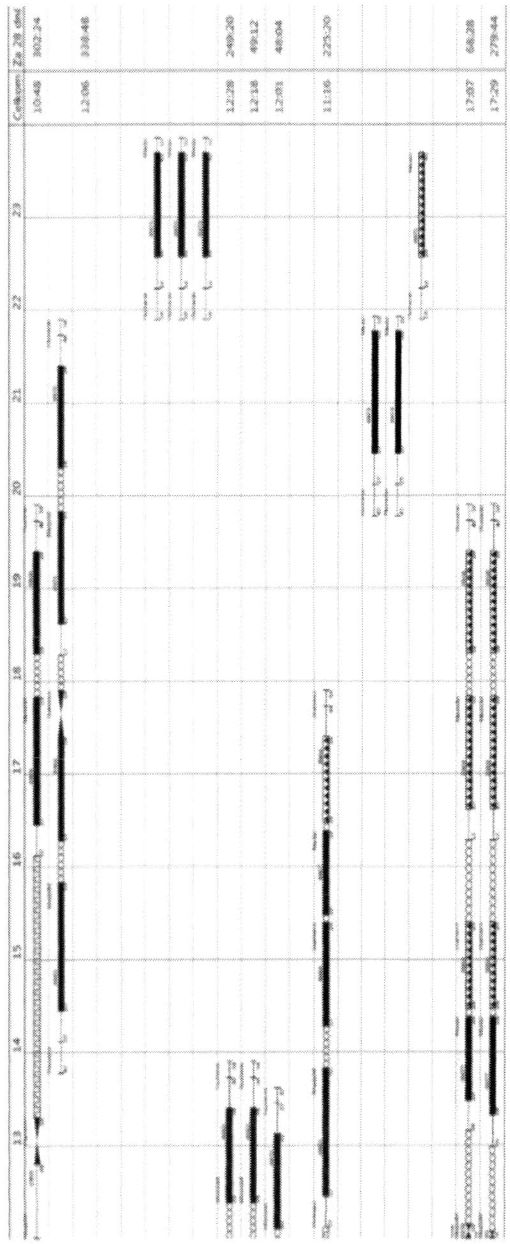

Figure 7 Design of turns group for the train crew 2 (starting point Humenné)

*** Formatting Issue - Best Available Paper/Graphic ***

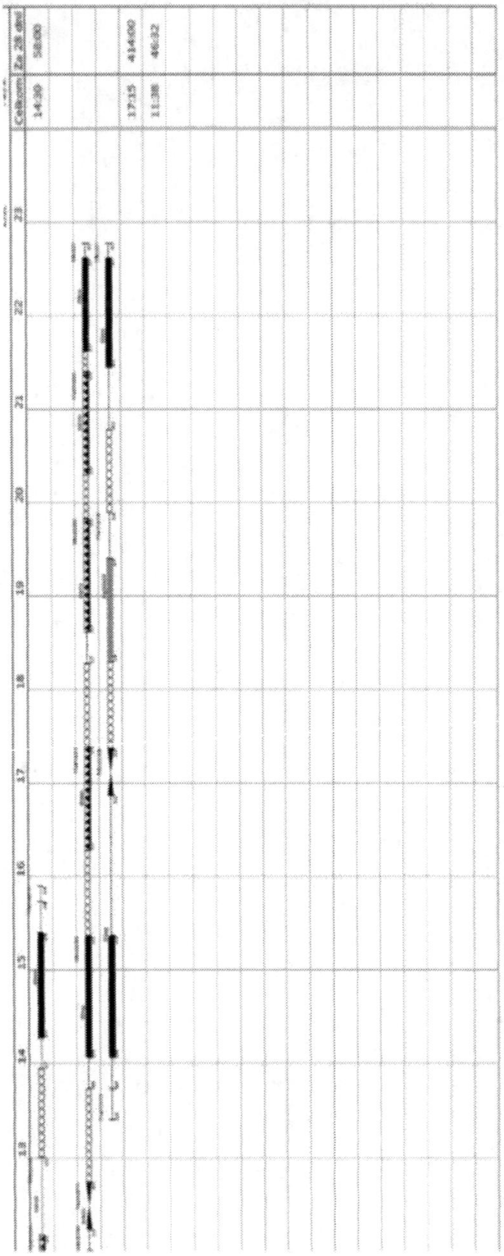

Figure 8 Design of turns group for the engine drivers (starting point Humenné)

Conclusion

Insufficient occupancy of connections outside the traffic peak causes a high cost of rail traffic and at the same time low return of invested funds. The aim of this study was to find free spaces in the current TTD for new connections and this can increase the interest of public and use of rail transport. Another possible solution is also the modernization of railway infrastructure from single-track to a double-track and increases the service speed to 160 km/h. This modernization needs significant expenses for realization. Similar designs could be applied to other, less attractive routes and lines. Investments to the modernization of railway infrastructure in Slovakia can bring desired effect in long-term aspect because benefits of rail transport in terms of safety, ecology, and economics are increasingly desirable for the public, especially in comparison with road transport. Increasing interest in rail transport and integrated transport system can in part eliminate the negative impact of road transport on the public.

VEGA 1/0063/16, VEGA 1/403/18, KEGA 018TUKE-4/2016.

References

1. D. Malindžák et al., *Theory of logistics* (TU Košice, 2007)
2. J. Gašparík, P. Blaho, D. Lichner, *Railway transport operation* (ŽU in Žilina, 2015)
3. J. Gašparík, Z. Pečený, *Train traffic diagram and transmittance of networks* (ŽU in Žilina, 2009)
4. R. Kampf, S. Lorincova, M. Hitka, O. Stopka. Sustainability, **9**. (2017) DOI: 10.3390/su9091561
5. Annual report of Slovak rails 2016. Available on: http://www.slovakrail.sk/fileadmin/dokumenty/2017_pdf/vyrocna_sprava_zssk_2016.pdf
6. Z. Záhumenská, J. Gasparík. Procedia Engineering. **192**, 976-981. (2017). DOI: 10.1016/j.proeng.2017.06.168
7. J. Gašparík, V. Ľupták, P. Meško. *Proceedings of the 3rd International Conference on Traffic and Transport Engineering (ICTTE)*, 388-392, (2016)
8. J. Čamaj, J. Daniš, A. Dolinayová. *Globalization and its Socio-Economic Consequences, 16th International Scientific Conference proceedings*, PTS I-V. 383-390. (2016)
9. P. Meško, J. Gašparík, V. Gáborová. *Proceedings of the 20th International Scientific Conference Transport Means* 2016. 224-229 (2016)
10. A. Dolinayova, J. Čamaj, J. Kanis. Transport problems. **12, 1**. 139-150 (2017)
11. M. Drábek, V. Jánoš, Z. Michl. Scientific Student Conference Interoperability of Railway Transport - (IRICON 2016). **5**. 12-16. (2016). DOI: 10.14311/APP.2016.5.0012
12. J. Gašparík, B. Abramović, V. Zitrický. Technicki vjesnik- Technical Gazette. **25,4**. 1190-1195. (2018). DOI: 10.17559/TV-20160917192247
13. J. Gašparík, V. Zitrický. Transport **25,4**. 397-393. (2010) DOI: 10.3846/transport.2010.48
14. M. Kendra, J. Mašek, J. Ponický, T. Skrúcaný. Transportation Research Procedia, **14**, 1977-1986. (2016) DOI: 10.1016/j.trpro.2016.05.165
15. J. Široký. MATEC Web of Conferences, **134**, 00052, (2017). DOI: 10.1051/matecconf/201713400052

Proposal of stiffness of triple spring of a passenger railway vehicle

Andrej Suchánek[1,*], *Jozef* Harušinec[1], *Mária* Loulová[1] and *Pavol* Kurčík[1]

[1]University of Žilina, Faculty of Mechanical Engineering, Department of Transport and Handling Machines, Univerzitná 8215/1, 010 26 Žilina, Slovak Republic

Abstract. The article deals with the calculation of stiffness of a secondary suspension spring built in a bogie of a rail vehicle with a tilting car body. The vertical stiffness of the springs was calculated using the ANSYS program. The results were compared with calculated values afterwards. The lateral stiffness was evaluated in a similar manner. Analyt-ical method by Gross, Wahl, Budrick, Timoshenko and Ponomarieva was used for comparison with numerical values. The ANSYS simulation was performed for calculating the vertical stiffness of the triple springs. The most suitable analytical method is a method by Timoshenko and Ponomarieva, where the percentage difference was the smallest. The obtained data will be used as an input for the design of coil springs which will be implemented in a model of a vehicle with a tilting car body, for which the comfort values during transition in curve will eventually be determined.

1 Introduction

Spring makes an important part of complex mechanical systems [1, 2, 5, 9, 10, 13]. By choosing a proper shape and material it is able to accumulate deformation energy. In mechanical engineering, they serve mostly to cushion a part of a tool or to produce pressure [6, 12]. Because of these properties, coil springs are used in construction of rail vehicle bogies as well. In case of using the coil springs in rail vehicle with tilting car body, the main function is to increase the tilt during transition in curve and therefore to decrease the impact of unbalanced lateral acceleration on passengers [3, 4, 8, 11, 14].

When modeling the bogie in CAD software CATIA it was necessary to determine its parameters needed for its later analysis in SIMPACK 9.8 program. In the article we will therefore deal with coil springs of the secondary suspension. We will set the force element in SIMPACK program, used for modeling springs and focus on determination of the spring stiffness. Spring valuse have been defined using analytical methods and compared with numerical results taken from the simulation program ANSYS R15.0 afterwards. The bogie model, imported to SIMPACK, is displayed in the Fig. 1.

2 Spring parameters necessary for the calculation

SIMPACK allows generation of a spring along the positive direction of the x-axis of a reference marker, which is located in the centre of the ground end of the spring. Stiffness of the coil springs is given by wire diameter d, number of active coils n, number of end coils n_0, free spring length h_0, pitch s. The free spring length is to be chosen in such manner, that even by its maximal compression z_{max} (restricted by bump stop) the single coils would not touch each other, but a clearance of cca 10 to 15 % of a wire diameter is to be held [11, 12].

The basic input parameters in SIMPACK program for the Spring-Damper Parallel Cmp Force element are [15]:

- nominal forces F_{nom_x}, F_{nom_y}, F_{nom_z} (Nominal force f_nom_x/y/z),

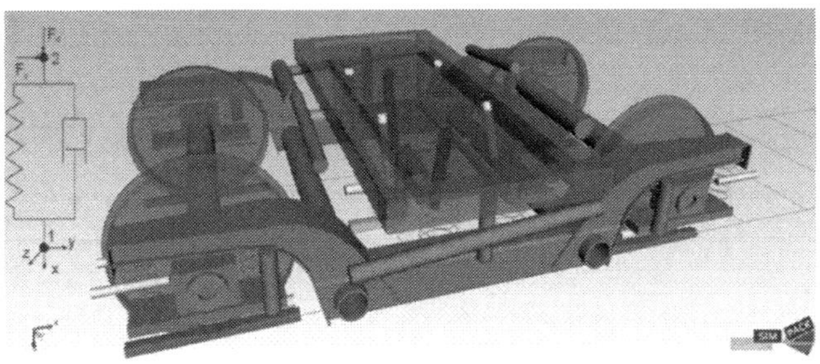

Fig. 1 Location of the suspension in a bogie.

* Corresponding author: andrej.suchanek@fstroj.uniza.sk

- stiffness k_x, k_y, k_z (Stiffness k_x/y/z),
- damping d_x, d_y, d_z (Damping d_x/y/z),
- imput function F_{kx}, F_{ky}, F_{kz}, F_{dx}, F_{dy}, F_{dz} (Input Function F_k(x/y/z), Input Function F_d(xd/yd/zd),
- clearance in direction x, y, z, s_x, s_y, s_z,
- root function - used for detection of the clearance constraints,
- reference marker for measurements and calculation - all the measurements of distance and speed between markers as well as all calculated forces use directions defined by the axis of the reference marker instead of From Marker,
- forces F_x, F_y, F_z which describe the forces in three directions.

3 Determination of vertical stiffness of the springs

Coil springs represent the most proper steel suspension element in the suspension system of rail vehicles. They are favourable in the means of dimensions and mass. They are used in systems along with dampers, because they do not have the ability to absorb the energy of oscillating motion of parts.

For the coil spring stiffness calculation, the formula (1) applies, which was used for analytical determination of individual vertical stiffness values. These were compared with values taken from ANSYS afterwards (Tab. 2):

$$k_z = \frac{G \cdot d^4}{64 \cdot D^3 \cdot n}\left[N \cdot mm^{-1}\right], \qquad (1)$$

Where:
G...Shear modulus [Pa],
d...Wire diameter [mm],
D...Diameter of spring [mm],
n...number of active coils [-].
In the following Tab. 1, basic parameters of coil springs used in the bogie model are displayed.

Table 1 Parameters of the coil springs.

Spring	d [mm]	D [mm]	n [-]	s [mm]	n_0 [-]	h [mm]	F_0 [N]	E [MPa]	G [MPa]
A	35	280	6	56.67	0.75	375	5707.10	$2.06.10^5$	$8.15.10^4$
B	25	210	7	50.00	0.75	375	3021.10	$2.06.10^5$	$8.15.10^4$
C	20	150	8	44.38	0.75	375	2983.70	$2.06.10^5$	$8.15.10^4$

Table 2 Coil springs vertical stiffness comparison.

Spring	Analytically determined stiffness - k_z [N.mm^{-1}]	Stiffness determined in ANSYS - k_z [N.mm^{-1}]	Difference [%]
A	103.17	114.50	9.90
B	55.45	59.74	7.19
C	54.57	60.89	10.37

Geometrical model of individual springs was created in CATIA and imported to ANSYS afterwards. As a boundary condition, a vertical displacement (along z axis) value of 50 mm was set. The manner of end coils placing is displayed in Fig. 2. After carrying out simulations and displaying results for individual springs A (outer), B (middle) and C (inner), we found the value of vertical force acting on the spring. From the rate of acting force and spring compression we determined the result value of vertical spring stiffness.

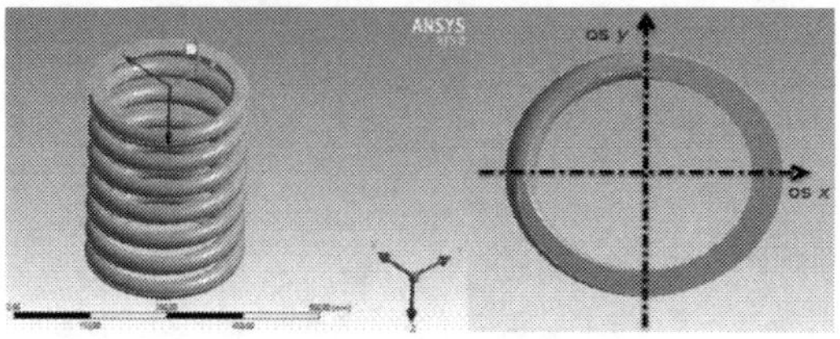

Fig. 2 Model of loaded coilspring A in ANSYS.

4 Determination of lateral spring stiffness

For the analytical calculation of lateral stiffness, many formulas from various authors were derived, but they are only approximate and do not apply in general, because they do not regard all the affecting factors [11]. For the analytical determination of the spring stiffness, formulas by Gross, by Wahl, by Budrick and by Tymoshenko and Ponomariov have been used.

- Calculation by Gross:

Stiffness of a spring under load in lateral direction:

$$k_y = \frac{1}{\frac{1}{F_0} \cdot \left[\frac{2}{\alpha} \cdot tg\left(\alpha \cdot \frac{h}{2}\right) - h\right] + \frac{h}{k_s}}\left[N \cdot mm^{-1}\right], \text{ where:}$$

$$\alpha = \sqrt{\frac{F_0}{k_0\left(1 - \frac{F_0}{k_s}\right)}}\left[-\right] \text{ is a constant.} \qquad (2)$$

Where:
k_0...Bending stiffness [N.mm^{-1}],
k_s...Shear stiffness [N.mm^{-1}].

Bending stiffness:

$$k_0 = \frac{h}{\pi \cdot n \cdot \frac{D}{2} \cdot \left(\frac{1}{E \cdot I_1} + \frac{1}{G \cdot I_p} \right)} \left[N \cdot mm^{-1} \right] \quad (3)$$

Where:
E...spring length [mm],
D...Wire diameter [mm],
π...mathematical constant [-],
I_1...Moment of inertia [kg.m^2],
I_p...Polar moment of area [kg.m^2].
Shear stiffness:

$$k_s = \frac{E \cdot h \cdot I_1}{\pi \cdot n \cdot \left(\frac{D}{2} \right)^3} \left[N \cdot mm^{-1} \right]. \quad (4)$$

Moment of inertia:

$$I_1 = \frac{\pi . d^4}{64} \left[kg.mm^2 \right] \quad (5)$$

Polar moment of area:

$$I_p = \frac{\pi . d^4}{32} \left[kg.mm^2 \right] \quad (6)$$

- Calculation by *Wahl*:

Stiffness of a spring under load in lateral direction:

$$k_y = \frac{2.6 \cdot k_z}{1 + 0.77 \cdot \beta^2} \cdot \left(1 - \frac{F_0}{U \cdot h_0 \cdot k_z} \right) \left[N.mm^{-1} \right],$$

where: $\beta = \frac{h}{D}$ is a constant. (7)

Table 3 Dependency of the U coefficient from β.

β	1.5	2	2.5	3	3.5	4	4.5	5
U	0.69	0.63	0.53	0.39	0.27	0.2	0.14	0.11

- Calculation by *Budrick*:

Stiffness of a spring under load in lateral direction:

$$k_y = k_z \cdot \frac{G}{E} \left(1 + \frac{2 + \mu}{3} \cdot \beta^2 \right) \left[N \cdot mm^{-1} \right]. \quad (8)$$

Where:

$$k_z = \frac{2}{\pi \cdot D \cdot n \cdot \left(\frac{h^2}{12} \cdot \left(\frac{1}{G \cdot I_p} + \frac{1}{E \cdot I_1} \right) + \frac{D^2}{4 \cdot E \cdot I_1} \right)} \quad (9)$$

$$\left[N \cdot mm^{-1} \right]$$

- Calculation by *Tymoshenko and Ponomariov*:

Stiffness of a spring under load in lateral direction:

$$k_y = k_z \cdot \frac{D^2 \cdot (1 - \gamma)}{\frac{0.2936 \cdot (h - \chi \cdot d)^3}{(h - 1.5 \cdot h \cdot d)} + 0.381 \cdot D^2} \left[N \cdot mm^{-1} \right], \quad (10)$$

Where:
γ...variable quantity [-],
χ...Constant [-],

for $\beta_0 < 2{,}62$: $\gamma = 0.375 \cdot \frac{F_0}{k_z \cdot h} \cdot \beta \cdot \left(\beta - 1.5 \cdot \frac{d}{D} \right)$, (11)

for $\beta_0 > 2{,}62$: $\gamma = \dfrac{\frac{F_0}{k_z \cdot h} \cdot \beta}{0.813 \cdot \left(\beta_0 - \sqrt{\beta_0^2 - 6.87} \right)}$. (12)

In formula (10) there is non-dimensional variable quantity γ, which is dependant on slenderness ratio of a loaded spring. Its value can be calculated from formulas (11) and (12). The constant χ is an auxiliary quantity, which regards the manner of mounting the end coils of the springs (joint or rigid mounting). For the analysed springs, the constant is equal 0.5 [5, 8]. We defined the values of lateral stiffness analytically, based on formulas according to individual methods (2-12). These values were afterwards compared with the values obtained from ANSYS. Input parameters of the individual springs are given in Tab.4.

Table 4 Calculated values and values obtained from ANSYS.

Method	A	B	C
ANSYS k_y [Nmm^{-1}]	172.12	38.28	35.47
ANSYS k_x [Nmm^{-1}]	142.99	50.36	25.60
Grossa [Nmm^{-1}]	100.36	34.18	15.90
Wahla [Nmm^{-1}]	88.57	32.10	17.70
Budricka [Nmm^{-1}]	179.49	78,99	66.85
Timošenka and Ponomarieva [Nmm^{-1}]	109.40	37.13	17.60
Percentage of difference between results from individual methods and ANSYS			
G-A-k_y [%]	41.69	10.70	55.18
G-A-k_x [%]	29.81	32.13	37.89
W-A-k_y [%]	48.54	16.14	50.11
W-A-k_x [%]	38.06	36.26	**30.88**
B-A-k_y [%]	**4.28**	106.37	88.46
B-A-k_x [%]	25.53	56.85	161.13
TaP-A-k_y [%]	36.44	**2.99**	50.37
TaP-A-k_x [%]	23.49	26.26	31.23

In ANSYS, we determined the lateral stiffness values similarly to vertical stiffness, but besides the 50 mm vertical displacement (compression) in z-axis direction, a lateral displacement of 1 – 5 mm in the x- and y-axis direction was set for the individual springs. The calculation was performed separately for the displacement in x-axis and in y-axis direction. The result values of the vertical force F_0 were used in analytical determination of stiffness. The lateral stiffness obtained from ANSYS was defined as proportion of the obtained lateral force to the lateral displacement. The calculated values and values obtained from ANSYS are in Tab. 4.

5 Determination of the triple spring stiffness

The secondary suspension of the analysed bogie is consists of three coil springs (Fig. 5). Lateral forces in springs F_p have been calculated using the ANSYS program for 8 different mounting positions of the end

 https://doi.org/10.1051/matecconf/201823500032

coil (Fig. 4). The maximal vertical displacement was set to the value of 50 mm and the lateral displacement was in the range of 0 – 4 mm. In Fig. 3 we can see the characteristic of the triple spring lateral forces in dependency on lateral displacement. If we turn the spring system around the vertical axis (45 degrees) the characteristic is shifting. As can be seen in Fig. 3, the mounting positions of the end coil P0 and P180 appear to be most suitable (Fig. 4). The resulting vertical stiffness of the triple spring is constant with a value of 221.88 Nmm^{-1} with an applied vertical force of 11 091.22 N. The calculated values of lateral forces F_p are in the following Tab. 5. We determined the lateral stiffness for the position of the end coils P180 (Tab. 6), its value is linear and rises with increasing lateral displacement.

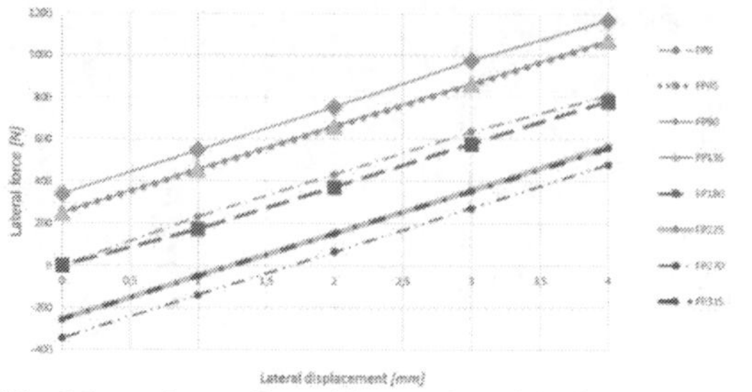

Fig. 3 Dependency of lateral force on lateral displacement.

Table 5 Calculated values of the lateral forces.

Lateral dispacement [mm]	F_{P0} [N]	F_{P45} [N]	F_{P90} [N]	F_{P135} [N]	F_{P180} [N]	F_{P225} [N]	F_{P270} [N]	F_{P315} [N]
4	809.21	1066.50	1165.30	1071.80	780.43	569.23	481,09	559.77
3	636.54	862.84	975.62	866.24	578.80	359.50	274.78	355.94
2	434.30	659.07	753.84	661.52	374.85	155.58	69.04	155.44
1	232.00	454.90	547.72	457.33	174.96	-47.52	-136.86	-48.64
0	2.99	251.72	342.89	253.00	2.99	-251.72	-342.89	-253.00

Table 6 Calculated values of secondary suspension lateral stiffness for the spring mounting position P180.

Lateral dispacement [mm]	F_{P180} [N]	Lateral stiffness k_y [Nmm^{-1}]
4	780.43	195.11
3	578.80	192.93
2	374.85	187.43
1	174.96	174.96

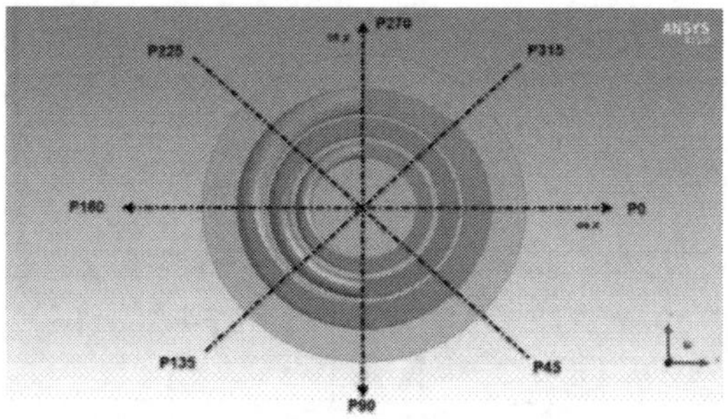

Fig. 4 Spring end coil mounting positions.

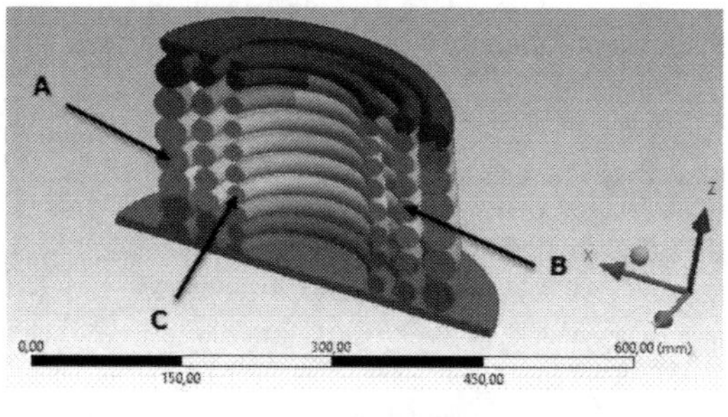

Fig. 5 Model of a loaded coil spring in program ANSYS.

6 Conclusion

Suspension is one of the most important parts of a bogie. In this article we focused on the secondary suspension, consisting of three coil springs. Stiffness of the individual springs was examined separately using analytical methods and numerical calculation.

When determing the vertical stiffness, the difference

between calculated values and values obtained from ANSYS was about 10 % in average, which is sufficient for use in SIMPACK program. After determination of lateral stiffness using individual analytical methods and consequential comparison with values obtained using numerical method we discovered that the method by Tymoshenko and Ponomariev, where the percentage difference was the smallest, is the most suitable. From the above can be concluded, that not every analytical method is suitable for determination of lateral stiffness of coil springs.

The advantage of using numerical method for spring stiffness determination and using simulation program ANSYS is the possibility of solving problems as a whole and the parametrisation of the models. Lateral stiffness of a triple spring is dependent on the end coils mounting position. The results will be further used in simulation of the whole bogie and of its transition in curve using SIMPACK program.

This work was supported by the Cultural and Educational Grant Agency of the Ministry of Education of the Slovak Republic in project No. KEGA 077ŽU-4/2017: Modernization of the Vehicles and engines study program. The work was also supported by the project No. APVV-0842-11: Equivalent railway operation load simulator on the roller rig and VEGA No. 1/0558/18: Research of the interaction of a braked railway wheelset and track in simulated operational conditions of a vehicle running in a track on the test stand.
Research-Educational Center of Rail Vehicles (VVCKV)

References

1. J. Gerlici, T. Lack Applied mechanics and materials. Vol. **486**, pp. 396-405 (2014).
2. T. Lack, J. Gerlici, *9th International Conference on Contact Mechanics and Wear of Rail/Wheel System*, 87-89 (2012).
3. T. Lack, J. Gerlici, Applied mechanics and materials. **486**. 359-370 (2014).
4. T. Lack, J. Gerlici, Applied mechanics and materials. **486**. 371-378 (2014).
5. T. Lack, J. Gerlici, *Communications – Scientific Letters of the University of Žilina*. **7**, 38-45 (2005).
6. T. Lack, J. Gerlici, *Communications – Scientific Letters of the University of Žilina*. **15**, 126-132 (2013).
7. J. Gerlici, T. Lack, An international journal on the science and technology of friction, lubrication and wear. **271**, Sp. iss., 246-258 (2011).
8. J. Gerlici, T. Lack, Procedia Engineering. . **2**, 1,. 2249-2257 (2010).
9. J. Gerlici, T. Lack, Applied mechanics and materials. **486**, pp. 379-386 (2014).
10. P. Baran, M. Brezáni, P. Kukuča, P. Šťastniak, Procedia Engineering. **192**. 34-39 (2017).
11. V. Hauser, et. al., Procedia Engineering. **192**, 289-294 (2017).
12. V. Hauser, et. al., Procedia Engineering, **192**: 295-300 (2017).
13. Simpack 9.9.2 Documentation. 4556p (2016).
14. G. Fedorko, V. Molnár, J. Strohmandl et al. *Transport Means 2015 – Proceedings of the International Conference*. 466-+ (2015)
15. P. Baran, P. Šťastniak, P. Kukuča, M. Brezáni, MATEC Web of conferences. **157**, 02003 (2018). DOI: https://doi.org/10.1051/matecconf/20181570200

Reduction of Kinematic Resistance To Movement Of the Railway Vehicles

E. Mikhailov[1], *S.* Semenov[1], *V.* Tkachenko[2] and *S.* Sapronova[2]

[1] East-Ukrainian National University named after Vladimir Dal,Central ave., 59-a/303, 93400, Severodonetsk, Ukraine,
[2] State University of Infrastructure and Technology, Kirilivska, 9/3, 04080, Kiev, Ukraine

Abstract. The aim of the work is to reduce the train rolling resistance to movement the railway vehicles. The article proves the expediency of changing the constructive scheme of railway vehicle wheel, which allows independent rotation their wheel mounting surface rolling and the guide surface (flange). There has been carried out the classification and analysis of the advantages and disadvantages of various structural schemes wheels of the railway vehicles. There has been analyzed the Features resistance to movement wheels of traditional and perspective constructive scheme. The obtained results allow drawing a conclusion about the possibility of a significant reduction in train rolling resistance to movement of railway vehicles with wheels perspective constructive scheme.

1 Introduction

The problem of reducing the level of force interaction between the vehicle and rail track is one of the most important problems of the rail transport.

The main share of energy recourses in rail transport (up to 75% of consumed electrical energy and fuel) is attributed to the haulage of train.

The expenditure of energy resources on haulage of train is directly related to the level of train rolling resistance to movement its locomotives and carriages.

Known theoretical and experimental studies in this direction are mainly related to the definition and refinement of empirical formulas for calculating the resistance to movement of certain types crews, the advantages of rotating bearings in front of plain friction bearings, and the coefficients of air resistance.

Analysis the components of total resistance to movement rail vehicles shows that the most promising direction of its reduction is the reduction of those components that are generated as a result of friction interaction of wheels and rails (kinematic resistance to movement) [1–6].

The research direction related to the determination of the possibility of influencing the level resistance to movement of the rail vehicles to structural improvements in their running parts (the radial of mounted wheels and bodies, the use of independently rotating wheels, the active control of guiding, etc.) is quite popular [7-12].

A number of scientific researches and developmental designs are known [1, 2, 12, 13], the results of which show the greater efficiency of rail vehicles in operation due to the fact that the designs of their running parts differed from traditional ones.

From the standpoint of generation of train rolling resistance to movement forces of the vehicle, this traditional mounted wheels design has a number of significant drawbacks that have been sufficiently fully disclosed in the works of a number of researchers [1, 2].

In case of two-point contacting wheel with rail (in the presence of the main and flange wheel contact), as a result of the parasitic power circulation in a closed loop, the «wheel flange – lateral face of top of rail – the running surface of rail-wheels thread» appears so-called differential kinematic resistance to movement.

Circular kinematic resistance to movement is the result of a group multi-contact interaction of the wheel system and mounted wheels with track in the course of guided movement in the rail track by circulating parasitic power within one mounted wheels or a group of mounted wheels united by a single frame.

A radical way to reduce the kinematic resistance to movement of rail vehicles is to reduce the level of circulation of parasitic losses by introducing decoupling points into the kinematic chains, which can be achieved through the use of new engineering proposal for mounted wheels and individual wheels.

Reduction of kinematic resistance to movement is possible with a decrease in the circulation of parasitic losses in a closed loop «wheel flange – lateral face of top of rail – the running surface of rail-wheels thread». Radical reduction of this circulation can be achieved only by changing the constructive scheme of a single wheel that allows independent rotation their mounting surface and guiding surface (flange).

It seems expedient to analyze the potential of using various structural schemes of the rail carriages in terms opportunities of reducing the kinematic resistance to movement.

* Corresponding author: mihaylov.evv@gmail.com

2 The Analysis of Constructive Schemes of Railway Carriages

When considering the known constructive schemes of the wheels of rail vehicles, they all were conventionally divided into six classification groups according to the criteria for ensuring the corresponding degrees of freedom in the coordinate of rotation wheel and the guide surface (flange) relatively to the wheel mounting surface around a common axis of rotation wheel.

The presence or absence of these degrees of freedom largely determines the level of kinematic running resistance to movement wheels on the rail [21].

A conditional classification of structural schemes of wheels of rail vehicles is shown in Fig 1.

In the above conditional classification, the wheels of the traditional constructive scheme are referred to group A, in which the wheel mounting surface thread and its guide surface (flange) are made monolithically.

The level of kinematic resistance to movement of rail vehicles with such wheels is taken as 100%.

Group B includes structural schemes of wheels in which, due to the use of elastomeric elements, there is the possibility of slight elastic displacements of the wheel mounting surface (tire) relative to the wheel centre.

The range of such constructive solutions is quite wide. In this direction, it should be especially noted a large complex of works on the creation of rubberized wheels, conducted by a team of researchers from the East Ukrainian National University named after V. Dahl, headed by prof. A. L. Golubenko for improving the dynamic and tractive characteristics of the vehicle, reducing the resistance to movement and reducing fretting wear of wheels and rails [1].

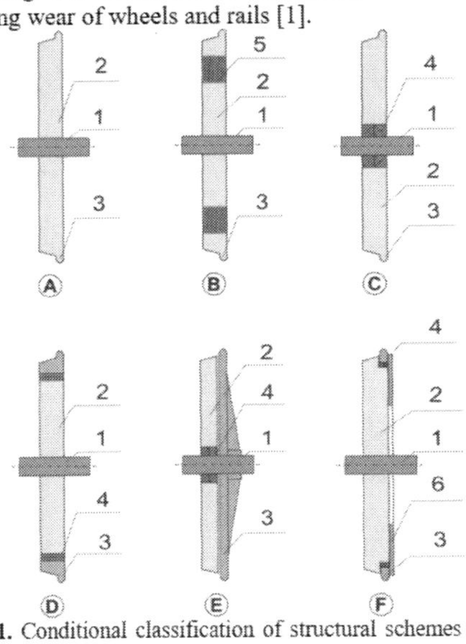

Fig. 1. Conditional classification of structural schemes of rail carriage wheels: 1 – axle; 2 – wheel center; 3 – flange; 4 – bearing bush; 5 – elastomeric element; 6 – pressure disk

By reducing the roughness of the torsion coupling of the wheel mounting surface (tire) with wheel centers, it is possible to provide a certain reduction in the wobble frequency when driving in straight sections of the track, improve the dynamical and tractive characteristics of the vehicle.

However, the use of such wheels does not make it possible to compensate for the range difference traveled by the left and right wheel in the mounted wheels when operation in curves, to eliminate associated wheel sliding along the rails and additional resistance to movement.

According to the data [1], the use of the wheels of this group allows to reduce the circulation component of the kinematic resistance to movement of rail vehicles by 5 ... 8%.

The use of the wheels of group C (so-called independently rotating wheels) in the truck of rail vehicles is a fairly well-known technical solution aimed at improving the dynamics and reducing the resistance to movement.

However, in the operation of rail vehicles with wheels of this constructive scheme, there are problems associated with insufficient guiding action of mounted wheels with such wheels when driving in a rail track.

This movement usually occurs with a prolonged two-point contact of one of the wheels with a rail significant angle of attack and negative consequences in the form of increased wear of the ridges and resistance to movement.

One of the technical solutions implemented in real operation that eliminate to some extent this disadvantage is the Spanish technology "TALGO", which makes it possible to reduce the angles of the attack of independently rotating wheels due to the forced radial of the single-axle truck in which they are installed.

However, studies show that during the exploitation of the TALGO trains there are facts of heavy wear of the ridges due to the special character of the front uniaxial body entry in the curve with the realization of increased directional forces and, correspondingly, increased running resistance to movement.

Thus, in work [13] in analyzing the wear characteristics of the flanges of wheels, it was established that the flanges of the wheels of the end and inter-wagon bogies had wear, the average intensity being 2.01 and 0.95 mm, respectively, for 20,000 km.

In this case, the use of such structures can reduce the circulation component of the kinematic resistance to movement of rail vehicles up to 50%.

So-called "wheels of differential rotation" [2, 14-17] developed by L.V.Vinnik can be attributed to a separate group of conditional classification (group D). The tire of such a wheel is mounted on the wheel centre by a free fit, which is actually a plain friction bearing.

Extensive theoretical studies of the features of the movement of rail vehicles with wheels of the proposed constructive scheme have shown the possibility of obtaining a positive effect in the form of reducing the resistance to movement and wheel thread and rails.

The disadvantage of the proposed technical solutions of the wheels of this group, despite the possibility of reducing the direct creep forces in the range of 28 ... 30% and the resistance to movement up to 12%, is a wide variation under various conditions of force of friction between the moving bandage and the wheel center.

At the same time, the claimed differential design effect may be appearing not in full.

In addition, there was severe wear on the contact surfaces of the wheel center and the tire.

This problem requires additional research, which, however, is recognized by the author their constructive scheme.

In the group E of conditional classification, constructive schemes of wheels are included, the feature of which is the possibility of independent rotation the wheel mounting surface and its guide surface (flange) relative to their common axis rotation.

One of the most successful designs of this group is development of the so-called «flexible wheels", conducted under the leadership of V.V. Shiller (Omsk State Transport University) [18, 19].

As a result of theoretical studies and tests on models, it is concluded that the use wheels of such constructive scheme can reduce the resistance to movement and reduce the cost of haulage of train up to 30%, and in some cases-up to 50%, while the wear of rails can be reduced to 30 %.At present, full-scale tests of the wagon with such wheels are carried out.

The group F of the conditional classification includes constructive schemes of the wheels, in which an independent rotation of the guide surface of wheel (its flange) is provided with respect to the wheel mounting surface, when the movable flange is mounted directly on the wheel.

When using wheels of such a constructive scheme in the composition of the carriage part of the railway rolling stock, the reduction of the differential component of the kinematic resistance to movement due to the reduction of parasitic slip in the flange contact can amount to 40%.

Thus, the conducted analysis showed that the greatest decrease in the kinematic resistance to movement can be achieved by using the structural schemes of the wheels of groups E and F of the proposed conditional classification by reducing the parasitic slip and reducing decrease the power of creep forces in the flange contact of the wheel with the rail.

3 The Features of Motion by Rail of Wheels of Different Constructive Schemes

Obviously, in the case of a fundamental change in the constructive of the wheel (allowing the possibility of turning the guide surface (flange) to the wheel mounting surface around their common axle - then the wheel of the perspective constructive scheme - the PKS wheel), the character of its movement along the rail should change in comparison with the wheel of the traditional constructive scheme (hereinafter the wheel TKS).

The study of the peculiarities of movement along the wheel of traditional and perspective wheels of the constructive scheme for the case of point-to-point contact with the rail showed significant differences in the character of this movement.

This can be judged by considering the calculation schemes shown in Fig. 2, which show the directions of the vectors of the components of the slip speed and creep forces in the flange contact in the longitudinal vertical plane for the TKS wheel (Fig. 2a) and PKS (Fig. 2, b).

The analysis of theses schemes made it possible to establish the analytical dependencies of the module and the direction of the slip speed in the flange contact of the PKS wheel on the speed of wheel and the angle of attack to the rail.

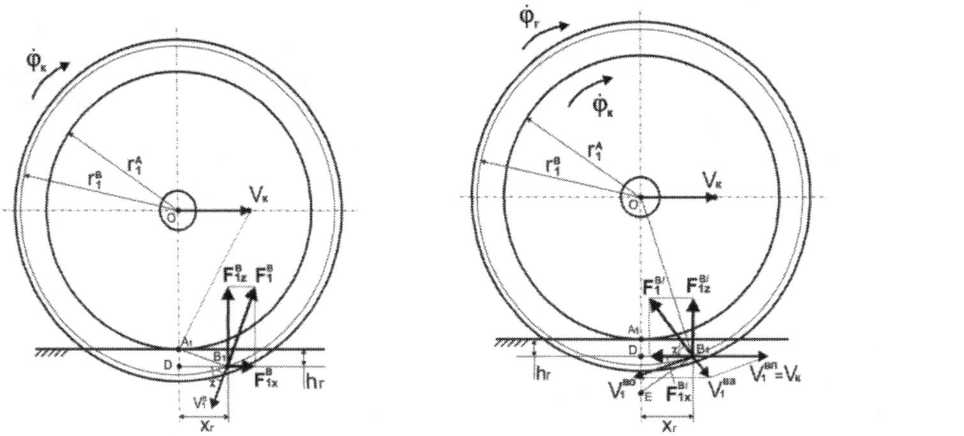

Fig. 2. Calculation scheme: a) wheel TKS; b) wheel PKS k

For the TKS wheel the module and the direction of the slip speed vector in the flange contact are uniquely determined by the geometric characteristics of the wheel and the rail and the angular speed of the wheel rotation.

When the wheel PKS moves, the direction and modulus of the slippage speed vector along the rail depend not only on the geometric parameters of the wheel and rail, but also on the ratio of the angular speeds of rotation the wheel mounting surface and guide surface of wheel (its flange) around a common axis of rotation wheel. Thus, for the calculation scheme of Fig. 2, in

$$\chi' = arcsin(\frac{(r_1^A + h_r)\cdot tg\psi \cdot tg\beta}{\sqrt{((K_w - 1)\cdot r_1^A - h_r)^2 + ((r_1^A + h_r)\cdot tg\psi \cdot tg\beta)^2}}) \quad (1)$$

$$V_1^{BA} = \dot{\varphi}_k \cdot \frac{\sqrt{((K_w - 1)\cdot r_1^A - h_r)^2 + ((r_1^A + h_r)\cdot tg\psi \cdot tg\beta)^2}}{K_w} \quad (2)$$

$$\chi' = arcsin(\frac{(r_1^A + h_r)\cdot tg\psi \cdot tg\beta}{\sqrt{((K_w - 1)\cdot r_1^A - h_r)^2 + ((r_1^A + h_r)\cdot tg\psi \cdot tg\beta)^2}}) \quad (3)$$

$$V_1^{BA} = \dot{\varphi}_k \cdot \frac{\sqrt{((K_w - 1)\cdot r_1^A - h_r)^2 + ((r_1^A + h_r)\cdot tg\psi \cdot tg\beta)^2}}{K_w} \quad (4)$$

$$\chi' = arcsin(\frac{(r_1^A + h_r)\cdot tg\psi \cdot tg\beta}{\sqrt{((K_w - 1)\cdot r_1^A - h_r)^2 + ((r_1^A + h_r)\cdot tg\psi \cdot tg\beta)^2}}) \quad (5)$$

$$\chi' = \arcsin\frac{(r_1^A + h_r)\times tg\psi \times tg\beta}{\sqrt{((K_w - 1)\;\; r_1^A \;\; h_r)^2 + ((r_1^A + h_r)\;\; tg\psi \;\; tg\beta)^2}} \quad (6)$$

$$V_1^{BA} = \dot{}_k \; \frac{\sqrt{((K_w \;\; 1)\;\; r_1^A \;\; h_r)^2 + ((r_1^A + h_r)\;\; tg\psi \;\; tg\beta)^2}}{K_w} \quad (7)$$

$$\chi' = arcsin\frac{(r_1^A + h_r)\;\; tg\psi \;\; tg\beta}{\sqrt{[(K_W \;\; 1)\;\; r_1^A \;\; h_r]^2}} \quad (8)$$

$$\chi' = \arcsin(\frac{(r_1^A + h_r)\;\; tg\psi \;\; tg\beta}{\sqrt{((K_w \;\; 1)\;\; r_1^A \;\; h_r)^2 + ((r_1^A + h_r)\;\; tg\psi \;\; tg\beta)^2}}) \quad (9)$$

$$V_1^{BA} = \dot{\varphi}_k \cdot \frac{\sqrt{((K_w - 1)\cdot r_1^A - h_r)^2 + ((r_1^A + h_r)\cdot tg\psi \cdot tg\beta)^2}}{K_w} \quad (10)$$

$$n^2 = \frac{3P'}{4\pi} \quad (11)$$

$$n^2 = \frac{3P'}{4\pi}\;\left(\frac{L_i^2 - L_{i+1}^2}{F_i \;\; F_{i+1}}\right) \quad (12)$$

$$b^2 = \frac{4\pi F_i n^2 \;\; 3P' L_i^2}{P'}$$

where $K_W = \dot{\varphi}_K / \dot{\varphi}_r$; ψ - the angle of attack wheel to the rail; β - the angle of inclination the generatrix the flange of wheel to the horizontal.

The ideal possibility of a complete absence of differential kinematic slip in the flange contact for such a wheel can be realized only if the following conditions are satisfied: the angle of wheel run-on to the rail ( ) is zero and a certain ratio of the angular speeds of the wheel and the wheel flange PKS $\dot{\varphi}_K / \dot{\varphi}_r = K_W$, where K_W is a coefficient depending on the geometric parameters contacting the wheel and rail.

In the absence of a rigid connection between the values of the angular speed of the wheel mounting surface PKS and its mounting surface (flange), to determine the stable state of the system under consideration we use the well-known principle of minimum energy dissipation.

This principle says that if not a single state of the system is admissible but a certain set of states that are consistent with conservation laws and constraints imposed on the system, then its state is realized which corresponds to a minimal increase in the entropy of the system or, what is the same, minimal energy dispersal.

Taking this into account, we can consider the quasistationary (stable) state of the system under consideration to be its state, in which the energy dispersal in the flange contact is minimal.

This state corresponds to the movement of the flange of the wheel PKS with the angular speed

$$\dot{\varphi}_r \to \dot{\varphi}_k / K_W \quad (13)$$

To assess the effectiveness of using the perspective constructive scheme of the wheel in terms of reducing the differential component of its kinematic resistance to movement, we determine how much the power of the creep forces in the flange contact can be reduced.

In order to do this, let's imagine the momentary high-power effect of creep forces as a dot product of vectors of the creep force vector in the center of the flanges contact B_1 and the vector of the corresponding instantaneous sliding speed of the point B_1 of the flange along the rail

$$N_1^B = \vec{F}_1^B \cdot \vec{V}_1^B = F_1^B \cdot V_1^B \cdot \cos\delta \quad (14)$$

$$N_1^{B/} = \vec{F}_1^{B/} \cdot \vec{V}_1^{BA} = F_1^{B/} \cdot V_1^{BA} \cdot \cos\delta \quad (15)$$

where δ - the angle between the lines of action of the corresponding vectors.

With the glance of the fact that the force vector of the creep is directed opposite to the vector of the corresponding instantaneous speed, i.e. $\delta = \pi$ and $\cos\pi = -1$, in our case

$$|N_1^B| = F_1^B \cdot V_1^B, \qquad |N_1^{B/}| = F_1^{B/} \cdot V_1^{BA}. \quad (16)$$

Let us estimate the possible relative decrease in the power of the creep forces in the flange contact for the wheel PKS by the coefficient K_N:

$$K_N = \frac{N_1^B - N_1^{B/}}{N_1^{B/}} \cdot 100\% \quad (17)$$

where N_1^B and $N_1^{B/}$ - the power of creep forces in the flanges contact when moving, respectively wheel TKS and wheel PKS.

The general character of the dependence of the coefficient on the main influencing factors-the angular speed of the movable flange $\dot{\varphi}_r$ (at a fixed angular speed the wheel mounting surface $\dot{\varphi}_K$) and angle of attack the wheel on the rail ψ is shown in Fig. 3.

The calculation is made for the following traffic conditions: linear speed of the wheel V_K = 20 m/s, wheel radius 0,475 m, magnitude the frictional force of creep in flange contact $F_1^B = 12500$ N, $K_W = 1{,}021$.

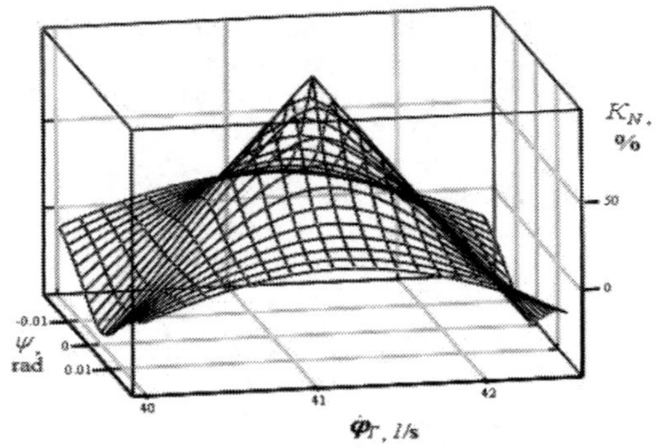

Fig. 3. Dependency graph $K_N = f(\dot\varphi_T, \psi)$ (V_K =20 m/s, $F_1^B = 12500$ N)

The analysis of the graph in Fig. 3 shows that for ideal conditions for the movement of the wheel ($\dot\varphi_T \to \dot\varphi_K / K_w$ and $\psi \to 0$), a virtually 100% reduction in the power of the creep forces in the flange contact for the PKS wheel in comparison with the TKS wheel is possible.

In real conditions the motion of wheel, due to the action of inertial forces and the moment of resistance to movement at the node of its interlinkage the wheel mounting surface and guide surface of wheel (its flange), and also with the practical impossibility of constant and strict observance of the condition , the resulting effect in the form of a decrease in the power of the creep forces in the flange contact will obviously be somewhat less.

Studies to assess the safety of the derailment of the wheel from the rail while rolling the ridge onto the rail made it possible to establish that the conditions for the implementation of the beginning of the process of derailing the wheel from the rail when rolling in the ridge are almost identical for the wheels of both structural schemes.

A significant difference is that the rolling of the wheel PKS on the head of the rail with the detachment of the rolling surface of the wheel from the surface of the rolling of the top of rail at the angles of attack the wheel on the rail more than 0.1 deg. is almost impossible, because under the influence of a system of applied forces and moments, the guide surface of wheel (its flange) will tend to turn of wheel mounting surface in the opposite side.

4 The Investigation of Resistance to Movement the Motion of the Fourth Gondola Wagon with Wheels of Different Constructive Schemes

In order to study the possibilities of reducing the resistance to movement of rail vehicles by using a promising constructive scheme in their trucks, a mathematical model [20] was simulated that imitates the movement along the railroad of a four-axle freight gondola wagon on a three-piece truck.

Satisfactory qualitative and quantitative coincidence of estimated and experimental data confirmed the suitability of the mathematical model not only for determining the dynamic index of the wagon, but also for the resistance to movement, because the work of creep forces at points of contact of wheels with rails cannot be measured directly.

There are results of calculations on mathematical models of some indicators of resistance to movement of four-axle gondola wagons for various modes of their movement below.

In Fig. 4-7 there are graphs of the dependence of the creep power M^g_{imj} in the flange contact of the approaching wheels (1st and 3rd mounted wheels of wagon) with wheels of different constructive schemes from the speed of movement in the loaded and empty states in the straight and curved sections of the curved track.

The figures are: TKS1, TKS3 - data for the 1st and 3rd mounted wheels with wheels of the traditional constructive scheme, PKS1, PKS3 – data for the 1st and 3rd wheel sets with the wheels of the perspective constructive scheme.

In the curved track, the value of the indicator M^g_{imj} was determined as the average power level of the corresponding creep forces in the flange contacts when the motion was steady.

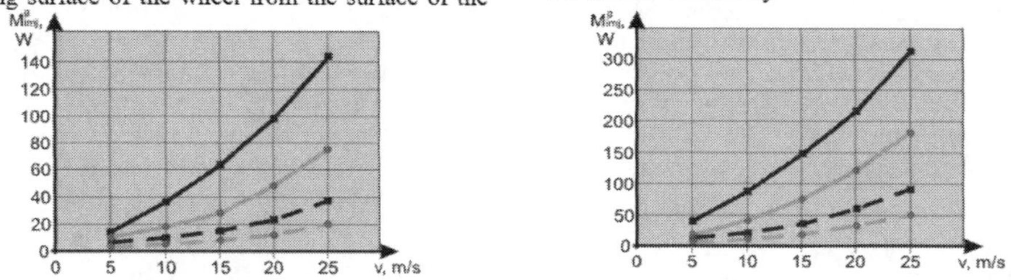

a) empty wagon b) loaded wagon

- TKS1; - TKSC3; - PKS1; - PKS3.

Fig.4. Dependences of the creep forces power M^g_{imj} in the flange contact of the foul wheels from the speed of the wagon V (straight track)

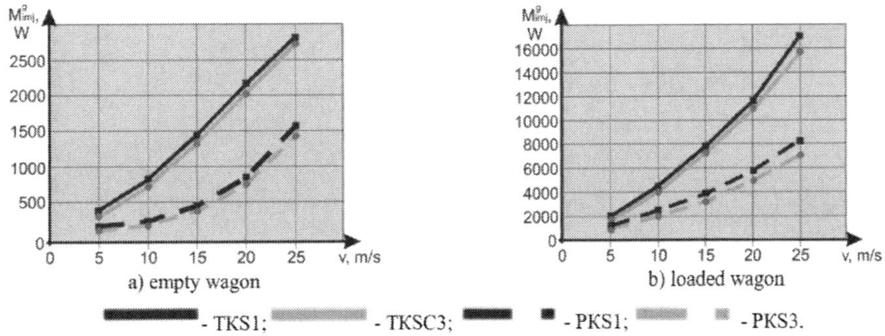

Fig.5. Dependences of the creep forces power M^{g}_{imj} in the flange contact of the foul wheels on the speed of the wagon V (curve with a radius of 350 m)

Since in the straight section of the path the effect of creep forces power in flanges contact due to the wobble of the mounted wheels was periodic, so the corresponding values of the power level of the creep forces for this mode of motion were averaged over by the maximum values of this power for the periods of contact of the corresponding flanges with the side faces the top of rail.

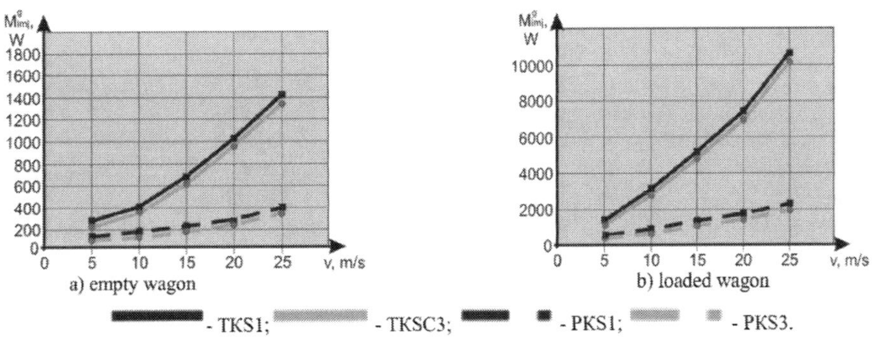

Fig.6. Dependences of the creep forces power M^{g}_{imj} in the flange contact of the foul wheels on the speed of the wagon V (curve with a radius of 750 m)

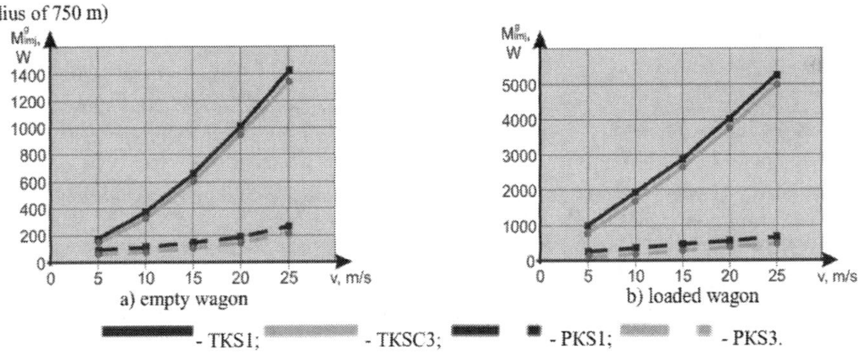

Fig.7. Dependences of the creep forces power M^{g}_{imj} in the flange contact of the foul wheels on the speed of the wagon V (curve with a radius of 1200 m)

Analysis of the results of calculations showed the possibility of a significant reduction in the of creep forces power in the flange contacts, which determines one of the most important components of total resistance to movement of the wagon, when using the PKS wheels in the truck of a four-axle freight gondola wagon compared to using TKS wheels.

The calculated values of the total resistivity for the movement of an empty and loaded car with PKS wheels in different driving modes are obtained lower than the values of this index for a wagon with TKS wheels under the same conditions.

In the straight sections of the track, this decrease is 10...12% for a loaded wagon and 13...15% for an empty wagon depending on the speed of movement.

In curves with a radius of 350 m, 750 m and 1200 m, respectively 20...22% and 23...25%, 16...18% and 17...19%, 13...15% and 14...16%.

In determining the possible savings in working expenditure due to the introduction of the wheels of the perspective constructive scheme in the truck of the four-axle general-purpose gondola cars, in comparison with traditional wheels, according to the real operating conditions of the South-Western railway, the projected forecasted value of the net discounted income from the introduction of mounted wheels with wheels of the perspective constructive scheme in the amount of $E = 14{,}85$ UAH million, which in terms of one wagon makes up $\Delta E = 29{,}7$ UAH thousand. At the same time, the estimated payback period of the project is $T_{OK} = 4{,}1$ year.

5 Conclusion

The generalization of the obtained results allows us to conclude that a radical decrease in the creep forces power in the flange contact of the foul wheels during point-to-point contacting may be achieved by changing the structural scheme of the wheel of the rail carriage allowing independent rotation of its bearing support surface and the director plane (flange).

The use of wheels of such a constructive scheme in the trucks of rail vehicles can allow reducing the total resistance to movement by minimizing the differential component of the kinematic resistance to movement with a decrease in slippage of the wheel flanges along the rails, which will make it possible to significantly reduce the power consumption for hauling operations with such rolling stock.

References

1. M. Mitsche, *Theory of Vehicle. Dynamic of Vehicle. Tom 2 Vibrations.*, 266 p. (1987)
2. J. Reimpell, *Podwozia Samochodow: Podstawy Konstrukcji.*, 411 s. (1997).
3. A. Rutka, J. Sapragonas *Mechanika* **1 (33)** pages 41-48. (2002)
4. J.L. Swayze, B.I. Bachrach, S.R. Shankar Suspension Force Optimization Using Quarter-Car Model with Elastomeric Elements. **Vol. 2**: 785-795. (1999)
5. V. Tkachenko, S. Sapronova, I. Kulbovskyi, O. Fomin, Eastern-European journal of enterprise technologies. **5/7 (89)**. 65–72 (2017)
6. S. Sapronova, V. Tkachenko, O. Fomin, V. Gatchenko, S. Maliuk, Eastern-European journal of enterprise technologies, **6/7(90)**. 19–25 (2017)
7. P. Lukaszewicz, Doctoral Thesis. Stockholm. 154 p. (2001)
8. H. Kobayashi. Quarterly Report of RTRI. **Nr. 1:** 16–20. (2000)
9. R. Upadhyay. Reduced wear wheels and rails. International Railway Journal. **7.** 33-34. (2000)
10. M.I. Spiryagin, V.I. Spiryagin, A.S. Klyuev, S.A. Klyuev, V.A. Ulshin TEKA Commission of Motorization and Energetics in Agriculture. **8a.** 142–149. (2008)
11. A. Golubenko, A. Malohatko, S. Klyuev, TEKA Commission of Motorization and Energetics in Agriculture.**11a.** 5–12. (2011)
12. V.A. Ulshin, S.A. Klyuyev, TEKA Commission of Motorization and Energetics in Agriculture. **14, 1.** 320–331. (2014)
13. M.C. Duffy, *Electric Railways.* 452 p. (2003)
14. A.M. Brzhezovsky, A.G. Parchevsky, M.S. Tikhov, Dynamics and impact on the way of passenger wagons of the company «PATENTES TALGO S.A.» with passive body inclination in curves. Bulletin of VNIIZhT. **2.** 10-18 (2007)
15. G.P. Bourtchak, L.V. Vinnik,. *Euromech colloquium* **409**. 34-35. (2000)
16. G.P. Bourtchak, L.V. Vinnik, *Proceedings of the 7th Miniconference on Vehicle System Dynamics Identification and Anomalies* (VSDIA 2000). 185-193. (2001)
17. L. Vinnik, G. Bourtchak, D. Pogorelov, *Proceedings of the 18th IAVSD Symposium held in Kanagawa.* **41.**. 607-616. (2003)
18. L.V. Vinnik, *Second International workshop on freight vehicle design.* 118-124. (2001)
19. A. Shiler, *Proceedings of the 25th DAAAM International Symposium Procedia Engineering.* 1714-1723. (2014)
20. I. Galiev, A. Shiler, A. Kildibekov, I. Kukushkin *Proceedings of the 25th DAAAM International Symposium Procedia Engineering.* 1724-1731. (2014)
21. A.G. Reidemeister *PhD thesis.* 185 p (2000)

MATEC Web of Conferences 235, 00034 (2018) https://doi.org/10.1051/matecconf/201823500034
Horizons of Railway Transport 2018

Information Technologies for Risk Management of Transportation – Logistics Branch of the "Russian Railways"

Galina V. Bubnova[1], *Olga V.* Efimova[1], *Irina V.* Karapetyants[1] and *Petr V.* Kurenkov[1,*]

[1]Doctor of Economics, Professor, Russian University of Transport (RUT), Moscow, Russian Federation

Abstract. The increased competition between modes of transport, the growing uncertainty of the business environment, the diversity of goals of the business units of the holding "Russian Railways" and the participants of the transport market inevitably increase the level of risks that can threaten the efficient development of the company. The development of information technologies is the base of risk management and higher competitiveness. The growth in uncertainty and speed of changes in the business environment leads to the necessity of implementation of digital technologies for risk management, influencing the added value of transportation-logistic services.

1 Introduction

The part of the management treatment is to appreciate the risk management. The development of information technologies is the base of risk management and higher competitiveness. The growth in uncertainty and speed of changes in the business environment leads to the necessity of implementation of digital technologies for risk management, influencing the added value of transportation-logistic services.
Special application field of this management is in the transport sector. The increased competition between transport modes, the growing uncertainty of the business environment, the diversity of goals of the business holding "Russian Railways" and the subjects of the transport market inevitably increase the level of risks that can threaten the efficient development of the company.

2 Strategy of Risk Management

The functional strategy of risk management of the holding "Russian Railways" has formed the methodical and regulatory framework of the risk management system, provided for the beginning of a change in attitudes towards risks in the railway transport, which resulted in awareness and acceptance of risks in all types of activities. Increased competition between modes of transport, growing uncertainty of the business environment, the diversity of the goals of the business units of the Russian Railways Holding and participants in the transport market inevitably increases the level of risks that may threaten the effective development of the company. The development of information technologies in the holding formed knowledge bases about risks and their consequences, therefore the information available in them allows to forecast various types of threats and manage risk factors that affect their consequences.
The interrelation of the main terms and definitions - "source of risk", "risk factor", "risk event" and "damage" form the logic of risk management. Sources of risk with a certain probability become risk factors, which in turn also with a certain probability can lead to risk-events, accompanied by damage. Such an approach to risk management dictates the need to formulate a list of sources of risk, group risk factors for each source, create a risk-event map and evaluate the damage for each risk event. Getting the maximum amount of meaningful information should curtail the sources of risk and reduce the likelihood of their becoming a risk factor. Management decisions should be aimed at reducing the likelihood of occurrence of a risk of events and the occurrence of damage.

3 Concept of Risk Management with use of Information Technologies

The proposed conceptual apparatus should be used both in the process of top-level risk management and in the standard information technology for risk management of business unit structural divisions involved in providing and providing transport and logistics services. To generate knowledge about risk factors, it is proposed to use the company's information resources, which allow them to predict risk factors aggregated by sources of risk, degree of control and degree of influence on business based on mathematical models of trend analysis and device and process characteristics (Fig. 1).
Information technologies when creating functional risk management strategies for business blocks of a transport holding company can be used to solve the following tasks:

* Corresponding author: petrkurenkov@mail.ru

- Formation of a register of typical risks (matrix of sources and risk factors).
- Establishment of quantitative values of risk assessment criteria (metrics).
- Assessment of the impact of identified risks on the target indicators of the business unit.
- Determination of the economic results of eliminating / reducing the consequences of the occurrence of events and incidents.
- Development of models for assessing the permissible level of risks for groups of typical risks.
- Monitoring of risks and activities for their identification and management, the structure of information support for the risk management system of the business unit.

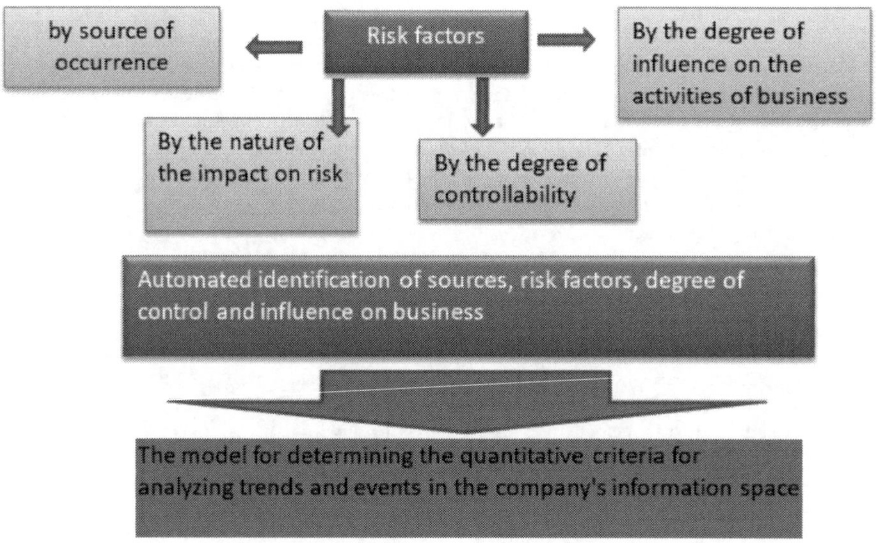

Figure 1. Structuring of risk factors on the basis of information technology

In the development of information technology, risk management should consider the appropriateness of accounting for and identification of individual risk factors based on the cost assessment of the costs of monitoring and managing each type of risk. Otherwise, an uncontrollable set of factors and parameters fall into the sphere of risk management. The cost of risk is not only damage, but also the costs of monitoring sources and risk factors, the implementation of information technology for the risk management process, insurance and hedging activities.

The magnitude of the cost of risk, which the company is ready to neglect and not build a system for monitoring factors and sources, is determined by its risk appetite - the willingness to accept certain risks. However, for a transport company only the valuation is not enough, here the safety and reliability of the transportation process have priority, therefore the corresponding risks should be identified as the most critical or significant.

4 Categorisation of Risks Management

Modern approaches to risk management consider risk not only as a threat, but also as a possibility, therefore information technologies for risk management should include processes of searching for new business prospects for the main activities in the business unit.

Information technologies that support the risk management strategy and policy are an important tool for ensuring the safety and reliability of the transportation process, and a set of principles and methods built into the processes and organizational culture of the Company ensures sustainable development, performance of work and rendering services to customers with the least risks and losses high quality of service delivery

In 2015, Ernst & Young conducted a study of the best risk management practices of more than 1,000 companies around the world. This study showed that the key task of risk management is the creation of value added, as well as the formation of integrated solutions based on a preventive response to risks and improving performance, which is impossible without the use of information technology.

For successful risk management it is proposed to divide them into three categories:

- strategic risks that must be taken, as they carry the possibility of obtaining benefits;
- avoidable risks, which should be avoided, as they have a negative impact;
- external risks that the organization does not control and that can negatively or positively influence the business.
- Identification and elimination of potential threats in the sphere of basic cargo transportation services are aimed at reducing damage or preventing the loss of financial and economic benefits of the Russian Railways holding, providing customer orientation, improving the quality of services, increasing the scale of the transport and logistics business, organizing effective interaction of business units in perimeter of the transport and logistics business block (Table 1).

Conclusion

The main goals of the risk management system are higher efficiency of management system due to the managers' awareness of the company's risks, higher quality of business planning and activities indicators monitoring, compliance with the regulatory requirements (both external and internal). Achieving these aims provides for sustainable income and investment attractiveness of the company (lower costs of funds attraction).

Till lately the attention was focused on risks managed by control means but having no or little potential for added value. As the requirements of the parties concerned grow and the business environment constantly changes the leading organizations start to pay more attention to risks, which create added value.

Table 1. Register of risks of transport and logistics business.

Name of risk	Significant signs
Pre-completion	Is characterized by lower production output in industry and agriculture
Macroeconomic	Is connected with the changes in taxation system, inflation, changes in transportation tariffs and rates, national currency rate fluctuations
Competitive	Is expressed in the growth of commercial efforts of competitors and relative decline of competitiveness of transportation-logistic services of the company including high-profit sectors, as well as shift of freight transportation to other transport means
Technogenic and environmental	Is caused by external sources, independent of the company's activity
Productional	Is connected with the violation of agreed transportation conditions, violations of technological transportation process which lead to lower revenue than planned
Commercial	Expressed in the failure of the task of revenue from providing customers with additional services
Application form (Interactions with customers)	Non-fulfillment of applications in domestic and international communication, as a consequence, reduced loading
Staffing	Increased staff turnover, insufficient staffing, reduced competence
Operating	Mistakes and delays in transportation documents, violations of technological process of arranging the transportation services, not fulfilling the technical requirements of freight placement and fixing which leads to lower loading
Planning quality	Inaccurate loading forecast and freight base determination, thus lower revenue than planned
Financial	Not completing the plan of accured revenue from transportation activities; Incomplete or delayed revenue return from transportation activities
Economic	Higher expenses due to increase of prices, insufficient managerial control of expenses elements
Managerial	Inefficient employment, arrangement and management of human resources, inefficient communications within the company
Technogenic	Is explained by internal factors (fire, disaster)
Reputational	Lower transportation services quality which leads to lower loading and lower revenue than planned

References

1. O.V. Efimova *Methods of justifying management decisions when introducing new information technologies*, 56, (2009).
2. Shiryaev, V.I. *Models of financial markets: Optimal portfolios, management of finances and risks*. 216. (2015)
3. T.N Kuzminova, E.Z. Makeeva. Economics of Railways, MFCED, 6, (2016)
4. N.A. Serebryakova, T.I. Ovchinnikova, I.N. Bulgakova, S.V. Sviridova, T.O. Tolstykh. European Research Studies Journal. **20**, **3**. p. 556-569. (2017)
5. E.N. Barikayev, M. Eriashvili. *.Management of business risks in the system of economic security. Theoretical aspect*: 159. (2015).
6. Efimova OV, Baboshin EB. Economics of railways, MFCED, **2**. (2016),
7. R. Gibson, *Formation of the investment portfolio: Financial risk management*, 274 (2016).
8. A.Yu. Egorov, A.A. Safronova. S.V. Tovsultanova Transport business in Russia .**5**. 290-298. (2013)
9. S.V. Emelyanov, *Proceedings of the ISA RAS: Algorithms. Solutions*, 102, (2014).
10. MJ. Piekarz, IS. Jenkins, PF. Mills. *Management and Control of Risks*. 89-116, (2015)
11. AA. Safronova, AM. Atayev. Innovative economy: information, analytics, forecasts. **1-2**. 3-9. (2012)
12. Desai, Bina Sarmiento, Juan Pablo. International Journal of Disaster Risk Reduction. **14**, 203-204, (2015).
13. P. Kurenkov, A. Astafiev, O. Kaplina. Komunikacie K. **18**, **2**. 148-152. (2016)
14. DG. Kagrimanova, EV. Borzova. Innovative economics and modern management. **1 (1).** 31-37. (2014)
15. SG. Blokhin, PV. Kurenkov, AS. Kornev Economy of railways. **11**. 40 (2009)

Implementation Of The Technical Interoperability For TAF TSI Subsystem In Freight Operation

Peter Ihnat[1,*]

[1]Železničná spoločnosť Cargo Slovakia, 040 01 Košice, Slovakia

Abstract. The article describes the cooperation between freight railway undertaking and carriers with regard to implementation of the TAF TSI into operation. It suggests for interfaces between railway undertaking/carrier which are important for success application of the TAF TSI and defines limitations which must be eliminated during the implementation of the TAF TSI into operation. Technical specification of interoperability for the telematics applications for freight is one of the subsystems of railway interoperability of which implementation into operation is mandatory in accordance with Directive of the European parliament (62/2006 and 328/2012).

1 Introduction

Interoperability of the rail sector is a European initiative designed to improve the competitive position of the rail sector towards other forms of transport. It does this through a regulatory framework that assuring the technical harmonisation and standardisation of Europe's railways.

The rail interoperability consist of a number of technical specifications for different kinds of the rails subsystems:[1,2]

Subsystems of the Structural areas:

a) Infrastructure – consist of the track, points, engineering structures (bridges, tunnels, etc.), associated station infrastructure (platforms, zones of access, including the needs of persons with reduced mobility, etc.), safety and protective equipment,
b) Energy – consist of the electrification system, including overhead lines and on-board parts of the electric consumptions measuring equipment,
c) Control-command and signalling – consist of the all equipment necessary to ensure safety and to command control movements of trains authorised to travel on the network, [3]
d) Rolling stock – defined structure, command and control system for all train equipment, current-collection devices traction and energy conversion units, braking, coupling and running gear (bogies, axles, etc.) and suspension, doors, man/machine interfaces (driver, on-board staff and passengers, including the needs of person with reduced mobility), passive or active safety devices and requisites for the health of passengers and on-board staff.

a) **Subsystems of the Functional areas:**
b) e. Traffic operation and management – consist of the procedures and related equipment enabling a coherent operation of the different structural subsystems, both during normal and degraded operation, including in particular training and train driving, traffic planning and management,
c) f. Maintenance – defined the procedures, associated equipment, logistic centres for maintenance work and reserves allowing the mandatory corrective and preventive maintenance to ensure the interoperability of the rail system and guarantee the performance required,
d) g. Telematics applications for passenger services, including systems providing passengers with information before and during the journey, reservation and payment systems, luggage management and management of conditions between trains and with other modes of transport [4]
e) h. Telematics applications for freight services (for detailed description see article 1.1).

The first series of TSI's on trans-European high - speed rail system were drawn up by AEIF (European Association for Railway Interoperability) in 2002 and most of them were subsequently revised in 2008. AEIF was acted as the joint representative body defined in the directive, bringing together representatives of the infrastructure managers, railway companies and industry (the AEIF was replaced by ERA in 2004). The second series of TSI's for European conventional rail system was published between 2006 and 2011. All of these TSI's were drawn up accordance with Council Directive 96/48/EC on the interoperability of the trans-European high - speed rail system and Council Directive 2001/16/EC on the interoperability of the European conventional rail system. A little difference between high-speed rail system and conventional rail system flow into consolidation these two directives of one Council

* Corresponding author: ihnat.peter@zscargo.sk

Directive 2008/57/EC on the interoperability of the rail system within the Community. [5]

As already foreseen in the interoperability Directives currently in force, this legal framework is begin further developed by correcting errors and close open points in TSI´s, and extending the geographical scope of the TSI´s to the whole European Union´s rail system.

On this basis, the Agency (ERA – European Rail Agency) is currently revising the following TSI´s: [6]

- Operation and traffic management (OPE),
- Freights wagons (WAG),
- Locomotives and passenger rolling stock (LOC and PAS),
- Control – command and signalling (CCS),
- Telematics applications for passenger services (TAP),
- Telematics applications for freight services (TAF),
- Accessibility for persons with reduced mobility (PRM),
- Safety in railway tunnels (SRT),
- Infrastructure (INF),
- Energy (ENE).

The dependence between TSI´s is described on the figure 1.

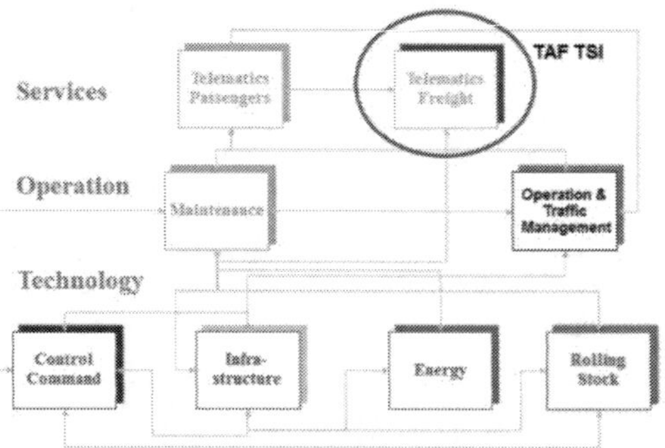

Fig. 1. TSI´s structure

1.1 Telematics applications for freight (TAF)

The TAF TSI is a one of the most important technical specifications of the interoperability for creation the integrated rail system for freight within Community. The implementation of these TSI´s is mandatory for all of the member states of EU. [7,8]

TAF TSI defined procedures, messages and content of these messages, which must be transmitted between subjects involved in the transport chain (railway undertakings, infrastructure managers, logistic bodies, customers, keepers, etc.). All of the relevant messages are split into next groups: [9]

- Consignment note data (the consignment note has to be send by the customer to the Lead RU. It must show all the information needed to carry a consignment from the consignor to the consignee,
- Wagon orders (the wagon order I primarily a subset of consignment note information. It must be forwarded to the RUs involved in the transport chain,
- Path request (defines the requested, accepted and actual data to be stored concerning the path of a train and the characteristics of the train for each segment of that path,
- Train preparation (consist of the Train Composition, Train Accepted, Train Not Suitable, Train Ready, Train Position, Train at Start and Train Running Information messages),
- Train running forecast (specifies the messages which must be exchanged during the normal running of a train without any interruption – Train Running Forecast and Train Running Information messages),
- Service disruption information,
- Train location (specifies the tracing possibility to get information about train location),
- Shipment ETI/ETA (the ETI (Estimated Time of Interchange) /ETA (Estimated Time of Arrival) calculation is based on the information from the infrastructure managers in charge, which sends, within the Train Running Forecast message, the TETA (Train Estimated Time of Arrival for defined reporting points.)
- Wagon movement (specifies the package of the messages which must be stored and electronically accessible – Wagon Release Notice, Wagon Departure Notice, etc.…),
- Interchange reporting (describes the messages attached to the transfer of responsibility for a wagon between two railway undertakings, which occurs at interchange points),
- Data exchange for quality improvement (a measurement process in an essential post trip process to support quality improvements. Defined the messages which must be transmitted between LRU´s, RU´s and IM´s),
- The main reference data (contain the Infrastructure Data and Rolling Stock Data which are most important for the operation of freight trains),
- Various reference files and databases (for the operation of freight trains on the network must be available an accessible also additional data, which are locally and centrally stored),
- Electronic transmission of documents (presents the communication network to be used for data exchange),
- Networking and communication,

The figure 2 reflected a share of single groups on to whole TAF TSI.

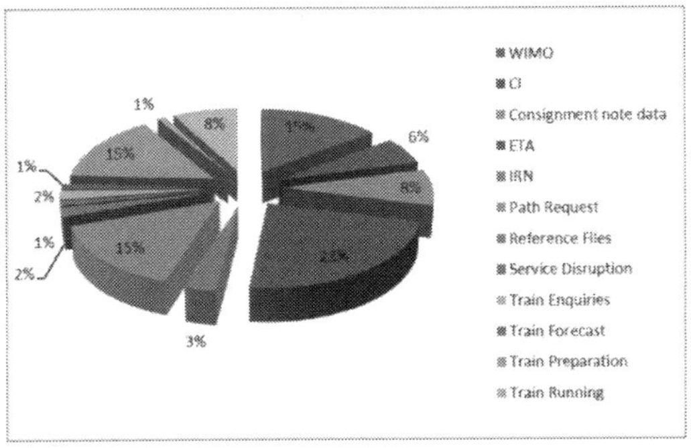

Fig. 2. Share of the TAF TSI

2 Restrictions of the implementation the TAF TSI into operation

The TAF TSI describes only baseline rail processes, which are important for realisation of the transport between two or more railway stations. This view limits to a great extent a successfully implementation within European rail network. [10]

The main problem is that into the transport process entry most of the railway and non-railway subjects with or without operational information systems, that it makes a process considerably complicated (see figure 3).

TAF TSI not allowance the electronic data exchange (EDI) between rail and non-rail subjects (railway undertaking and intermodal terminal, railway undertaking and road carrier, etc.), so in case of intermodal transport will be problematic to inform railways undertaking, consignee, consignors and other subjects involved in the transport chain about the transport by the road, river, sea. [5]

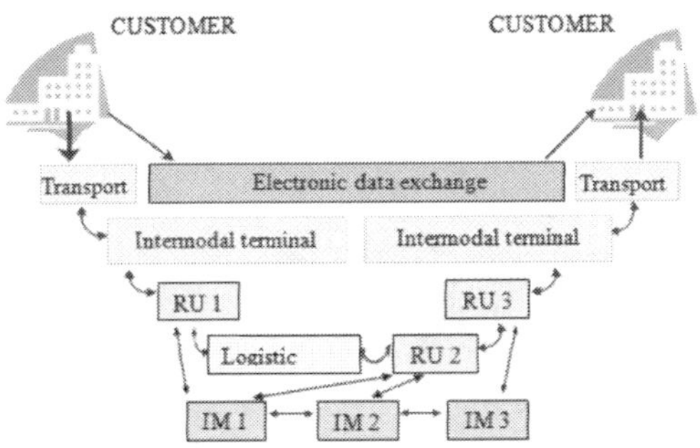

Fig. 3. Freight railway process map

The second problem of the implementation of the TAF TSI is, that no all railway subjects to own information system, which covered operational process described in regulation and related documents on information interoperability within European rail network. Because of this situation is much information into the information system typing manually by the RU´s staff, what is a potentially place of the errors creation.

TAF TSI defined Consignment note data as first of all actions of the rail transport, but consignment note data must to precede wagon order send from consignee to RU, or to LRU. (this type of message and EDI process are not described in TAF TSI). In case of various RU´s may be various structures of the wagon order messages, which must be used by the consignors.

In the figure 4 point 2 and 3 are illustrated the main areas of the cooperation between RU´s and carriers (consignor and consignee) in context of the TAF TSI. [4]

The main problem of this cooperation been described in previous text is EDI. In case of absence information systems on the customer sides, must be information collected RU´s by the own staff. But it is not correctly, because the costs of data collection and responsibility of the data quality are bared by the RU´s. Next problem of this, are limited possibilities to transmitting the information about the transport between RU´s and carriers. [9]

The same restrictions are created in case of cooperation between two RU´s without reciprocally EDI.

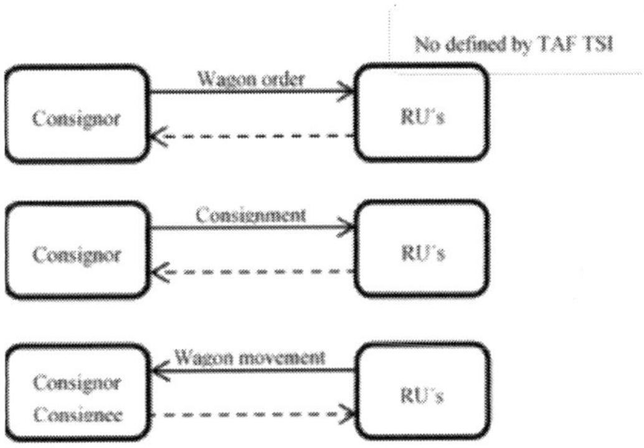

Fig. 4. Cooperation between RU´s and carriers

3 Conclusion

The progress made thanks to the EU regulatory framework for railway interoperability will encourage the further development of the internal rail market, helping new business to emerge, cutting entry costs and, ultimately, improving the competitiveness of rail compared with other modes of transport.

But without regulatory framework which force to subject implementing own information systems in context of TAF TSI, won´t be possible to encourage the further development of the internal rail market.

The one of the arrangements towards to successful implementation of the TAF TSI into operation, can be price policy of the access onto rail track charge. The new price policy would by make provision for total access onto rail track charge for the subjects with EDI and subject without EDI.

References

1. A. Mecke, I. Lee, J.R. Baker jr., M.M. Banaszak Holl, B.G. Orr, Eur. Phys. J. E **14**, 7 (2004)
2. A. Dolinayová, M. Ľoch, J. Čamaj. TRA2016. Transportation Research Procedia. **14**. 916-925. (2016).

3. European Commission: Commission Regulation (EC) No 62/2006 concerning the technical specification for interoperability relating to the telematics applications for freight subsystem of the trans-European conventional rail system: Brussels 2005
4. J. Mašek, P. Kolarovszki, J. Čamaj. Proceedia Engineering. **134**. 231-236. (2018)
5. Z. Zahumenská, J. Gašparik. Procedia Engineering. **192**. 976-981 (2017)
6. J. Klimková, J. Čamaj, J. Gašparík. Transport means 2016 : proceedings of the 20th international scientific conference. 891-896. (2016)
7. J. Gašparik, V. Gaborova, V. Ľuptak. Transport Means – Proceedings of the International Conference. 245-249. (2016)
8. European Commission: Report from the commission to the European parliament, the council, the European economic and social committee and the committee of the regions on the progress made towards achieving interoperability of the rail system, Brussels 2013
9. A. Dolinayová, J. Čamaj, J. Danis. 16th International Scientific Conference on Globalization and its Socio-Economic Consequences. 383-390. (2016)
10. S. Stoilova, R. Nikolova. 15th International Scientific Conference on Engineering for Rural Development. 586 – 596. (2016).

Investigation Of Methanol Utilization In Ci Engines

Ákos Bereczky[1*], *Kristóf Lukács*[1] *and Zoraze Khan*[1]

[1] Budapest University of Technology and Economics (BME), Department of Energy Engineering 1111 Műegyetem rkp. 3-7, Budapest, Hungary

Abstract. The decrease in the fossil fuel resources and more rigid emission standards make the necessity to look for new fuels and technologies. One of the possible fuels can be the different alcohols. The oxygen content of the alcohols can improve the combustion process and reduce the emission rates of incomplete combustion products. The blending methanol (M) and ethanol (E) with diesel fuel is limited by separation. This article shows the research results and analysis of two typical solutions. First are blends of stabilized rape seed methyl ester (RME) to prevent the phase separation. The tests are carried out on a conventional TDI engine (VW 1Z) at different speeds and loads. Experimental results indicate that the addition of 10% methanol to 80% Diesel and 30% of FAME blend increased the unburned hydrocarbon (THC) with 60%, decreased CO emission with near 4% and the particulate emission (PM) with near 30%, the NOx emissions were near the same. The other technology is the dual fuel method. In this case, different amount of methanol is injected into the intake manifold. The tests are carried out with a conventional DI engine (IVECO AIFO 8031) at constant speeds and different loads and energy ratios. Experimental results indicate that the increase of energy ratio of the methanol at full load decreased NOx emission with up to 60% and the particulate emission (PM) up to 90%, while the CO and THC emissions increased extremely mainly at part loads.

1. Introduction

The oxygen content of the alcohols can improve the combustion process and reduce emission rates of incomplete combustion products. In addition, they modify the physical properties of the blends, decrease the density, cetane number, lower heating value (LHV), boiling point and the viscosity, but the oxygen contents of the blend increases [17, 19, 21]. The higher alcohols like different isomers of propanol and butanol can make stable blends [8, 12]. The blending methanol and ethanol with diesel fuel is limited by separation. One solution can be to utilize to blend stabilized methyl esters to prevent the phase separation [3, 7, 9, 10, 13] or surfactant additive like tetrahydrofuran (THF) [2].

Other possibility is the injection of the alcohol in the intake manifold or in the combustion chamber. To inject the alcohol in the combustion chamber a high-pressure injector is needed, but the lubrication of the high-pressure injector and the atomiser is problematic. Furthermore, another atomiser system is needed in the combustion chamber which is expensive [5]. According to these considerations, the most effective solution is the multipoint injection in the intake manifold. In case of dual fuel engine, the alcohol as a primary fuel is injected in the intake manifold of the engine and the produced homogenous mixture is ignited in the combustion chamber by the injected diesel fuel as secondary fuel [5, 15, 16]. In this case the main parameter is the energy ratio:

$$E_r = \frac{LHV_p * \dot{B}_P}{LHV_p * \dot{B}_P + LHV_{D2} * \dot{B}_{D2}} \qquad (1)$$

2. Engine and test regimes

The tests are carried out with the AVL-7030 fuel balance as fuel consumption measuring system, HORIBA MEXA-8120 F exhaust gas analyser system, and AVL 415 smoke meter.

Engine tests were carried out in Jendrassik György laboratory of Budapest University of Technology and Economics. VW- Audi diesel engine 1.9 TDI 1Z was used for r triple blend tests (D2-M-RME). Engine with an open combustion chamber (direct injection), containing a turbo compressor, an exhaust gas recirculation system (to achieve more accurate heat release results, the system was off during tests) and an electronic control unit (ECU).

* Corresponding author: bereczky@energia.bme.hu

https://doi.org/10.1051/matecconf/201823500036

In the first step of the test, at the chosen speeds (2000, 2500, 3000 and 3500 RPM) the maximum loads were determined and the 25%, 50% and 75% of the maximum torque with diesel fuel (D2) were determined.

1.9 TDI engine model was created by the software AVL Boost Burn and used for analysis of heat release ($dQ/d\varphi_{max}$) and pressure rise ($dp/d\varphi_{max}$) characteristics [4]. Having the required experimental data, the software AVL Boost Burn makes these calculations relatively easy and quick.

The dual fuel tests the developed system, a M8B 160 generator is driven by an IVECO AIFO 2.9 diesel engine with direct injection. The alcohol is injected in the intake manifold of each cylinder. The installation of the injection lines needed a new intake manifold system, which was made possible by welding from patent boards and cut pieces which were suitable for the receiving of the nozzles. The fuel was atomised by the nozzles on the intake valves, avoid the condensation on wall of the intake manifold to. To regulate the combustion process the timing of the injection was resolved by modification of injection pump.

3. Results with triple blends [results based on 18]

Practically in the whole engine load range the BTE increased in the case of B30, 1–2% and with B30+10M 2–2.5% respectively. The NOx increased with the B30 and B30+M10 blends, the highest change was at 3000 RPM at 25 % and 50 % load, 4.5–6 % at full load, 0–2 % at 75% load, the EDC increased the pre-injection angle with the fuel mass flow, therefore there was intensive increase (Fig. 1):

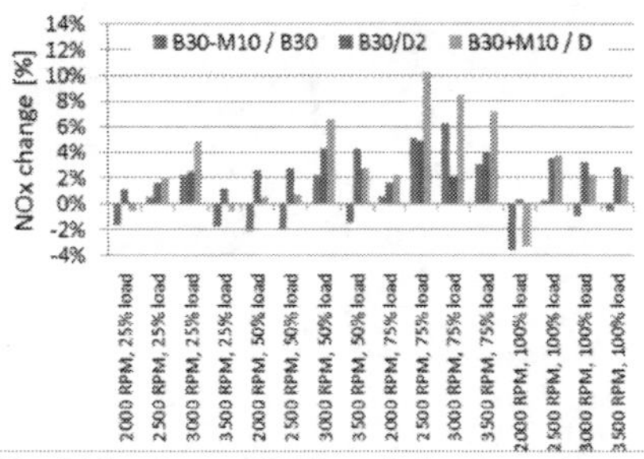

Fig. 1. The change of NOx emission comparing to D2 and B30 at different loads and speeds (results based on 18)

The distribution of the incomplete combustion products' (CO and HC) concentration values is ambiguous: in case of using B30, the concentration of CO was decreased mainly at 50 %-100 % loads, with methanol blend at low load (25 %) increased 10-20 %, at high loads deceased (Fig. 2):

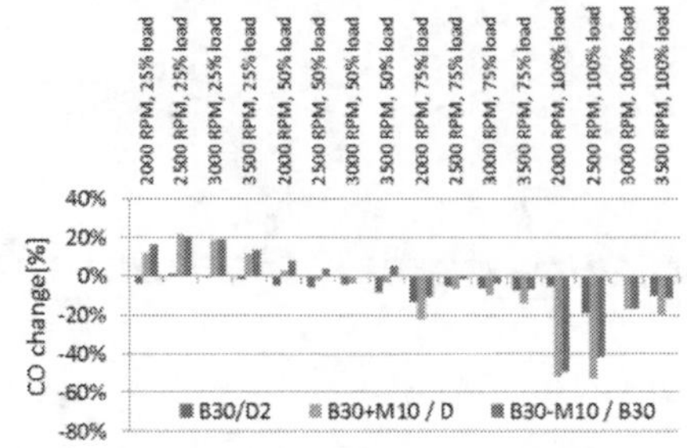

Fig. 2. The change of CO emission comparing to D2 and B30 at different loads and speeds (results based on 18)

The THC concentration values reduced by 32 % in the whole load range; in case of B30+10M, THC values were varying in the range of -8 to +18 % through all experimental load points (Fig. 3):

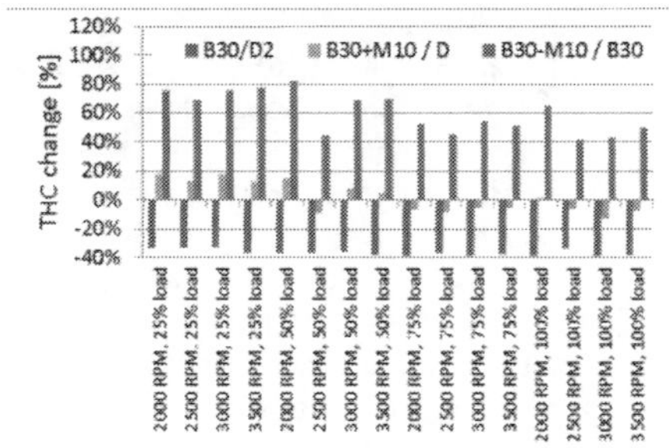

Fig. 3. The change of THC emission comparing to D2 and B30 at different loads and speeds (results based on 18)

The PM concentration values were reduced by 12–22% (B30) and 28–54% (B30+ 10M) compared to D2 (Fig. 4):

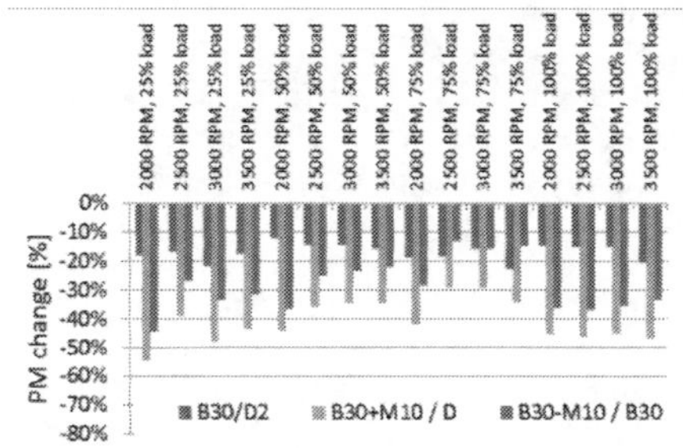

Fig. 4. The change of PM emission comparing to D2 and B30 at different loads and speeds (results based on 18)

Table 1 presents the averaged change of the emissions. In the case of NOx, the averages do not include the 75 % load points as mentioned the influence of the modified pre-injection was high. The methanol had low influence on the NO_x [14] emission but the THC emissions increased showing the methanol influence was higher, whereas with the RME the THC emissions were decreased. Soot concentration values were reduced by more than

16 % with biodiesel and 40.7 % with RME methanol blend compared to D2:

	NO_x change [%]	THC change [%]	CO change [%]	PM change [%]
B30/ D2	2.5%	-36%	-5.8	-16.9
B30+ M10/ D2	1.8%	1,9	-8.6%	-40.7
B30+ M10/ B30	-0.7%	60.4%	-3.5	-28.5

Table 1. The average change of the different emission components comparing with D2 and B30

4. Results with dual fuel system

At 24 kW loads, the maximum pressure increased from 0 % to 75 % Er, the maximum values were at 75% Er, but the maximum appeared later (Fig. 5). In the case of 90% Er, the pressure was significantly reduced and the maximum values occurred significantly later (Fig.5):

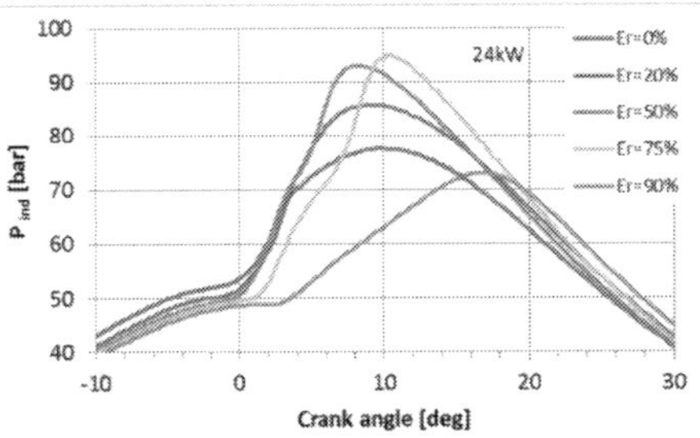

Fig. 5. The indicated pressures in the function of the crank angle at different energy rations at 24 kW load [16]

The effects are well visible on HRR functions (Fig.6):

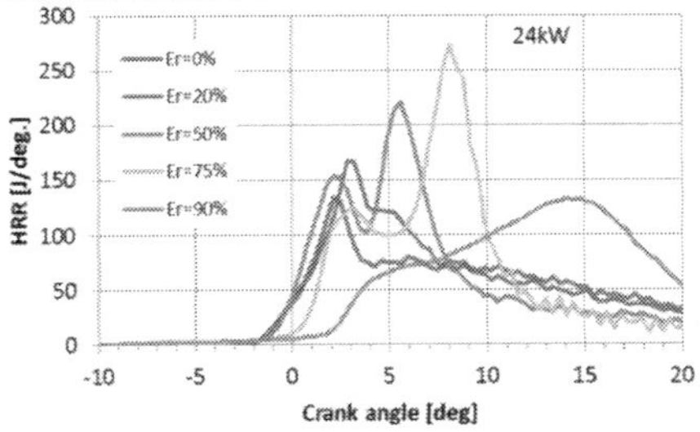

Fig. 6. The heat release rate (HRR) in the function of the crank angle at different energy rations at 24 kW load [16]

In the case of pure gas oil, conventional pre-mixed - diffuse combustion phenomena can be observed, at 20 % Er the pre-mixed section increased to the same extent as the diffuse part of the combustion. With the further increase in methanol content (50 % Er), combustion has been significantly transformed, intense combustion of methanol started at the beginning of the pre-mixed phase and continued in the diffuse stage and reached the maximum of HRR. By further increase in methanol content (75 % Er), ignition delay is increased, as D2 fuel was significantly reduced, the pre-mixed phase could ignite a small amount of methanol, therefore the intense burning of methanol started later and reached its maximum in the diffusion phase. In case of 90 % Er, combustion changes significantly. The inflammation delay increased significantly due to the heavy overflow of methanol in the combustion chamber.The NO_x emissions at full load (24 kW) were maximum at 20 % Er, in this case the premixed combustion was maximum and the NO_x emission increased compare to D2 (Fig. 7):

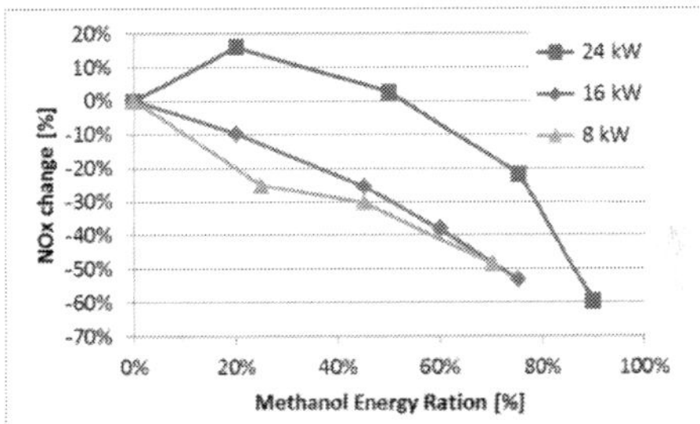

Fig. 7. The change of NO_x emissions in the function methanol energy ratio (Er) at different loads (results based on 16)

With the growth of Er, the intensity of the pre-mixed stage appeared later and the evaporation of Methanol decreased NO_x emission. At part loads, the decreased fuel and increased air-to fuel ratio decreased NO_x emission. The distribution of incomplete combustion products' (CO and THC) concentration values increased with Er, mainly at part loads (16 kW and 8 kW). This indicates that the quality of the methanol burning process has decreased (Fig.8-9):

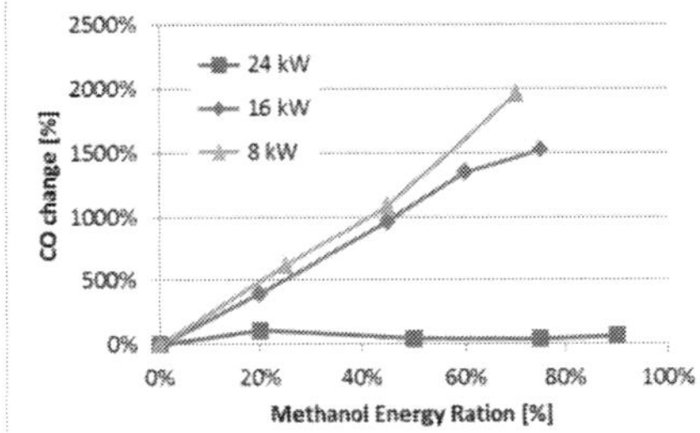

Fig. 8. The change of CO emissions in the function methanol energy ratio (Er)at different loads (results based on 16)

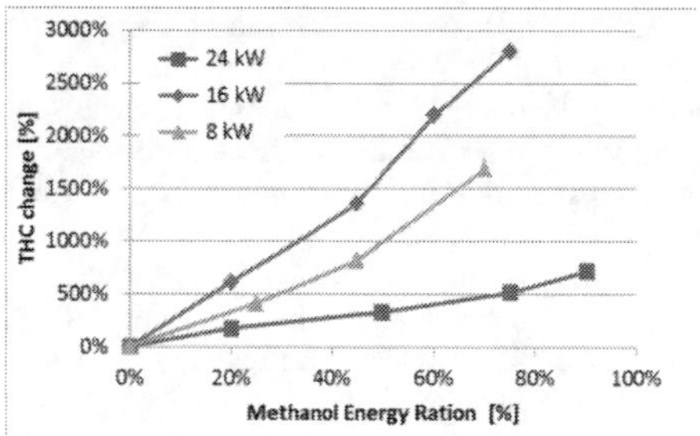

Fig. 9. The change THC emissions in the function methanol energy ratio (Er)at different loads (results based on 16)

Particle matter (PM) emissions, with increased Er, drops considerably, at full load at 90% Er, it is reduced to just 4 % compared to D2, the primary reason for this being the decreased diesel fuel (D2) content (Fig. 10):

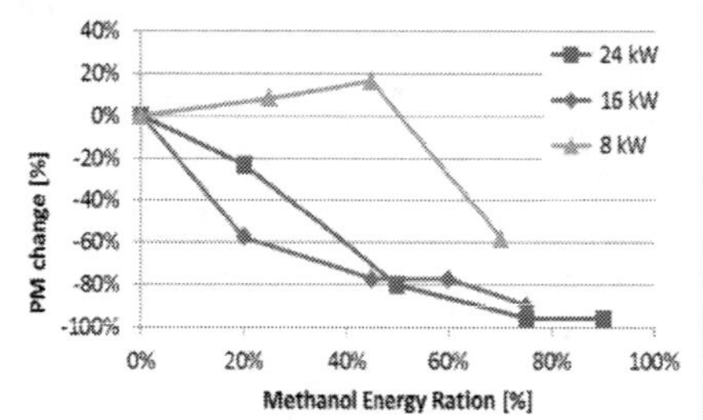

Fig. 10. The change of PM emissions in the function methanol energy ratio (Er)at different loads (results based on 16)

5. Modelling

Modelling of the working process is very important for educational and development purposes. For problems solving we use a one dimensional system, like the AVL Boost system and the AVL FIRE CFD software package for CFD calculations. There are many possibilities for modelling the different fuel blends and dual fuel system with CFD methods [11, 16].

5.1 CFD simulations

Numerical modelling of the in-cylinder processes is low cost and less rigorous method for the analysis of performance and emission characteristics. The model studies are based on ECFM combustion model. This used code is developed based on the FV approach with the pressure based segregated solution algorithm and the SIMPLE algorithm is used for velocity and pressure coupling. ECFM model is founded on the concept of laminar flame propagation and the flame velocity and thickness are taken to be the average flame front values. Various other combinations of models are used. Turbulence of the charge is modelled by K-zeta-f model, combustion for diesel by ECFM-3Z, and for dual fuel, ECFM by combining it with the diesel ignited gas engine ignition model, spray atomization by WAVE breakup and multicomponent evaporation model for spray evaporation, Extended Zeldowich and Lund flamelet models are used for modelling NOx and soot emissions respectively. With this CFD software package the combustion process of different methanol energy rations investigated (Fig. 11.).

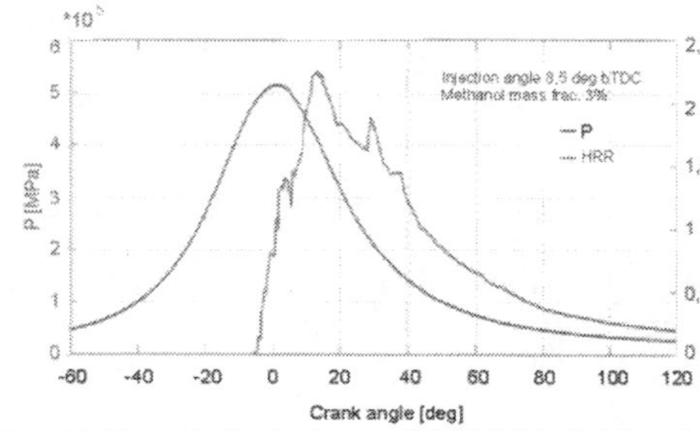

Fig. 11. Numerical calculation of HRR [J/deg*100] and pressure in the function of crank angle at 3% methanol mass fraction

The main target of present simulation work to simulate PPCI and HCCI combustion. Because at low-mid engine load range these methods can be successful to achieve low NO_x and PM emissions. [20].

6. Conclusions

The triple blend give a possibility to decrease PM up to 40 % and CO emission up to 9 %, but the NO_x emission increase (1.8%), when the BTE increased and with B30 + 10M, 2–2.5 %.

The dual fuel technology give opportunity the decrease the PM and NO_x emission together, but the low load has problems with incomplete combustion products (CO and THC).

Authors are grateful to AVL GmbH Company for the possibilities to use their simulation software (AVL Boost).

REFERENCES

1. AVL: AVL FIRE® VERSION 2011. Combustion and Emission module. Edition 10/2011.
2. I. Barabás, A. Molea., R. Suciu. *Proceedings of the 4th International Congress of Automotive and Transport Engineering (AMMA 2018)*. Proceedings in Automotive Engineering. Springer, (2018)

3. I. Barabás, M. Zöldy, I. A. *Todorut Oil and Natural Gas*. 644: Studium Press LLC, 2015. pp. 577-606.
4. Á. Bereczky, K. Lukács, Á. Sipos, *OGÉT 2015: XXIII. Nemzetközi Gépészeti Találkozó*, 32-36, (2015)
5. Emőd - Tölgyesi – Zöldy: Alternatív járműhajtások, Maróti kiadó (2006)
6. G. Szabados, Á. Bereczky: Polytechnica-Mechanical Engineering **59.** (3) 120-125. (2015)
7. A. Jamrozik, W. Tutak, M. Gruca, M. Pyrc, Journal of Mechanical Science and Technology. **32, 6,** 2947-2957, (2018).
8. A. Jamrozik, W. Tutak, M. Pyrc, M. Gruca, M. Kočiško, Fuel, **221**, 329-345, 2018.
9. A. Jamrozik W. Tuta, M. Pyrc, M. Sobiepański, Environmental Progress & Sustainable Energy, **36**, pp. 1151-1163
10. T. Laza: Polytechnica-Mechanical Engineering **52. (2**). 67-70. (2008)
11. K. Lukács, Z. Szurdoki, Á. Bereczky, *ENELKO2013 XIV*. 95-100, (2013).
12. M Zoldy, A Hollo, A Thernesz: *SAE Technical Paper*, 2010-01-0481
13. M Zöldy: Transport **26 (3),** 303-309
14. G. Szabados, Á. Bereczky, Periodica Polytechnica Mechanical Engineering, **59(3),** 120-125. DOI: 10.3311/PPme.7989. (2015)
15. W. Tutak A. Jamrozik R. Gnatowska, Thermal Science, **22, 3**, 1191-1203(2018).
16. W. Tutak, A. Jamrozik, Á. Bereczky, K. Lukacs, Transport, **33(3),** 633-646. (2018)
17. J. Žaglinskis, "Periodica Polytechnica Mechanical Engineering, **59(4),** 169-175. DOI: https://doi.org/10.3311/PPme.8216. (2015)
18. J. Žaglinskis, K. Lukács, Á. Fuel **170.** 245-253. (2016)
19. M. Zoldy, A. Torok Period. Polytech. Transp. Eng., **43, 4**. 172-176 (2015),
20. H. Xiaoye, Y. Zhenyi, W. Meiping, T. Jimi, Z. Ming. Applied Energy **198,** 347–359, (2017)
21. A. Rimkus, J. Matijošius, M. Bogdevičius, Á. Bereczky, Á. Török, Energy, **152**, 640-651. (2018)

Modelling fuel consumption and refuelling of autonomous vehicles

Máté Zöldy* *and Imre* Zsombók[1,*]

[1]BME Budapest University of Technology and Economics, Department of Automotive Technologies, Budapest Stoczek u. 6.

Abstract. In our research we highlighted the problematic of the refuelling of autonomous vehicles. During the way to be full autonomous, the vehicles take over more and more driving function from the driver. It is lot of focus on automotive cyber security or trajectory following, but refuelling is not in the main researches. After reviewing the vehicle drivetrains, it was specified which to focus for further testing. In the second part of the article the main influencers of fuel consumption was listed based on a literature overview. Main aspects of the carried out test is listed and results helped us to refine the research direction. For modelling the refuelling of the vehicle it is an enabler to set realistic models for fuel consumption. Relevant researches were studied and these was extended with overview of refuelling studies. Based on these the main direction of our research was set.

Introduction

Autonomous vehicles are more than more part of our everyday life in Europe. It has been started with driving and brake assistants, but nowadays it is common that a vehicle can manage the parking alone or helps to solve a takeover situation [17, 22]. As it is in technical and tabloid news as well, manufacturer's after each other announcing more and more feature containing vehicles. Some of these vehicles are on traditional internal combustion engine basis, some others are using hybrid technology and certainly full electric cars are available as well.

In our research our aim is to work out, simulate and test a decision making system supported with artificial intelligence to optimize autonomous vehicles refuelling. In this paper the first problem definition is presented.

1 Vehicle propulsions and refuelling

For long time it was not to mention the drivetrain system, the main difference was about the fuel: vehicle was operated with diesel or gasoline. Nowadays the situation is more complex, first hybrids, than plug-in-hybrids latest electric drivetrains arrived.

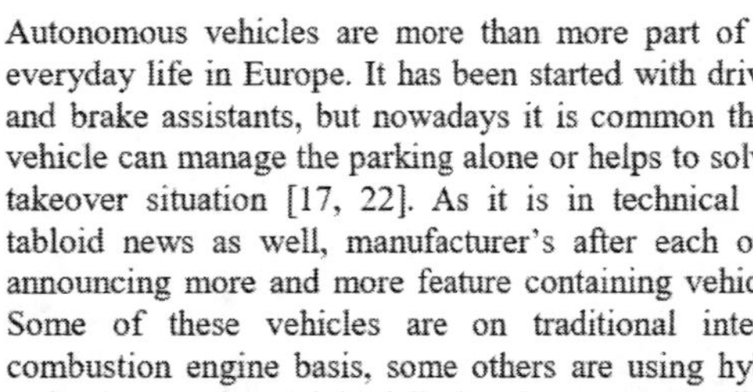

1. Figure New automotive drivetrains [7]

As Figure 1 shows parallel and series hybrids are more closer to the traditional internal combustion engine (ICE). With orange it is to see the refuelling with conventional fuel. In case of Plug-in Hybrid Electric Vehicles (Plug-in HEV or PHEV) first appears the external electric recharging option. These vehicles, as it can be followed at Figure 1 can be refuelled with conventional fuel and parallel to that with electricity. Battery electric vehicles (BEV) are rechargeable only with electricity, they are disconnected to fossil fuels.

1.1 Internal combustion engine

Conventional spark ignition and compression ignition engines, as called together internal combustion engines are part of mobility since 1876. In this year invented Nicolaus Otto four-stroke spark ignition engine. First engines was constructed for alternative fuels: gases and vegetable oils. They had four strokes: it carries out four piston strokes during a combustion cycle. Design of spark ignition engines was changed a lot during the years.

History of diesel started a bit later and the main difference is in combustion start: diesel engines uses the compression to start the combustion in the chamber. Daimler designed the first engine according to this principle.

Traditional combustion engines, independently if it is fuelled by diesel, gasoline, liquid or even gaseous alternative fuels. Most of the vehicles on the road today are very similar from drivetrain point of view. Lot of development happened during the decades and thanks for that today's internal combustion engines have considerably better efficiency, have better emission characteristic than those from decades ago, the basic

* Corresponding author: mate.zoldy@auto.bme.hu

theory to combust fuel to create power remains the same [20, 21].

Refuelling of internal combustion engine is managed most of the cases in refuelling stations. They have fixed locations along the roads, they are mainly settled according to customer demand. Near to fuel they could offer extra services, as food, store, relax zone etc. Vehicles has a fuel sensor that measures the remaining fuel level in the tank. The signal of this sensor is placed into the driver's dashboard and gives virtual feedback. In older models driver see the remaining fuel level a based on that the decision was made to stop for refuelling. In modern vehicles there is a calculation based on the last mileages driving behaviour and its predicts how many kilometres can be driven with the car until the tank will be empty.

1.2 Hybrid Drivetrains

In first view hybrids seems to be the first and strongest alternatives of internal combustion engines. After a Japanese manufacturer puts his first hybrid vehicles into series, most of the manufacturers have hybrid models. If it is checked in detail, hybrids are more a good tool to reduce fuel consumption of traditional engines instead of replace them. Mechanical setup of hybrid electric vehicles could differ a lot, fundamentally all hybrids have a mixed electric motor and internal combustion engine drivetrain and an electric battery, where either or both combustion engine and the electric motor could power the drivetrain. Batteries could be recharged in several ways, as be spanned by a generator when traditional engine is operating or in some cases, kinetic energy of the vehicle can be converted into electric energy with the help of systems like regenerative brakes.

The refuelling of these hybrids are very similar to ICE vehicles, while they are using as primarily energy source mainly gasoline and or diesel. Electricity is generated on board so it is not needed to refuelled outside. Optimization of utilization of conventional fuel and electricity is decided in motor management system.

1.3 Plug-in hybrid electric vehicles

Plug-in hybrid vehicles have a lot of similarities with hybrid electric vehicles as their drivetrain is hybrid: both have a traditional combustion engine and electric motor. The difference to hybrid vehicles, that these batteries can be recharged by plugging it in to a power source. When the battery is depleted, the plug-in hybrid starts acting as a regular hybrid, with the combustion engine taking the role of primary power source.

Energy management of plug-in-hybrid electric vehicles enables a much broader optimization scale. It includes the traditional opportunities and setup of traditional engines and has the option to refuel the battery. Battery refuelling nowadays has three main options. First is the charging facilities placed into traditional fuel stations. These are easy to find, loading level can be various, depending on the generation of the charger. There are standalone chargers as well mostly in city centres in parking places or park zones of hypermarkets. Most of the cases the waiting for free charger is not solved. Third option is charging home at residential areas or some workplaces are also offering such option. The public availability of both these charging types are limited.

1.4 Battery electric vehicles

This type of vehicles are the first in this summary, those do not have internal combustion engine. They run on a fully electric with rechargeable battery. These vehicles can be recharged by be plugged into a power source. Recharging time and driving ranges are varying depending on the type of the vehicle.

Recharging options of battery electric vehicles are covered with the options of plug-in-hybrid electric vehicles.

Table 1. Refueling possibilities of different drivetrains

refuel	ICE	HEV	PHEV	BEV
fuel	yes	yes	yes	no
electricity	no	no	yes	no

Based on the complexity of the refuelling of plug-in-hybrid technology, as it summarized in Table 1, it is worth to put it in the focus of our research. Our modelling will focus on this drivetrain and as it contains the solutions for pure ICE and pure electric vehicles as well, results will be easily convertible to these.

2 Fuel consumption Influencers

Increasing fuel prices growing need for more efficient vehicles resulted that fuel consumption has become one of the key challenges of the vehicle industry. Control of fuel consumption and its maintenance an average level requested lot of new solutions to be introduced. These solutions are covering energy optimized planning of routes, support and training of drivers, autonomous or partially autonomous vehicles, intelligent fleet management and platooning. To able to develop highly automated self-driving vehicles the basics of fuel consumption have to be understood. It is also a great opportunity to understand and utilized in an increased way the driver assist systems for eco-driving. This also provides a great help in assisting drivers in eco-driving training. Vehicles fuel- and energy consumption could depend on various different reasons and that are to be categorized as,

1 vehicle driven parameters,
2 road related parameters,
3 usage (driver) related parameters,
4 ambient parameters [1].

2.1 Vehicle driven parameters

Vehicle driven parameters are the physical attributes of vehicle like its length and weight, gearbox, type of the engine and motor, vehicle type, etc.

Vehicle fuel efficiency related questions as if an engine is not maintenance properly are also incorporated in this, as they could influence engine fuel efficiency, it will lead to a decline in the engine's fuel efficiency.

Simple maintenance activities such as regular, in time changes of oil and oil filters, change of air-filters, and in-time replacement of spark plug will increase vehicle lifespan and parallel they will improve vehicle fuel economy. In contrary missing to change worn out spark plugs, elements of air cleaner system or fuel filters as well as improper maintenance of throttle valves or oxygen sensors will effect fuel consumption increase.

Engine oil quality is also having an impact on fuel economy of the vehicle. In case of non-standard engine oil utilisation, at certain loads and temperatures, pistons of the engine and other moving parts freedom of moving will be lower. It increases the load on engine and it results increased fuel consumption.

Condition of tires is an important factor to manage the fuel consumption. If the tire pressure is too low, then vehicle needs more energy from the engine for efficient move around. It is important to keep the pressure barriers proposed by manufacturers.

2.2 Road related parameters

Road has several parameters that have influence on fuel consumption. Several test was conducted to investigate relationship between road surface type and the rolling resistance of vehicles. A excessive knowledge database has been created on the low-noise road surfaces interconnection with a lower rolling resistance and how it can be a good tool to decreasing vehicles fuel consumption and carbon-dioxide emissions [2, 23]. Research results demonstrated for lightweight vehicles a rolling resistance decrease up to 10% in case when fuel economy of driving on a double-layered highly porous asphalt road surface was compared to a regular highly porous surface. This 10% decrease in rolling resistance is equivalent to a 2 - 3% reduction fuel consumption and as well in carbon-dioxide emissions. It doesn't seems to be a huge amount but if it is multiplied with the total distance of the vehicles, than it is easy to calculate that it represents an amazing amount of fuel saving per annum [2]. Not only the road surface but construction material of the road have effect on fuel consumption. New materials, as rubber bitumen have a positive effect on vehicles fuel consumption [3].

2.3 Usage related parameters

Vehicle utilisation parameters or operational is nowadays stored in fleet management systems and it gives an easy opportunity to reach and to analyse. Database could contain different data types related to the vehicle as id of the vehicle, odometer actual value, total runtime, vehicle speed, severe accelerations, number and intensity of brakes, over speeding, etc.

Driving behaviour or style of the driver also have huge effect on fuel consumption. Aggressive driving and inappropriate gear handling are the most common ways to keep engine in high revolution areas that results extra fuel consumption. Keeping the right distance between the vehicles result lower number of unnecessary brakes and

through that it improves fuel consumption. Automatic transmission systems could compensate this for a defined range, while these are quicker in the shift up. This mostly applies to manual cars because automatic transmissions shift up more quickly. Definite speeds are proposed to be driven in given gears. For example, if vehicle runs at 90 km/h in fourth gear, then vehicle will need more energy to be moved in comparison to driving at same speed in fifth gear, and it will result extra fuel consumption.

Planning and optimising routes has effect fuel consumption. From fuel efficiency point of view, the two neuralgic point are the start-up of the vehicle and short trips. One of the reasons behind this phenomenon is that catalytic converters (that are used to decrease the emitted air pollutants) cannot be operated in a proper way until they reach the utilization temperature. The border is around 5-6 km; under this distance the internal combustion engine cannot reach the operating temperature.

Over speeding is also a bad actor in fuel consumption. Vehicles are constructed to be utilized up to a certain speed, that is depending on their engine and power output. Using vehicles on too high speed results in all cases extra fuel consumption.

2.4 Ambient parameters

Ambient parameters have effect on fuel consumption as well. Intake air temperature could one parameter, but in the time of turbocharged and with intercooler equipped engines, its role is marginal. High external temperatures are mostly followed with the start-up of the cooling system that has a measureable effect on fuel consumption. In another end, if auxiliary heating is necessary, while ambient temperature is cold, then it has also an increasing effect on fuel consumption. Driving with the AC on generally adds extra load to the engine hence it consumes more fuel when operating. However, if driving at speed of over 80 km/h, use of air-conditioning is better for fuel consumption compared to an open window as the latter creates aerodynamic drag, which is bad for fuel economy.

Table 2. Main influencers of vehicle fuel consumption

Fuel consmuption influencers	
Vehicle driven parameters	**Usage related parameters**
weight	total runtime
gearbox	vehicle speed
engine type	severity of accelerations
electric motor type	number of brakings
vehicle type	over speeding
maintenance	driver style
engine oil	engine start-up
tires	short trips
Road related parameters	**Ambient parameters**
surface type	Intake air temperature
material	cold temperature
	extra warm temperature
	landscape profile

Landscape profile for any trip has huge impact in the achievable fuel economy of the vehicles. A test was conducted to measure the landscape effect on fuel consumption, it results [6] that there is measurable connection between the road landscape and the fuel consumption.

Table 2 summarizes main influencers of fuel consumption. As it is easy to recognize, vehicle related parameters are influencing the fuel consumption on long term mostly.

Road related parameters and ambient parameters are external from the vehicle point of view, they have to be recognized and if possible a decision should be made between alternatives. Usage related parameters are those where driver or automated vehicle can influence fuel consumption.

3. Predicting of fuel consumption

Ability to model and predict the fuel consumption is vital in enhancing fuel economy of vehicles and preventing fraudulent activities in fleet management. Fuel consumption of a vehicle depends on several internal factors such as distance, load, vehicle characteristics, and driver behaviour, as well as external factors such as road conditions, traffic, and weather. However, not all these factors may be measured or available for the fuel consumption analysis.

Predicting methods for fuel consumption are mainly focusing on vehicle data or handling transportations as a flow. Utilization of fleet management system interface [1], is an easy way to get data from the vehicle. This study utilizes various vehicle parameters like speed, total fuel consumption, distance travelled, total runtime for example.

Another approach is to measure vehicles on test bench and make predictive figures based on the experiences. A passenger car with diesel engine was measured according to the UDC test cycle. Results of this work shows that fuel consumption can measured with UDC cycle only with huge simplifications. A research was carried out to creating a simulation for prediction of mileage fuel consumption based on engine operational parameters, such as fuel consumption and effective power [4]. With a self-developed algorithm, the main goal was to determine instantaneous fuel consumption not only for constant but for variable velocities as well, and based on these simulate mileage fuel consumption. Measurement uncertainties was taken into consideration, but with them as well the fuel consumption of the measured vehicles was higher by 34.8 to 46 % than the manufacturer given value.

Utilization of big data and data mining techniques is a utilized solution as well. In a research (5) the data acquisition was focusing on GPS data, authors proposed a solution that can estimate on a more accurate way and it is able to visualize energy consumption of vehicle based on analysis of vehicles' mobile and stationary activities. According to their proposal by adopting analytical concept of space-time path in time geography, they recommend methodology with that vehicle energy consumption and emissions can be estimated more accurately

Direction of vehicle routes can be a basis of fuel consumption estimation as well. In a case study [8], a method is presented that deal with fuel consumption and vehicle emissions of a single vehicle combined with a road network. Both are estimated and visualized with help of GPS data. Estimation accuracy of the utilized method reaches nearly 89%. Different road segments typical fuel consumption is analysed as well to able to determine patterns and internal mechanisms.

During the investigation five machine-learning methods were compared to able to build up the best model based on historical data analysis and prediction. From the five investigated techniques the quantile regression was found as best suitable. It is a classical least square estimation extension [8].

An investigation [9] puts sampling time into the fuel consumption centre. Results of this study clearly shows that increasing sampling rate to 10 minutes results better fuel consumption prediction as a sampling rate of 1 minutes. Main influencers of fuel consumption is road slope, weight and speed of vehicle is determined. Random forest, supporting vector model and artificial networks are compared and according this study the differences in punctuality are negligible.

Most of the methodological approaches are investigating a part of the fuel consumption elements and do not handling the issue as a whole. Consumption predictions are reduced to conventional fuel management. In this approach focus is on the most complex vehicles: plug-in-hybrids and as method the aim is to combine all elements that have predictable and measurable effect on fuel economy [9].

A research focused on three different machine learning techniques to see their usability in fuel consumption prediction [10]. In their research paper it is presented the application of the chosen three techniques to as a fuel consumption predicting tool. The model was set on trucks for a large dataset. Random Forest (RF), Support Vector Machine (SVM), and Artificial Neural Network (ANN) models have been developed and their prediction was compared. In most of the cases the telematics data is the basis of fleet performance evaluation for fleet managers, and maintenance related decisions or drivers' education based on these as well.

Available vehicle data is standardized according to SAE J1939 regulation. Road characteristic data for the test was given by local authority database. Based on those data it was possible to develop a new fuel consumption model, that incorporates road parameters into the fuel consumption prediction, and in this way influence fleet managers. Ten-times cross-validation was carried out to let learn the neural network, random forest and support vector machine models. The methodological results of the research were that random forest outperforms slightly support vector machine and artificial neural networks with higher R2 and lower error level [11].

Artificial neural network was used for personal car fuel consumption prediction in a research [13]. As carbon-dioxide emission is in predict direct relationship with vehicles fuel consumption with fuel prediction the emission cold as well. Vehicles have to face with most of influencing parameters in urban areas. Based on these facts the test was carried out in an urban area. Main factors that influenced fuel consumption were speed of the vehicle, city zone and time within the day and within the week. Satisfactory prediction accuracy was reached with a single hidden layer artificial neural network having 10 neurons [13].

As it is summarized in Table 3, fuel consumption prediction methods are tested mainly random forest, support vector machine and artificial neural network for prediction, but no clear preference is to see. In most of the cases GPS coordinates and fuel consumption are basis of the calculations but several other data is used as input. To special aspects is worth to mention: 10-minute sampling time is proposed and road segmentation could be a value added input data for the prediction.

Table 3. Fuel Consumption prediction summary

Nr.	Artificial Intelligence used	Engine type	Vehicle type	Data	Speciality
1		diesel	HD	speed, fuel consumption, distance, runtime	
4		diesel	PC		
5	data mining	diesel		GPS, energy consumption	
8	quantile regression	gasoline		GPS, fuel consumption	road segmentation
9	RF, SVM, ANN	hybrid			sampling time
10	RF, ANN, SVM	diesel			
11	ANN, SVM, RF				
13	ANN				

4 Refuelling

Prediction of fuel consumption is the first step to calculate the refuelling need of the vehicle and as next step extend it to refuelling optimization. There ae only a few researches that deals with the problems of refuelling.

One of the researches that aims to increase the effectiveness of refuelling decision instead of handling the refuelling as a technical prediction. As it is in other sources as [15, 16, 24] it is mentioned that pricing has an important role in refuelling decision problems. This study puts in the focus of the examination the viability of utilization a refuelling plan that aims to keep fuel expenses as low as possible for heavy-duty vehicles. They proofing the idea, that if refuelling happens more often but in cheaper fuel stations and with this method not only the lower price is beneficial but the lower weight of the vehicle causes lower fuel consumption.

In the current study, when the optimized refuelling strategy is calculated, the weight of the fuel is taken into consideration which leads to more frequent stops but also a lower fuel consumption. To test this strategy, a fuel consumption model was created with dynamic programming algorithm. Fuel consumption of the heavy-duty vehicle is calculated with the help of a recorded drive cycle. Dynamic programming algorithm is the responsible to decides whether to refuel at a specific fuel station in the function of fuel price of the station and the distance to upcoming stations. Main statement of the research is in one side there is a small amount of both money and fuel saving in most of the cases, but the in many cases, the predictability of the model makes its utilization at least questionable [12].

Refuelling prediction of electric vehicles is a much sharper topic, as the full load of a battery is around in that distance with which with a traditional internal combustion engine the driver starts to think that it is high time to refuel the vehicle. A research presents results of route planning for electric vehicles in a road network [14]. Main driver of route planning is in most of the cases to find the fastest way. It is ordinary for traditional vehicles with fuel, while the refuelling time is negligible compared to travelling time. The situation differs in case of electric vehicles: this energy is enough for shorter trips and refuelling time could take much longer. Furthermore, recharging prices of electric vehicles follows different standards than the pricing for conventional vehicles. The research propose a solution how to find balance between travel time and electric vehicle price of travel.

A study was focusing on the special requisites of battery electric vehicles, as they require different driver information systems. According to the anticipation of the researchers, a new highly integrated and networked information system is vital for electric vehicles that combines central car systems, drivers' behavioural data and parameters of the environment. Proposal of the research is to introduce an OEM-independent cruising range prediction system, which combines measurement and GIS-system based calculations on the road. The proposed model uses as main input parameters the most relevant range-affecting variables as driving style of the driver, ambient temperatures, battery charge status, navigational data, plan of energy consumption, and prediction of the route distance. Based on this data the optimal range can be managed.

Vehicle system used in the research consists of an optional on-board unit with data acquisition and measurement opportunity and communication device that offers navigation, support the power management. A self-learning algorithm is the heart of the model, as an aggregation feature that collect all data from environment and vehicle. It is responsible for propose charging or battery swapping stations on the planned route, which will then be passed on to a reservation management system, making sure the driver can enjoy a maximum of comfort and peace-of-mind on his or her journey. [18]

One of the neuralgic points of electric vehicles is the range uncertainty. It hindrance the extensive dissemination of this vehicles. Electric vehicles owners and driver stress on driving range can be reduced if vehicle's energy consumption is better understood. A research work [19] propose a predictive strategy for energy needs estimation of an electric vehicle. During the estimation vehicle related parameters and the topology data of the roads, and actual traffic conditions were used. In the model the energy consumption was represented with a macroscopic model. Basis of that was aggregated real-time data from web providers' maps. Model of the road network was an adaptation of a weighted direct graph.

Conclusions

As conclusion of this paper is can be stated, that plug-in-hybrid vehicles are the most complex ones at this stage of the technology. Our research aims to build fuel economy and refuelling prediction of these complex vehicles. Our goal is to incorporate all external and internal parameters that could have measurable and predictive effect on fuel consumption and refuelling of the vehicle. Vehicle related parameters have the lowest prompt effect on fuel consumption but they could be featured with tendencies. Road and ambient related parameters can be detected with different sensors, and signals should be processed within the vehicle. Usage related parameters could be reached via vehicle CAN system. A good mix of these signals and data should be the basis of the refuelling decision extended with refuelling stations data.

The research reported in this paper was supported by the Higher Education Excellence Program of the Ministry of Human Capacities in the frame of Artificial Intelligence research area of Budapest University of Technology and Economics (BME FIKP-MI/FM).

References

1. A. Viswanathan, 'Data driven analysis of usage and driving parameters that affect fuel consumption of heavy vehicles', *Dissertation*, http://www.diva-portal.org/smash/record.jsf?pid=diva2%3A634988&dswid=851 (2013)
2. K. Barry, Rigid Roads Can Reduce Fuel Consumption, VIRED, Gear (2012)
3. A. Geiger A. Holló, A Thernesz, R. Durgo S. Czibor, L. Bartha, P. Gergó Chemically Stabilized Rubber Bitumen, EEC, 1-8. (2012).
4. W. Gołębiewski., T. Stoeck TEKA. Commission of Mot. & Energ. in Agriculture – 2014. 14, 3, 17–24 (2014)
5. Z. Kan, L. Tang, M.P. Kwan, X. Zhang. Int. J. Environ. Res. Public Health 2018, 15, 566; (2018) DOI:10.3390/ijerph15040566 (2018)
6. O. Franzese, Freight Trucks, 8 (2011) https://info.ornl.gov/sites/publications/files/Pub33386.pdf
7. Hybrid and Battery Electric Vehicles, http://autocaat.org/Technologies/Hybrid_and_Battery_Electric_Vehicles/
8. X. Zhigang W. Tao, E. Said, Z. Xiangmo, Q. Xiaobo, Computer-Aided Civil and Infrastructure Engineering,. 33, pp. 209, (2018.)
9. H. Almér, Machine learning and statistical analysis in fuel consumption prediction for heavy vehicles in KTH, Sweden, (2015)
10. F. Perrotta,, T. Parry, L. Neves,.. IEEE Conference on Big Data (2017). DOI: 10.1109/BigData.2017.8258382.2017
11. S. Wickramanayake, D. Bandara, Fuel consumption prediction of fleet vehicles using Machine Learning: A comparative study. 90-95. 10.1109/MERCon.2016.7480121. (2016).
12. J. Telberg Fixed Route Optimal Refuelling Plan. KTH Royal Institute of Technology. www.diva-portal.org/smash/get/diva2:957311/FULLTEXT01.pdf
13. B. Predić, M. Madić, M. Roganović, M. Kovačević, D. Stojanović, Facta Universitatis, Series: Automatic Control and Robotics, 1(2), 105-116. Retrieved from http://casopisi.junis.ni.ac.rs/index.php/FUAutContRob/article/view/1597 (2016).
14. M. Højsleth, A. T. Lau, A. Strandfelt, Personalized Route Planning with Recharging for Electric Vehicles. https://projekter.aau.dk/projekter/files/239555254/article.pdf
15. A. Török, A. Török, F. Heinitz, Transport and Telecommunication Journal. 15(4), 292-298. https://doi.org/10.2478/ttj-2014-0025 (2014).
16. R. Ildarkhanov, Periodica Polytechnica Transportation Engineering. Onlinefirst, 7 DOI 10.3311/PPtr.10553 (2018)
17. Z. Szalay, T. Tettamanti, D- Esztergár-Kiss, I. Varga, C. Bartolini, Periodica Polytechnica Transportation Engineering, 46(1), 29-35. DOI: 10.3311/PPtr.10753. (2018)
18. P. Conradi, P. Bouteiller, S. Hanßen, Advanced Microsystemsfor Automotive Applications, Heidelberg:Springer,. 269-277, (2011)
19. G. De Nunzio, L. Thibault, A. Sciarretta, Intelligent Transportation Systems (ITSC) 2017 IEEE 20th International Conference, 1-7, (2017)
20. A. Rimkus, J. Matijošius, M. Bogdevičius, Á. Bereczky, Á. Török,. Energy, 152, 640-651. (2018)
21. I. Barabás. Fuel, 109, 563-574. (2013)
22. D. Dobrowolski, P. Droździel, R. Madleňák, I. K. Rybicka, D. Siłuch, Advances in Science and Technology. Research Journal, 12(3). (2018)
23. R. Madleňák, L. Madleňáková, D. Hošťáková, P. Drozdziel, Á. Török, Advances in Science and Technology. Research Journal, 12(2). (2018).
24. L. Hansut, A. David, J. Gasparik. *Bussines Logistic In Modern Management.* 29-42. (2017)

AUTHOR INDEX

EDP Sciences – Web of Conferences
17, Avenue du Hoggar
Parc d'Activité de Courtabouf
BP 112
F-91944 Les Ulis Cedex A
France

ISBN 978-1-5108-7489-3

9 781510 874893

CURRAN ASSOCIATES INC.
proceedings.com

Eighteenth International Conference on Emerging Security Information, Systems and Technologies (SECURWARE 2024)

Held at NetWare 2024

Nice, France
3-7 November 2024

IARIA